# Compendium of Plant Genomes

**Series Editor**

Chittaranjan Kole, Raja Ramanna Fellow, Government of India, ICAR-National Research Center on Plant Biotechnology, Pusa, New Delhi, India

Whole-genome sequencing is at the cutting edge of life sciences in the new millennium. Since the first genome sequencing of the model plant Arabidopsis thaliana in 2000, whole genomes of about 100 plant species have been sequenced and genome sequences of several other plants are in the pipeline. Research publications on these genome initiatives are scattered on dedicated web sites and in journals with all too brief descriptions. The individual volumes elucidate the background history of the national and international genome initiatives; public and private partners involved; strategies and genomic resources and tools utilized; enumeration on the sequences and their assembly; repetitive sequences; gene annotation and genome duplication. In addition, synteny with other sequences, comparison of gene families and most importantly potential of the genome sequence information for gene pool characterization and genetic improvement of crop plants are described.

**Interested in editing a volume on a crop or model plant?** Please contact Prof. C. Kole, Series Editor, at ckoleorg@gmail.com

More information about this series at http://www.springer.com/series/11805

Xuan Hieu Cao · Paul Fourounjian ·
Wenqin Wang
Editors

# The Duckweed Genomes

Springer

*Editors*
Xuan Hieu Cao
Department of Plant Physiology
Institute of Biology
Martin-Luther-University of
Halle-Wittenberg
Halle (Saale), Germany

Paul Fourounjian
Waksman Institute of Microbiology
Rutgers University
Piscataway, NJ, USA

Wenqin Wang
School of Agriculture and Biology
Shanghai Jiao Tong University
Shanghai, China

ISSN 2199-4781 ISSN 2199-479X (electronic)
Compendium of Plant Genomes
ISBN 978-3-030-11044-4 ISBN 978-3-030-11045-1 (eBook)
https://doi.org/10.1007/978-3-030-11045-1



This Springer imprint is published by the registered company Springer Nature Switzerland AG
The registered company address is: Gewerbestrasse 11, 6330 Cham, Switzerland

*This book series is dedicated to my wife Phullara and our children Sourav and Devleena*
Chittaranjan Kole

# Preface to the Series

Genome sequencing has emerged as the leading discipline in the plant sciences coinciding with the start of the new century. For much of the twentieth century, plant geneticists were only successful in delineating putative chromosomal location, function, and changes in genes indirectly through the use of a number of "markers" physically linked to them. These included visible or morphological, cytological, protein, and molecular or DNA markers. Among them, the first DNA marker, the RFLPs, introduced a revolutionary change in plant genetics and breeding in the mid-1980s, mainly because of their infinite number and thus potential to cover maximum chromosomal regions, phenotypic neutrality, absence of epistasis, and codominant nature. An array of other hybridization-based markers, PCR-based markers, and markers based on both facilitated construction of genetic linkage maps, mapping of genes controlling simply inherited traits, and even gene clusters (QTLs) controlling polygenic traits in a large number of model and crop plants. During this period, a number of new mapping populations beyond $F_2$ were utilized and a number of computer programs were developed for map construction, mapping of genes, and for mapping of polygenic clusters or QTLs. Molecular markers were also used in the studies of evolution and phylogenetic relationship, genetic diversity, DNA fingerprinting, and map-based cloning. Markers tightly linked to the genes were used in crop improvement employing the so-called marker-assisted selection. These strategies of molecular genetic mapping and molecular breeding made a spectacular impact during the last one and a half decades of the twentieth century. But still they remained "indirect" approaches for elucidation and utilization of plant genomes since much of the chromosomes remained unknown and the complete chemical depiction of them was yet to be unraveled.

Physical mapping of genomes was the obvious consequence that facilitated the development of the "genomic resources" including BAC and YAC libraries to develop physical maps in some plant genomes. Subsequently, integrated genetic–physical maps were also developed in many plants. This led to the concept of structural genomics. Later on, emphasis was laid on EST and transcriptome analysis to decipher the function of the active gene sequences leading to another concept defined as functional genomics. The advent of techniques of bacteriophage gene and DNA sequencing in the 1970s was extended to facilitate sequencing of these genomic resources in the last decade of the twentieth century.

As expected, sequencing of chromosomal regions would have led to too much data to store, characterize, and utilize with the-then available computer software could handle. But the development of information technology made the life of biologists easier by leading to a swift and sweet marriage of biology and informatics, and a new subject was born—bioinformatics.

Thus, the evolution of the concepts, strategies, and tools of sequencing and bioinformatics reinforced the subject of genomics—structural and functional. Today, genome sequencing has traveled much beyond biology and involves biophysics, biochemistry, and bioinformatics!

Thanks to the efforts of both public and private agencies, genome sequencing strategies are evolving very fast, leading to cheaper, quicker, and automated techniques right from clone-by-clone and whole-genome shotgun approaches to a succession of second-generation sequencing methods. The development of software of different generations facilitated this genome sequencing. At the same time, newer concepts and strategies were emerging to handle sequencing of the complex genomes, particularly the polyploids.

It became a reality to chemically—and so directly—define plant genomes, popularly called whole-genome sequencing or simply genome sequencing.

The history of plant genome sequencing will always cite the sequencing of the genome of the model plant Arabidopsis thaliana in 2000 that was followed by sequencing the genome of the crop and model plant rice in 2002. Since then, the number of sequenced genomes of higher plants has been increasing exponentially, mainly due to the development of cheaper and quicker genomic techniques and, most importantly, the development of collaborative platforms such as national and international consortia involving partners from public and/or private agencies.

As I write this preface for the first volume of the new series "Compendium of Plant Genomes," a net search tells me that complete or nearly complete whole-genome sequencing of 45 crop plants, eight crop and model plants, eight model plants, 15 crop progenitors and relatives, and three basal plants is accomplished, the majority of which are in the public domain. This means that we nowadays know many of our model and crop plants chemically, i.e., directly, and we may depict them and utilize them precisely better than ever. Genome sequencing has covered all groups of crop plants. Hence, information on the precise depiction of plant genomes and the scope of their utilization are growing rapidly every day. However, the information is scattered in research articles and review papers in journals and dedicated Web pages of the consortia and databases. There is no compilation of plant genomes and the opportunity of using the information in sequence-assisted breeding or further genomic studies. This is the underlying rationale for starting this book series, with each volume dedicated to a particular plant.

Plant genome science has emerged as an important subject in academia, and the present compendium of plant genomes will be highly useful to both students and teaching faculties. Most importantly, research scientists involved in genomics research will have access to systematic deliberations on the plant genomes of their interest. Elucidation of plant genomes is of interest not only for the geneticists and breeders, but also for practitioners of an array of plant science disciplines, such as taxonomy, evolution, cytology, physiology, pathology, entomology, nematology, crop production, biochemistry, and obviously bioinformatics. It must be mentioned that information regarding each plant genome is ever-growing. The contents of the volumes of this compendium are, therefore, focusing on the basic aspects of the genomes and their utility. They include information on the academic and/or economic importance of the plants, description of their genomes from a molecular genetic and cytogenetic point of view, and the genomic resources developed. Detailed deliberations focus on the background history of the national and international genome initiatives, public and private partners involved, strategies and genomic resources and tools utilized, enumeration on the sequences and their assembly, repetitive sequences, gene annotation, and genome duplication. In addition, synteny with other sequences, comparison of gene families, and, most importantly, the potential of the genome sequence information for gene pool characterization through genotyping by sequencing (GBS) and genetic improvement of crop plants have been described. As expected, there is a lot of variation of these topics in the volumes based on the information available on the crop, model, or reference plants.

I must confess that as the series editor, it has been a daunting task for me to work on such a huge and broad knowledge base that spans so many diverse plant species. However, pioneering scientists with lifetime experience and expertise on the particular crops did excellent jobs editing the respective volumes. I myself have been a small science worker on plant genomes since the mid-1980s and that provided me the opportunity to personally know several stalwarts of plant genomics from all over the globe. Most, if not all, of the volume editors are my longtime friends and colleagues. It has been highly comfortable and enriching for me to work with them on this book series. To be honest, while working on this series I have been and will remain a student first, a science worker second, and a series editor last. And I must express my gratitude to the volume editors and the chapter authors for providing me the opportunity to work with them on this compendium.

I also wish to mention here my thanks and gratitude to the Springer staff, particularly Dr. Christina Eckey and Dr. Jutta Lindenborn for the earlier set of volumes and presently Ing. Zuzana Bernhart for all their timely help and support.

I always had to set aside additional hours to edit books beside my professional and personal commitments—hours I could and should have given to my wife, Phullara, and our kids, Sourav and Devleena. I must mention that they not only allowed me the freedom to take away those hours from them but also offered their support in the editing job itself. I am really not sure whether my dedication of this compendium to them will suffice to do justice to their sacrifices for the interest of science and the science community.

Kalyani, India Chittaranjan Kole

# Preface

The duckweed or Lemnaceae family is a collection of five genera (*Spirodela*, *Landoltia*, *Lemna*, *Wolffiella*, and *Wolffia*) and 37 species of the smallest, fastest growing flowering plants living in aquatic environments. Many of these monocotyledonous plants can grow all over the world in a variety of climates. Provided their simplified and neotenous morphology, duckweeds have been researched for several decades as a model species for plant physiology and ecotoxicological research, contributing to the knowledge, e.g., about flowering response, plant circadian system, sulfur assimilation pathways, and auxin biosynthesis. In addition, duckweed-based water treatment has been proven as a feasible and inexpensive solution, especially within developing countries, to remove phosphorus and pharmaceutical chemicals from sewage and wastewater. With a dry mass yield per hectare per year up to 80 tonnes (equivalent to 10 tonnes of protein), duckweed is a promising aquatic crop in new modern and sustainable agriculture. Besides being an excellent primary or supplemental feedstock for production of livestock and fish, duckweed biomass can be utilized as a potential resource for human nutrition, biofuel, or bioplastics, depending on water quality as well as protein or starch accumulating procedures. Those academic and commercial interests led to the international effort to sequence the *Spirodela polyrhiza* genome, the smallest and most ancient genome in the family. *Spirodela* genomes reveal novel insights into the 158-Mbp genome size with less than three quarters number of *Arabidopsis thaliana* protein-coding genes and no signs of recent retrotranspositions.

In view of above, a total of 46 authors, representing 23 academic institutions or companies from five countries, have contributed 18 chapters for this book. This volume in the genome compendium series covers not only the latest findings in modern genetics, phylogenetics (Chaps. 2, 5), epigenetics, cytogenetics (Chap. 4), transcriptomics (Chaps. 12, 13, and 16), proteomics (Chap. 14), and genomics research in all five genera of duckweeds but also efforts toward transformation, genome editing and sequencing of the over one Gigabase *Wolffia* genomes (Chaps. 15, 17), with their large potential impacts on genome evolution and agricultural research. The introductory chapter stresses the importance of duckweeds as an aquatic plant model and as an extensive resource for biotechnological applications. The book tells the tale of the first *Spirodela* genome sequencing adventure (Chap. 7), details the nuclear (Chap. 9) and organelle (Chap. 10) genome sequences of *Spirodela polyrhiza*, which is the smallest, least methylated, and least transposon-rich monocot genome sequenced to date (Chap. 8). It describes the current genomics applications of these findings (genotyping by sequencing in Chap. 11; small RNA in Chap. 16) and the strategies to obtain new genome sequences within the family (Chap. 6). Finally, Chap. 18 is devoted to deeper insights and future perspective of using the duckweed genome information for duckweed research and applications.

It has been a great privilege to work with colleagues of the duckweed research community on this book. We are grateful to all the authors for their contribution in writing chapters of high quality. We are also thankful to the reviewers (Dr. Olaf Barth and Dr. Wiebke Zschiesche from Martin Luther University of Halle-Wittenberg, Germany and Dr. Hien Le Thu from Institute of Genome Research, Vietnam) for helping us in improving the quality of the chapters. The editors would like to express our sincere thanks to Prof. Chittaranjan Kole, Editor-in-Chief, of the Genome Compendium Series for cordial inviting us to contribute on this important masterpiece as well as to Springer, in general, Naresh Kumar Mani and Praveen Anand Sachidanandam, in particular, for constant help and support in publication and promotion of this book. We also appreciate and recognize cooperation and moral support from our family members for sparing us precious time for writing and editorial work.

We hope that our efforts in compiling the information on different aspects of duckweed will help the duckweed research and application community in enhancing better understanding about the duckweed biology and developing an extensive resource for biotechnological applications. This book will also benefit students, scientists both in academia and industry, and policy-makers in updating their knowledge on the importance and recent advances of duckweeds as an aquatic plant model.

Halle (Saale), Germany — Xuan Hieu Cao
Piscataway, USA — Paul Fourounjian
Shanghai, China — Wenqin Wang

# Contents

# Contributors

**Dong An** School of Agriculture and Biology, Shanghai Jiao Tong University, Shanghai, China

**K.-J. Appenroth** Matthias Schleiden Institute—Plant Physiology, Friedrich Schiller University of Jena, Jena, Germany

**Azizullah Azizullah** Department of Botany, Kohat University of Science and Technology, Kohat, Pakistan

**Gerrich Behrendt** Faculty of Natural Sciences, Institute of Plant Genetics, Leibniz University of Hannover, Hannover, Germany

**M. Bog** Institute of Botany and Landscape Ecology, University of Greifswald, Greifswald, Germany

**R. Brandt** Leibniz Institute of Plant Genetics and Crop Plant Research (IPK), Gatersleben, Germany

**Merlin Brychcy** Faculty of Natural Sciences, Institute of Plant Genetics, Leibniz University of Hannover, Hannover, Germany

**Xuan Hieu Cao** Institute of Biology/Plant Physiology, Martin-Luther-University of Halle-Wittenberg, Halle (Saale), Germany

**Yanqiang Ding** Key Laboratory of Environmental and Applied Microbiology, Chengdu Institute of Biology, Chinese Academy of Sciences, Chengdu, China;
Environmental Microbiology Key Laboratory of Sichuan Province, Chengdu, China;
University of Chinese Academy of Sciences, Beijing, China

**Anping Du** Key Laboratory of Environmental and Applied Microbiology, Chengdu Institute of Biology, Chinese Academy of Sciences, Chengdu, China;
Environmental Microbiology Key Laboratory of Sichuan Province, Chengdu, China

**Pengfei Duan** Collaborative Innovation Center of Water Security for Water Source Region of Mid-Line of South-to-North Diversion Project, Henan Province, Nanyang Normal University, Nanyang, China;

Key Laboratory of Ecological Security for Water Source Region of Mid-Line of South-to-North Diversion Project of Henan Province, Nanyang Normal University, Nanyang, China;
College of Agricultural Engineering, Nanyang Normal University, Nanyang, China

**Tamra Fakhoorian** The International Lemna Association, Mayfield, KY, USA

**Yang Fang** Key Laboratory of Environmental and Applied Microbiology, Chengdu Institute of Biology, Chinese Academy of Sciences, Chengdu, China;
Environmental Microbiology Key Laboratory of Sichuan Province, Chengdu, China

**Paul Fourounjian** Waksman Institute of Microbiology, Rutgers University, Piscataway, USA

**Lydia Gramzow** Matthias Schleiden Institute - Genetics, Friedrich Schiller University, Jena, Germany

**Lin Guo** Key Laboratory of Environmental and Applied Microbiology, Chengdu Institute of Biology, Chinese Academy of Sciences, Chengdu, China;
Environmental Microbiology Key Laboratory of Sichuan Province, Chengdu, China

**Kaize He** Key Laboratory of Environmental and Applied Microbiology, Chengdu Institute of Biology, Chinese Academy of Sciences, Chengdu, China;
Environmental Microbiology Key Laboratory of Sichuan Province, Chengdu, China

**A. Himmelbach** Leibniz Institute of Plant Genetics and Crop Plant Research (IPK), Gatersleben, Germany

**Hongwei Hou** The State Key Laboratory of Freshwater Ecology and Biotechnology, The Key Laboratory of Aquatic Biodiversity and Conservation of Chinese Academy of Sciences, Institute of Hydrobiology, Chinese Academy of Sciences, Wuhan, Hubei, People's Republic of China;
University of Chinese Academy of Sciences, Beijing, People's Republic of China

**Shiqi Hu** The State Key Laboratory of Freshwater Ecology and Biotechnology, The Key Laboratory of Aquatic Biodiversity and Conservation of Chinese Academy of Sciences, Institute of Hydrobiology, Chinese Academy of Sciences, Wuhan, Hubei, People's Republic of China;
University of Chinese Academy of Sciences, Beijing, People's Republic of China

**Yanling Jin** Key Laboratory of Environmental and Applied Microbiology, Chengdu Institute of Biology, Chinese Academy of Sciences, Chengdu, China;
Environmental Microbiology Key Laboratory of Sichuan Province, Chengdu, China

**Suliman Khan** The State Key Laboratory of Freshwater Ecology and Biotechnology, The Key Laboratory of Aquatic Biodiversity and Conservation of Chinese Academy of Sciences, Institute of Hydrobiology, Chinese Academy of Sciences, Wuhan, Hubei, People's Republic of China

**Sunjeet Kumar** The State Key Laboratory of Freshwater Ecology and Biotechnology, The Key Laboratory of Aquatic Biodiversity and Conservation of Chinese Academy of Sciences, Institute of Hydrobiology, Chinese Academy of Sciences, Wuhan, Hubei, People's Republic of China;
University of Chinese Academy of Sciences, Beijing, People's Republic of China

**Donald H. Les** Department of Ecology and Evolutionary Biology, University of Connecticut, Connecticut, USA

**Changsheng Li** Institute of Plant Physiology & Ecology, Shanghai Institutes for Biological Sciences, Chinese Academy of Sciences, Beijing, China

**Gaojie Li** The State Key Laboratory of Freshwater Ecology and Biotechnology, The Key Laboratory of Aquatic Biodiversity and Conservation of Chinese Academy of Sciences, Institute of Hydrobiology, Chinese Academy of Sciences, Wuhan, Hubei, People's Republic of China;
University of Chinese Academy of Sciences, Beijing, People's Republic of China

**Anke Londenberg** Faculty of Natural Sciences, Institute of Plant Genetics, Leibniz University of Hannover, Hannover, Germany

**Kim Lühmann** Faculty of Natural Sciences, Institute of Plant Genetics, Leibniz University of Hannover, Hannover, Germany

**Thomas Reinard** Faculty of Natural Sciences, Institute of Plant Genetics, Leibniz University of Hannover, Hannover, Germany

**K. S. Sree** Department of Environmental Science, Central University of Kerala, Periye, India

**Günter Theißen** Matthias Schleiden Institute - Genetics, Friedrich Schiller University, Jena, Germany

**Li Tan** Key Laboratory of Environmental and Applied Microbiology, Chengdu Institute of Biology, Chinese Academy of Sciences, Chengdu, China;
Environmental Microbiology Key Laboratory of Sichuan Province, Chengdu, China

**Xueping Tian** Key Laboratory of Environmental and Applied Microbiology, Chengdu Institute of Biology, Chinese Academy of Sciences, Chengdu, China;
Environmental Microbiology Key Laboratory of Sichuan Province, Chengdu, China;
University of Chinese Academy of Sciences, Beijing, China

**Nicholas P. Tippery** Department of Biological Sciences, University of Wisconsin, Whitewater, USA

**Giang T. H. Vu** Leibniz Institute of Plant Genetics and Crop Plant Research (IPK), Gatersleben, Germany

**F. Wagner** Institute of Plant Sciences, University of Regensburg, Regensburg, Germany

**Wenqin Wang** School of Agriculture and Biology, Shanghai Jiao Tong University, Shanghai, China

**Maren Wichmann** Faculty of Natural Sciences, Institute of Plant Genetics, Leibniz University of Hannover, Hannover, Germany

**Xiaoli Xiang** Institute of Biotechnology and Nuclear Technology, Sichuan Academy of Agricultural Sciences, Chengdu, China

**S. Xu** Institute for Evolution and Biodiversity, University of Münster, Münster, Germany

**Yaliang Xu** Key Laboratory of Environmental and Applied Microbiology, Chengdu Institute of Biology, Chinese Academy of Sciences, Chengdu, China; Environmental Microbiology Key Laboratory of Sichuan Province, Chengdu, China;
University of Chinese Academy of Sciences, Beijing, China

**Jingjing Yang** The State Key Laboratory of Freshwater Ecology and Biotechnology, The Key Laboratory of Aquatic Biodiversity and Conservation of Chinese Academy of Sciences, Institute of Hydrobiology, Chinese Academy of Sciences, Wuhan, Hubei, People's Republic of China;
University of Chinese Academy of Sciences, Beijing, People's Republic of China

**Lunguang Yao** Collaborative Innovation Center of Water Security for Water Source Region of Mid-Line of South-to-North Diversion Project, Henan Province, Nanyang Normal University, Nanyang, China;
Key Laboratory of Ecological Security for Water Source Region of Mid-Line of South-to-North Diversion Project of Henan Province, Nanyang Normal University, Nanyang, China;
College of Agricultural Engineering, Nanyang Normal University, Nanyang, China

**Jiaming Zhang** Institute of Tropical Bioscience and Biotechnology, Chinese Academy of Tropical Agricultural Sciences, Haikou, China

**Yating Zhang** School of Agriculture and Biology, Shanghai Jiao Tong University, Shanghai, China

**Hai Zhao** Key Laboratory of Environmental and Applied Microbiology, Chengdu Institute of Biology, Chinese Academy of Sciences, Chengdu, China;
Environmental Microbiology Key Laboratory of Sichuan Province, Chengdu, China

# Importance of Duckweeds in Basic Research and Their Industrial Applications

1

Paul Fourounjian, Tamra Fakhoorian and Xuan Hieu Cao

**Abstract**

The Lemnaceae family, commonly called duckweeds, is 37 species of the smallest and simplest flowering plants found floating on nutrient-rich waters worldwide. Their small size and rapid clonal growth in aseptic conditions made them a stable and simple model for plant research especially from 1950 to 1990, when they were used to study plant physiology and biochemistry including auxin synthesis and sulfur metabolism. Duckweed research then saw a resurgence in 2008 when global fuel prices rose and the US Department of Energy funded the sequencing of the *Spirodela polyrhiza* genome. This launched not only the genomic investigations detailed in this book, but the regrowth of duckweed industrial applications. Thanks to their ability to quickly absorb nitrogen, phosphorous, and other nutrients while removing pathogens and growing at a rate of 13–38 dry tons/hectare year in water treatment lagoons, scientists are currently exploring ways that duckweed can convert agricultural and municipal wastewater into clean water and a high-protein animal feed. The potential of these plants for phytoremediation of heavy metals and organic compounds also allows the possibility to clean the wastewater from heavy industry while providing biofuels and even plastics. Finally, thanks to their superb nutritional profile *Wolffia* species grown in clean conditions promise to become one of the healthiest and most environmentally friendly vegetables. Given the importance of these incredible plants, it is no wonder researchers are investigating the genetic mechanisms that make it all possible.

This chapter was revised and significantly expanded upon, with the guidance of T. F., from the chapter "The Importance and Potential of Duckweeds as a Model and Crop Plant for Biomass-Based Applications and Beyond," in the Handbook on Environmental Materials Management, which X. H. C. and P. F. wrote for Springer Nature a year ago (Cao et al. 2018). We hope this chapter thoroughly explains non-genomic research and application topics, especially for those who are unfamiliar with the family.

P. Fourounjian (✉)
Waksman Institute of Microbiology,
Rutgers University, Piscataway 08854, USA
e-mail: pjf99@scarletmail.rutgers.edu

T. Fakhoorian
The International Lemna Association,
Mayfield, KY 42066, USA

X. H. Cao
Institute of Biology/Plant Physiology,
University of Halle-Wittenberg
Martin-Luther-Weinbergweg,
10, 06120 Halle, Germany





X. H. Cao et al. (eds.), *The Duckweed Genomes*, Compendium of Plant Genomes,
https://doi.org/10.1007/978-3-030-11045-1_1

## 1.1 Introduction

Duckweed (known as monocotyledon family *Lemnaceae* or recently classified as subfamily *Lemnoideae* in the arum or aroid family *Araceae*) is a small group of aquatic plants with only five genera (*Spirodela*, *Landoltia*, *Lemna*, *Wolffia*, and *Wolffiella*) and 37 species (see Landolt 1986; Nauheimer et al. 2012; Sree et al. 2016). Except for *Wolffiella* (commonly named as bogmat) that is restricted to the Americas and Africa, species of other duckweed genera occur around the whole world. Although highly adaptable across a broad range of climates, most diverse species of duckweed appear in the subtropical or tropical zones. Duckweed species tend to be associated with nutrient-rich or eutrophic freshwater environments with quiet or slow-moving flow. However, they are extremely rare in deserts and are absent in the cold polar regions (Arctic and Antarctica).

Duckweed species are the smallest flowering plants with minute sizes from 0.5 mm to less than two cm (Landolt 1986). Species of duckweed can be easily distinguished morphologically from species of any other flowering plants, even closely related aquatic plants, due to their highly reduced body structure. The leaflike body of the duckweed species, sometimes called a frond or thallus, is a modified stem with only few cellular differentiations (Fig. 1.1). The growth of duckweed vegetatively occurs by budding within the pouches or cavities of the basal sections of the fronds. Each daughter frond emerging from the pouch of mother bud already contains two new generations of daughter fronds. Therefore, under optimal conditions, the growth rate of duckweed is nearly exponential. The frond number of fast-growing species (e.g., *Lemna aequinoctialis*, *Wolffiella hyalina*, and *Wolffia microscopica*) almost doubles within 24 h (Ziegler et al. 2015; Sree et al. 2015b), presenting the fastest growing flowering plants. With a miniaturized body plan and such rapid growth leading to maximum fitness, duckweed has arguably been interpreted as an example of the hypothetical Darwin–Wallace Demon for the lifetime reproductive success (Kutschera and Niklas 2015).

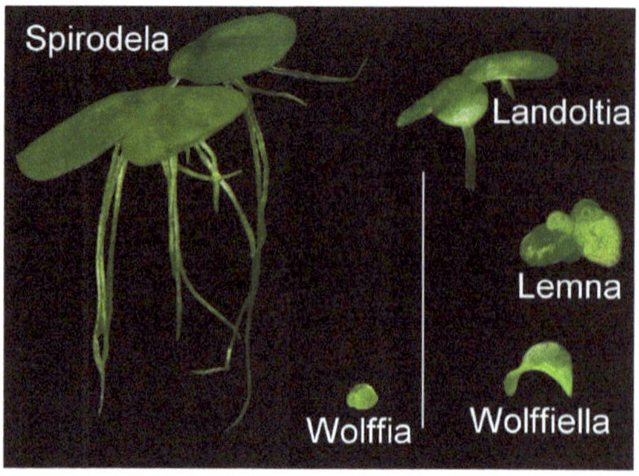


**Fig. 1.1** Morphology of five representative species for duckweed genera. Spirodela: *Spirodela polyrhiza*; Landoltia: *Landoltia punctata*; Lemna: *Lemna minor*; Wolffiella: *Wolffiella lingulata*; Wolffia: *Wolffia arrhiza*. Bar: 1 cm

Only occasionally or very rarely, several species of duckweeds produce microscopic flowers in nature as well as under in vitro conditions (Fu et al. 2017; Schmitz and Kelm 2017; Sree et al. 2015a). In *Spirodela* and *Lemna* (belonging to the subfamily *Lemnoideae*), the flowering organs (1 membranous scale, 2 stamens, and 1 pistil) originate in the same pouches in which the daughter fronds are normally formed. In the subfamily *Wolffioideae* (consisting of *Wolffiella* and *Wolffia*), generative and vegetative reproductions are spatially separated occupying the floral cavity on the upper surface of the frond and the budding pouch, respectively.

Duckweed fronds are free floating on or near the surface of the water, often forming dense mats in suitable climatic and nutrient conditions. In unfavorable weather, such as drought or freezing winter seasons, in addition to flowering, several duckweed species are able to form special "resting fronds" (in the dormant phase) to persist until conditions return that can support growth. In place of a frond, the greater duckweed (*Spirodela polyrhiza*) produces a starch-rich tissue called a turion, which sinks to the bottom of the water. Turion production has been reported also for *Lemna turionifera*, *L. aequinoctialis*, *Wolffia brasiliensis*, *Wolffia borealis*, *Wolffia angusta*, *Wolffia australiana*, *Wolffia arrhiza*, *Wolffia columbiana*, and *Wolffia globosa*. These turions do not grow any further but can germinate and start a new life cycle from the bottom of the water body or mud when the water temperature reaches about 15 °C. In addition, resting fronds of the ivy duckweed (*Lemna trisulca*) and *Wolffiella gladiata* with reduced air spaces can accumulate starch and still rather slowly grow on the bottom of the water, forming new but similar fronds. However, the common duckweed (*Lemna minor*), gibbous duckweed (*Lemna gibba*), *Lemna perpusilla*, and some strains of *Lemna japonica* produce starch-rich fronds that do not sink to the bottom of the water but are just pressed down under the ice cover during freezing temperatures. Interestingly, formation of turions as a survival and adaptive capacity of *S. polyrhiza* strains collected from a wide geographical range seems to be genetically determined and highly influenced by the mean annual temperature of habitats (Kuehdorf et al. 2013). Furthermore, the family displays significant inter- and intraspecies differences of cell physiology (e.g., starch, protein, and oil contents) together with duckweed potential for industrial applications (Alvarado et al. 2008; Appenroth et al. 2017; Hou et al. 2007; Mkandawire and Dudel 2005; Tang et al. 2015; Yan et al. 2013; Zhang et al. 2009).

Due to their small and abbreviated structures, morphological and physiological classification of the 37 duckweed species (*Spirodela*: 2 species; *Landoltia*: 1; *Lemna*: 13; *Wolffiella*: 10; *Wolffia*: 11) can be challenging. In the past decade, for species assignment as well as resolving intraspecies differences, several attempts have been carried out to employ molecular genotyping techniques, including random amplified polymorphic DNA (RAPD; Martirosyan et al. 2008), inter-simple sequence repeats (ISSR; Fu et al. 2017; Xue et al. 2012), simple sequence repeats (SSR; Feng et al. 2017), amplified fragment length polymorphism (AFLP; Bog et al. 2010, 2013), and DNA barcoding using plastid sequences (Borisjuk et al. 2015; Wang et al. 2010) or nuclear ribosomal sequences (Tippery et al. 2015). Although DNA barcoding using two plastidic barcodes aids in identifying most duckweed species (at least 30 among 37 species) in a quite simple and straight forward manner, combination of different techniques or using additional barcodes may help to unambiguously and economically assign remaining duckweed species.

The Lemnaceae family was one of the earliest model plants due to their ease of aseptic cultivation in the laboratory and simple morphology. The second volume of Landolt and Kandeler's 1987 monographic study contains 360 pages dedicated to the physiological research of the family in particular and plants as a whole (Landolt and Kandeler 1987). The professors who organized the first duckweed conference summed up the duckweed research stating that duckweeds were the main model for plant biology from 1950 to 1990, when Arabidopsis and rice were used for their sexual reproduction and applicability to terrestrial crops (Zhao et al. 2012). In that time, investigations of duckweeds revealed the tryptophan-independent synthesis of auxin (Baldi et al. 1991), translational regulation in eukaryotes (Slovin and Tobin 1982), and seven of the first stable plant mutants (Posner 1962). Today, physiological studies continue largely in the fields of circadian rhythm research, xenobiotic plant–microbe interactions, and phytoremediation and toxicology. Starting in 2011, a biannual series of international duckweed conferences in research and applications has connected and helped expand this research community and increased public awareness and recognition of duckweed economic and environmental importance (Zhao et al. 2012; Lam et al. 2014; Appenroth et al. 2015). Together with the completion of the *Spirodela* genome in the year 2014 and rapid advances in sequencing technologies, this resurgence of interest has resulted in a proliferation of genome and transcriptome sequences for duckweed species and ecotypes discussed in the remainder of this book.

One of the largest fields of duckweed research is ecotoxicology, where the widely distributed *Lemna* species *minor* and *gibba* serve as model plants to determine the effect of a compound on an ecosystem. These growth tests have been standardized in the International Organization for Standardization's protocol ISO 20079 which handles environmental samples and the Organisation for Economic Co-operation and Development's assay OECD 221, which was developed for specific chemicals and compounds (ISO 2005; OECD 2006). Both are seven-day growth rate tests, which use different media, to measure the effective concentration of the substance, or $EC_{50}$, where the growth rate by frond count or frond area is half of the control. These tests date back to the 1970s and have surveyed the effects of heavy metals, pharmaceuticals, various pesticides and organic compounds, and even radioactivity on Lemna growth rate and health, helping us quickly asses and monitor environmental safety.

In order to perform major gene function studies, as well as to improve duckweed agronomic performance (Cao et al. 2016), it is required to establish an efficient system for genetic manipulation and transformation. Several stable transformation protocols for *Lemna* (Chhabra et al. 2011; Yamamoto et al. 2001), *Landoltia* (*Spirodela oligorrhiza*; Vunsh et al. 2007), and *Wolffia* (Boehm et al. 2001; Khvatkov et al. 2015) using either *Agrobacterium*-mediated or biolistic gene transfer together with a recent gene-silencing platform in *L. minor* (Cantó-Pastor et al. 2015) have been described, providing the means to further develop gene/genome-edited duckweed as a powerful biomanufacturing platform.

## 1.2 Current State of Duckweed-Based Applications

### 1.2.1 Historical

For centuries, people have seen the role duckweed can play in their food production. Perhaps by observing their livestock consume duckweed species, small-scale farmers in Southern Asia started feeding duckweed, often fresh as a portion of the diet, to their fish, ducks, chickens, pigs, and goats. In addition to animal feed, the people of Thailand, Laos, and Cambodia have consumed wild harvested and farmed *Wolffia*, mainly *globosa*, rinsed, and then incorporated into soups, salads, sauces, and a wide variety of foods (Bhanthumnavin and Mcgarry 1971). If the *Wolffia* is not cooked in with other ingredients, it is generally briefly boiled, thereby selecting a duckweed species without harmful calcium oxalate crystals and killing potential pathogens. Recently, farmer education programs in Guatemala, Indonesia, and across the globe have improved the use of duckweed to treat manure while using it as a fertilizer and expanded the practice within Asia and around the world, especially in Central America where a consortium of ~200 small-scale farmers grows duckweed and tilapia. It is estimated by the executive director of the International Lemna Association that the thousands of small-scale farmers collecting wild duckweed or growing it on site for human or animal consumption are currently a greater part of the duckweed applications by volume than the large-scale, higher tech companies.

### 1.2.2 Water Treatment

As global population rises, so does demand in clean water supply and wastewater treatment systems. While developed nations have often relied on a combination of aerobic bacteria degradation and chemical treatment in activated sludge systems, a variety of natural treatment systems have been growing in popularity for their often 50% lower capital and operating costs, ability to recapture nitrogen, phosphorous, and other valuable nutrients, and in some cases convert them into appropriate products. The main downsides of these natural treatment systems are their larger land requirements (up to 5 $m^2$/person), poorer performance at cold temperatures, and the requirement of knowledgeable and diligent staff to manage ecosystems through toxic wastewater streams, harsh weather, etc. All this indicates that natural treatment systems such as constructed wetlands are ideal in rural locations, especially of tropical countries, precisely where many of the 2.5 billion people without access to sanitary wastewater treatment live (Zhang et al. 2014b).

While a variety of plants have been used effectively in constructed wetlands, we will focus here on the 37 species of the Lemnaceae family for their global distribution, tolerance of ammonia, heavy metals, other stresses, high yield of biomass (especially at 20–30 °C), ease of harvest, high protein and starch content, and range of uses. As seen in Fig. 1.2, duckweed can treat agricultural, municipal, and even industrial wastewater streams into clean non-potable water, and a biomass that can be used for feed applications, or fuel if it was used to treat harmful industrial wastewater.

The classic example of a duckweed treatment system and feed application would be the Mirzapur Bangladesh hospital wastewater facility, which was designed by the PRISM group, monitored from 1989 to 1991, and thoroughly described in the book "Duckweed Aquaculture: A New Aquatic Farming System for Developing Countries" (Skillicorn et al. 1993). The book describes a pilot plant facility with clean effluent water of 1 mg/l BOD (biological oxygen demand) and 0.03 mg/l of both $NH_3$ and P, an annual duckweed dry yield of 13–38 metric tonnes/hectare year (t/ha yr), carp production of 10–15 t/ha yr, and positive economic analysis of duckweed, duckweed-fed carp, and duckweed-fed tilapia farming. As of 2015, the Mirzapur facility was still operational, profitably treating wastewater above the standards of any US city, while providing fresh, pathogen-free, sustainably farmed fish. Professor Zhao Hai's group from Chengdu Institute of Biology, Chinese Academy of Sciences, also has extensive records from their pilot plant at Dianchi Lake, in subtropical Yunnan, China (Fig. 1.2). In a year-long comparison between duckweed and water hyacinth, they found a higher total yield for water hyacinth (55 compared to 26.5 t/ha yr) and a higher nitrogen removal rate, partially due to denitrifying bacteria. However, they chose to focus on duckweed for its consistent year-round production, $\sim$33% protein content, and biofuel potential as a low lignin, high starch ethanol feedstock (Zhao et al. 2014). In follow-up

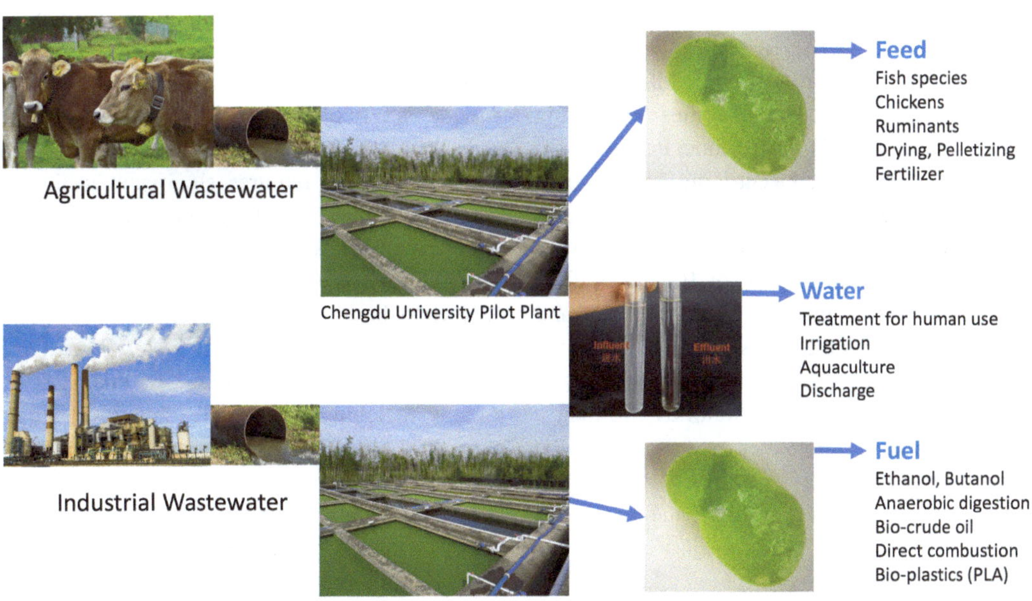


**Fig. 1.2** Flowchart of duckweed wastewater treatment and biomass application. Farm and factory examples, and the pilot plant at Chengdu University. Their influent agricultural wastewater and effluent water in the two test tubes. Mother and daughter frond of *Lemna minor*. While duckweed can be grown on agricultural or industrial wastewater and used for feed or fuel, the applications of the biomass are determined by the input water source and local regulations. *Source* Hai Zhao, Chengdu University, China

experiments, they found they could increase duckweed starch content from 9.5 to 40% through 11 days of growth on clean water, and that a hydraulic residence time (HRT) of 6 days achieved their treatment standards and optimized the *Landoltia punctata* starch yield above maize and wheat to 13.9 t/ha yr. Considering that these are experimental water treatment plants, their duckweed yield is expected to rise with further optimization, or in more intensive cultivation. For their size, length of study, and abundance of publicized information, these two facilities stand as prime examples to study duckweed's water treatment capabilities, yield, and applications in practice.

If a wastewater stream comes from an industrial point source or a large municipality, it likely has persistent chemical compounds, such as textile dyes and metalworking fluids, or bioaccumulating heavy metals in it. There is a large body of academic evidence illustrating the potential of duckweed and other plants to treat wastewater from cities, tanneries, mines, metalworking shops, and textile mills by degrading compounds like pharmaceuticals and antibiotics, and accumulating phenols along with heavy metals (van der Spiegel et al. 2013). Rezania et al. reviewed the heavy metal absorption of 5 different plant species and described 19 studies evaluating *Lemna minor* and *gibba* as moderate or hyperaccumulators, often concentrating metals over 400-fold, depending on the metal and circumstance; even when used as a dried powder (Rezania et al. 2016). A table of 10 studies illustrated removal efficiencies of Cu, Cd, Pb, Zn, and 9 other metals, with the lowest being 29% and the majority being over 70%. In these cases, duckweed and its microbial communities can treat a variety of harmful wastewater streams and then be utilized outside of the food supply for biofuel applications to further concentrate the metals.

### 1.2.3 Bioenergy

While these applications have been researched academically, few have been practiced in large scale. The simplest bioenergy application would be direct combustion of dried duckweed, possibly as a drop-in fuel for a trash incinerator or coal-fired power plant. This would concentrate heavy metals in the smoke, which could be scrubbed, and ash for proper disposal, or even encapsulated reuse in concrete or gypsum as coal ash is in the USA. A second relatively simple option would be anaerobic digestion to produce methane. Conveniently, many municipal wastewater treatment plants already have anaerobic digesters to treat sludge, and the liquid digestate has been well studied as a fertilizer for duckweed ponds. A duckweed and pig manure mixture increased gas production 41% in comparison with pig manure, while the increased production from cow manure tapered after a 2% inclusion of duckweed (Cui and Cheng 2015).

Another possibility is pyrolysis of dried biomass or hydrothermal liquefaction (HTL) of wet biomass. Both processes are similar, yet we will focus on HTL since it conveniently avoids drying the ~90% water content duckweed biomass. In HTL, biomass and water processed at 200–400 °C and 50–200 times atmospheric pressure for 10–90 min to create aqueous solutes, $H_2$, $CO_2$, and $CH_4$ gasses, high molecular weight bio-char, and bio-crude oil with 95% of the energy content of diesel (Zhang et al. 2014a). A wide variety of feedstocks from algae to wood and to sewage sludge can be used, separately, or mixed, and each requires significant testing to optimize, which is likely why there are no large-scale HTL operations at the present day. The algae can be converted to crude oil with a 26–68% yield depending on the conditions, yet all the crude oil tends to have a high water content and require hydro-deoxygenation to dewater it thereby matching the stability and viscosity of petroleum crude oil. A wide range of molecules can be created and isolated so there is petrochemical potential as well. This option is interesting for its theoretical ability to match the wide variety of the crude oil applications in a carbon neutral manner and the ability to produce in hours what naturally takes ~150 million years.

The most versatile and best studied application of potentially harmful duckweed is fermentation of the starch, which can be accumulated at rates varying from 46% after 5 days to 31% after 10 days of nutrient starvation and fermented at 95% efficiency after enzymatic pre-treatment. These fermentation processes also create protein-rich distiller's grains, which can be used as an animal feed supplement if they are not concentrating heavy metals. As the first commercially viable example of ethanol fermentation, the Andrew Young Foundation conducted a private research trial using the ecosystem technology, produced by resource recovery experts Greenbelt Resources Corporation, which was presented in a feasibility study report conducted by an independent party Agregy and submitted to the USDA in 2017. With successful feasibility determined, the foundation created a corporation called Duckweed Days LLC, which partnered with Greenbelt Resources to conduct a pilot system development project in Paso Robles, California, USA, in 2018. Leveraging its farming and agricultural expertise as well as its engineering prowess, Greenbelt has developed a species agnostic prototype cultivation, harvesting and processing system. For the biorefining of the cultivated duckweed, Greenbelt's proprietary, partially AI-operated modular machinery uses membrane filtration to produce anhydrous bioethanol that can be sold as a fuel or solvent, plus chemically safe distillers' grains that can be used as animal feed or a nutritious protein concentrate.

Ethanol is not the only fermentation product, since *Clostridium acetobutylicum* bacteria can convert the sugars of 32% starch content duckweed into a mixture of 68% butanol, with acetone and ethanol coproducts (Cui and Cheng 2015). Ethanol can of course be blended into gasoline at rates up to 10% or 85% for certain flex-fuel vehicles, while significantly more expensive butanol behaves very similarly to gasoline. Finally, the Argentinian company MamaGrande experimented with fermentation as a means to generate lactic acid for polymerization into PLA. Polylactic acid, or PLA, is a renewable and degradable plastic produced by enzymatically digesting starch to glucose, fermenting the glucose to lactic acid, and then purifying and polymerizing it. At the moment, anaerobic digestion and ethanol fermentation appear to be the best studied options, while fermentation is the only biofuel in full-scale commercial application.

### 1.2.4 Animal Feed

Most agricultural wastewater and certain domestic wastewater streams will have undetectable or legally permissible levels of heavy metals, enabling a design where duckweed can recycle nutrients back into the food supply, provided it is monitored for heavy metals and other hazards, and legally approved. Agricultural wastewater, which can come from greenhouses, livestock barns, anaerobic digesters, or even food processing facilities, is often heavy metal "free" and therefore diluted down to 20–50 mg/l total nitrogen for optimal duckweed growth. Considering the pilot plant examples above, and publicly posted information from Paul Skillicorn's Agriquatics Blog, we see the following steps for domestic wastewater treatment (Fig. 1.3).

First, solids will be removed by screening and then primary settling lagoons or laminar flow systems and hydrocyclones, possibly for anaerobic digestion. Secondly, there may be a buffer lagoon or lagoons, which treat soaps and other chemicals that may harm duckweed or its downstream applications. Third will be the duckweed farm portion, where a diluted influent with $NH_3$ concentrations of 10–30 mg/l, BOD of 15–30 mg/l, and pH from 6.0 to 7.0 will fertilize rapidly growing high-protein duckweed biomass. Fourth, ponds with slower growing, high starch content duckweed can polish wastewater as the final cleaning step. Here, once nitrogen has been depleted heavy metals will be accumulated, with the majority of municipal effluents producing duckweed passing US food and feed safety standards. HRT can vary from 6 to 15 days depending on environment, degree of effluent recirculation, and treatment standards. For example, the Mirzapur duckweed ponds reduced $NH_3$ from 32 to 0.03 mg/l. This high HRT increases the footprint compared to a conventional system, while providing resilience against heavy rains or community crashes that

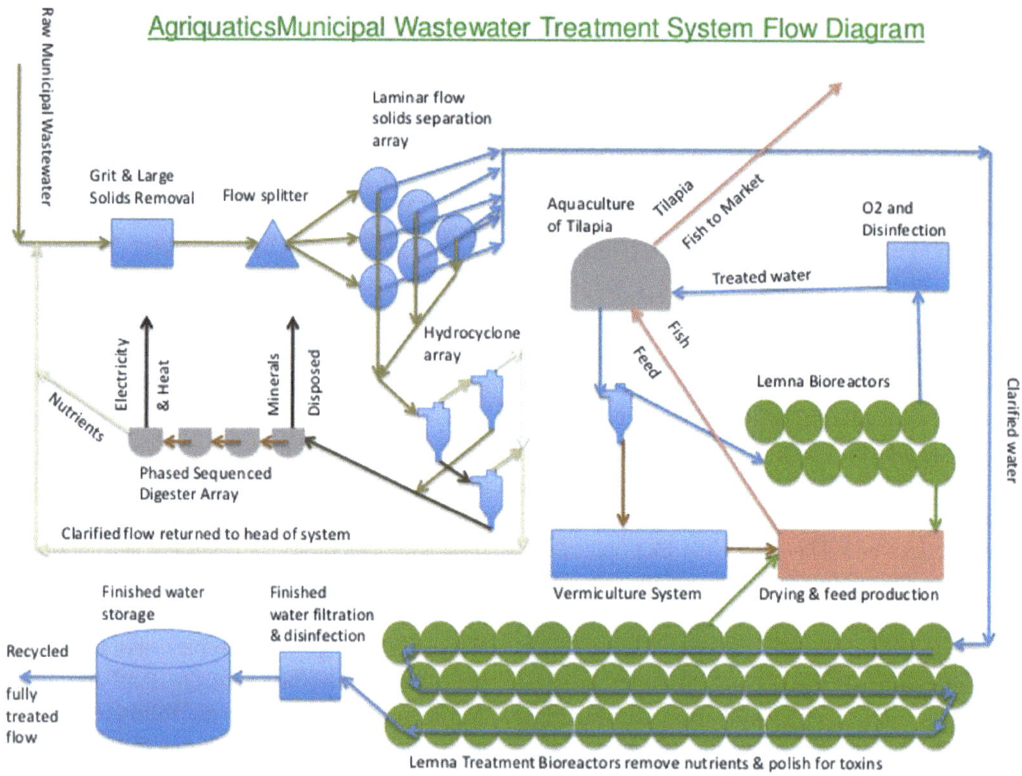


**Fig. 1.3** Agriquatics wastewater treatment and aquaculture diagram for Olmito, Texas. Proposed blueprint for a municipal treatment facility designed by Agriquatics. The systems start with solids removal through laminar flow separators and hydrocyclones, and sends solids to an array of bacterial digesters, which act as an improved anaerobic digester similar to conventional methods. A series of duckweed ponds remove solutes, and their circular shape facilitates central harvesting. Water is then filtered and disinfected with conventional methods. Duckweed biomass can be tested, sterilized, and converted to Tilapia feed, while aquatic worms and duckweed purify water and provide food for the Tilapia. *Source* Paul Skillicorn, Lyndon Water Limited, UK, https://paulskillicorn.wordpress.com/about/

occasionally overwhelm smaller systems. Throughout this process, pathogens are largely killed off, evaporation is reduced 33%, mosquito populations are reduced, and odors are partially suppressed by the duckweed mat (Goopy and Murray 2003; van der Spiegel et al. 2013). Finally, polished water and duckweed biomass can be sterilized and utilized. In a budget estimate for a medium-sized treatment plant in Texas, USA, Agriquatics illustrated that their treatment system would have 52% of the capital and 66% of the total annual costs of a conventional oxidative ditch system. This budget completely excluded the proposed tilapia aquaculture system that had been proved profitable in Mirzapur. To make larger duckweed treatment systems, even more cost-effective Agriquatics has positioned them on the outskirts of cities to benefit from rising real estate value as the city grows, while providing greenspace and reducing pipe distance.

Since duckweeds have been a traditional feed for fish and poultry in South East Asia for centuries, they are now being quantitatively investigated in a variety of feed trials. In many cases, $NH_3$-tolerant *Lemna* and *Spirodela* species are used and harvested with dry weight protein contents of 20–30%. To minimize pathogen transfer, feed trials often use effluent from one species to grow duckweed, which is then fed to a different species. While ozone and microwave disinfection were used in the long-term commercial operation of Mirzapur, many feed trials have simply washed with water, or just harvested duckweed, and have no report of pathogens (Goopy and Murray 2003; Skillicorn et al. 1993). Surprisingly, several studies have found duckweed, including samples from hospital wastewater to be safe as chicken and fish feed with regard to *E. coli* and *Salmonella*, with no significant differences in the quantity of five different pathogens in chickens fed on duckweed compared to control, presumably due to the severe pathogen reduction seen in wastewater treated by duckweed and its associated microbial communities (Goopy and Murray 2003; van der Spiegel et al. 2013). The feed trials often use dried duckweed as a percentage of complete commercial feed or substitute it for a percentage of the soybean or fishmeal component, with duckweed performing very similarly to soy in the case of chickens, ducks, and fish, up to a point where it is suspected that oxalates or other anti-nutritives inhibit growth (Goopy and Murray 2003; Skillicorn et al. 1993). For tilapia, inclusion rates of 30% were found equivalent to control, and 30% replacement of fishmeal component was seen as the most cost effective (Goopy and Murray 2003). An ecosystem of 5 different carp species or the grass, catla, and mirror carp and tilapia species individually can be fed on a pure duckweed diet, with a carp yield of 10–15 t/ha yr (Skillicorn et al. 1993). Duckweed was found to be beneficial in replacing ~15% of the soybean meal in the feed for chicks or broilers, and 40% in the case of laying hens (Goopy and Murray 2003; Skillicorn et al. 1993). In some cases, pig saw decreased growth in response to small inclusion rates of duckweed, while the Mong Cai piglets of Vietnam had higher growth rates than their Large White counterparts due to higher nitrogen digestibility (Goopy and Murray 2003; Gwaze and Mwale 2015). Finally, ruminants have shown promising results with high nitrogen digestibility in merino sheep, and cattle consuming and effectively digesting up to 10% of their weight in dried duckweed per day (Goopy and Murray 2003). Taken together, these results show the potential of duckweed to reduce the environmental impact of livestock by recycling nitrogen phosphorous and other nutrients that currently cause eutrophication, while partially replacing human edible soy and non-sustainable fishmeal. Furthermore, recycling wastewater to grow animal feed has been shown in several economic analyses to raise farmer income, especially in developing countries.

Considering the economic and environmental benefits, and the success of duckweed as feed for a variety of livestock species, there will likely be a rapid expansion of the duckweed agricultural sector and its use as a sustainable animal feed. In the FAO's 2012 estimates, global demand for non-fish animal protein is expected to increase at 1.3% per year till 2050, with the largest growth of 4.2% in South Asia, with similar numbers in the 2030 projection (Alexandratos and Bruinsma 2012). Roughly, half of this increase is expected to be as poultry (OECD/FAO 2017). Additionally, the largest increase in animal protein supply will be aquaculture, which was ~17% of global fish supply in 1990, grew largely in Asia between 4 and 10% per year, and is forecasted to exceed the global catch in 2020 (OECD/FAO 2017). The livestock sector is, however, very feed, land, and water intensive, and all reports stress the need to reduce the environmental impact particularly through improving the feed supply. With their ability to treat agricultural wastewater on non-arable land and provide an affordable protein-rich feed, a greater number of farmers are turning to duckweeds as a cheap sustainable feed source. There are currently several commercial ventures and hundreds of thousands of small-scale farmers growing duckweed primarily in Asia and Central America feeding tilapia, ducks, chicken, and pigs. Since they are sustainably feeding the livestock species in the regions where the FAO expects the largest growth in the world, it is natural to expect this industry to grow. While working with farmer education programs in Guatemala and Indonesia, the ILA, International Lemna Association, has seen an increase in educational activities for small-scale farmers and 20% more businesses seeking to enter the industry for the past 7 years (Table 1.1, Director of the ILA).

**Table 1.1** Summary of the duckweed applications in use or development and the major companies working on them

| Application | Company (if blank academic) | Genera |
|---|---|---|
| Human food | Hinoman, Green Onyx, Parabel | Wolffia, Lemna |
| Protein isolate | Plantible, Parabel, CAIS | Lemna |
| Livestock | Many small-scale farmers | Lemna, Spirodela, others |
| Conversion chemicals | MamaGrande | Lemna |
| Wastewater treatment | MamaGrande, CAIS | Mixture |
| Space life support | Space Lab Technologies | Lemna, Wolffia |
| Isolation chemicals | CAIS | Mixture |
| Transformation | | |
| Specialty (cosmetics, pets, tea) | | |
| Biofuels or energy | Greenbelt Resources | |

### 1.2.5 Human Nutrition

The high growth rate, protein content, and success in a variety of animal feed trials naturally beg the question of whether duckweeds could be a healthy and environmentally friendly food for humans? As previously stated, the *Wolffia* genus of the duckweed family has been traditional cuisine in Thailand, Burma, and Laos for centuries, since they lack the kidney stone forming calcium oxalate crystals found in the other genera. At the time of writing, there are three large companies producing *Wolffia* or *Lemna* for human consumption, namely Hinoman with greenhouse precision agriculture cultivation, Parabel with open pond *Lemna* cultivation and protein extraction, and Green Onyx, which has developed robotic farming systems that can dispense *Wolffia* on demand. Due to their successful scale-up since their founding in 2010, and abundant public information, we will focus on the Israeli company Hinoman here. They currently grow *Wolffia* (aka Mankai$^{TM}$) on formulated, clean water media in greenhouses with automated energy-efficient climate control and harvesting systems operated by their cultivation algorithm. Through this system, they are able to grow a pesticide- and herbicide-free vegetable year-round, with a fraction of the water used in cultivation of soy, spinach, or kale, (http://www.hinoman.biz/what-we-do/). Their product is stable with approximately 25% carbohydrate content, 45% protein content, and a complete and bioavailable amino acid profile such as egg or soy, with a higher PDCAAS than soy. They have currently conducted three publicly visible clinical trials demonstrating the protein and iron bioavailability, as well as the mitigating effect on Glycemic Index of their *Wolffia*, and posted multiple recipes for their product, which will soon be made available to consumers.

Furthermore, compared to kale *Wolffia* is more abundant in most minerals and vitamins A, B2, B12, and E, which survive the gentle drying process. An extensive academic investigation of the species *Wolffia microscopica* confirmed the high mineral content and that the protein ($\sim$25% of dry weight) exceeded WHO recommendations, while finding abundant antioxidants and a high omega-3 content ($\Omega6/\Omega3$ ratio is 0.61) for the relatively scarce lipids (Appenroth et al. 2017, 2018). Fresh, or dry powdered *Wolffia*, with a neutral taste, can be juiced, consumed fresh, or incorporated into breads, pastas, and sports nutrition products (Fig. 1.4). With supporting data from academic laboratories, records of historical consumption, and thorough testing of their product for harmful metals and oxalates, Hinoman and Green Onyx were able to achieve the generally recognized as safe (GRAS) status for the *Wolffia* species *arrhiza*

**Fig. 1.4** *Wolffia* fortified breads. Hinoman has tested the addition of *Wolffia* to multiple food and beverage products. Note the retention of the chlorophyll pigments throughout the baking process, and unchanged texture and leavening of the bread. *Source* http://www.hinoman.biz

and *globosa* in the USA in 2015 and 2016, respectively. Now, with South East Asia, Israel, and the USA recognizing select *Wolffia* species and *Lemna minor* as human food the crop and its producers have significant potential to grow and provide abundant plant protein for minimal land, water, and energy inputs.

With their small size, growth rate, aquatic lifestyle, and high protein content, the duckweeds provide a promising new crop to grow and an assortment of cultivation and preparation processes for human consumption. Given the growing consumer demand for novel vegetables and healthy leafy greens, companies like Hinoman and Green Onyx grow these tiny nutritious vegetables in clean environments with robotic systems and plan to bring them into our grocery stores and homes both frozen and fresh. The global market for plant-based protein (57% of total global protein supply; Henchion et al. 2017) has been growing at 12.3% per year from 2013 to 2016, and is anticipated to grow 6.7% annually from 2018 to 2021, when it is anticipated to exceed 1 billion USD. Seeing this demand for protein, Plantible Foods is developing a gentle protein isolation process using *Lemna* in order to create a colorless, tasteless protein isolate with the physical properties of egg whites to create a vegan product that can finally match the textures of many beloved foods. Additionally, Parabel has chosen to sell its duckweed product as a high protein powder. Given the expansion of the plant protein market in both whole and extract formats, and their current progress, we expect these and other companies to increase in size, dramatically, providing a healthy and environmentally friendly alternative to less efficient protein sources.

As seen above, duckweed wastewater treatment performs well in tropical and subtropical environments, requires more land, yet less funding to operate, and even has the potential to generate revenue if duckweed biomass and clean effluent are well utilized. Agricultural wastewater can be converted into animal feed supplements, while industrial effluents can be treated to degrade or accumulate harmful chemicals and heavy metals while producing bioenergy, according to the laws of the land. The duckweed has proven to be a suitable food source for both humans and livestock, and will likely play an expanding role in meeting future food demands. There is plenty more to learn at the International Lemna Association and The Charms of Duckweed Web sites, and in the Duckweed Forum newsletter. Given the tremendous diversity of species, strains, environments, and applications, along with the relatively recent commercial interest, duckweed researchers are continuously rediscovering what is possible and practical.

## 1.3 Future Prospects in Duckweed-Based Applications

The field of duckweed applications has made tremendous progress recently. For centuries, it was harvested from wild ponds and used as a vegetable or animal feed in certain parts of the world, and largely in the twenty-first century humans have recognized the potential of these tiny overlooked plants and started applying them to wastewater treatment, and larger-scale animal feed and human nutrition operations. While certain applications are mature enough for large-scale deployment, those discussed below include important developing technologies. In terms of scale and possibility of duckweed applications, we believe in 2019 we are still looking at the tip of the iceberg.

Due to the success and low prices of other crops, many companies growing duckweed are focused on high-tech, high-value applications to avoid commodity markets. Similar to protein extracts, several high-value products, like sugars, antioxidants, and oils, are being extracted from duckweed biomass in academic and commercial research laboratories. Appenroth et al. conducted a thorough investigation of *W. microscopica* and found a complete plant protein, roughly 150 mg carotenoids and 22 mg of tocopherols/gram dry weight, and an oil profile of 61% polyunsaturated fatty acids with a high content omega-3 s and a phytosterol content minimum fivefold higher than common plant oils, presenting several healthy, high-value compounds that may be extracted (Appenroth et al. 2017). After or without extraction of certain compounds or protein, biomass can be converted to other products, for example MamaGrande's research in converting starch to sugar, and then polylactic acid valued at ~$2000 USD/ton. After enzymatically converting starch to sugar, the sugars can be fractionated and sold, or converted to levulinic, formic, or succinic acid (Liu et al. 2018). Pyrolysis and HTL discussed above can be used to create bio-char, gases, and a bio-crude oil. A subset of a single sample of duckweed derived bio-crude oil contained over 100 distinct compounds, mainly ketones, alcohols, fatty acids, and cyclic compounds (Duan et al. 2013). When considering the variables of biomass, solvents, temperature, pressure, and time HTL, pyrolysis can be adjusted to offer countless compounds that can be created and fractionated. Finally, there are a variety of other high-value application niches that duckweed can be used for including tea, cosmetics, pet food, and aquarium plants, which have been tested on small scale and may develop further. Major crops such as corn and soy have been used as feedstocks for hundreds of uses including food-thickening agents, cosmetics, construction adhesives, and ink. It is therefore reasonable to expect that as duckweed abundance grows there will be a greater number and variety of applications.

Another sector where duckweed species will likely play an expanding role is water reclamation and supply. In 2018, the Duckweed forum issue 22 described 23 companies in 9 countries, with 4 each working in water quality testing and water treatment (Shoham 2018). Provided the perpetual rise of water pollution and increased testing, and the roughly 50% lower capital and operating costs of duckweed (Skillicorn 2013) and constructed wetland (Zhang et al. 2014b) treatment systems compared to their bacterial counterparts, these industries are expected to grow, likely more so in developing countries. Sadly, 14 years of satellite observations reveal decreasing clean water availability across the world and in heavily populated areas like California, the Middle East, Northern India, and Northern China where groundwater is being depleted (Rodell et al. 2018). Many regions suffer clean water scarcity for at least 1 month of the year resulting in inadequate supply for people as well as agricultural losses. Duckweed treatment systems to reclaim water, as well as water efficient duckweed crops, with many other measures, might be utilized in these and other regions to increase supply. Similar to water reclamation, there is a lesser known need for phosphorous reclamation, since our current practice is to mine and refine phosphorous deposits, fertilize our crops, and then let the phosphorous run directly off of fields and into the ocean, or through our wastewater treatment systems into the ocean where it causes eutrophication damage like the Gulf of Mexico hypoxic zone. Economically mineable, organically available phosphorous is expected to be scarce by 2050 or 2100, and production might decline by 2030 raising its price possibly beyond the reach of poorer farmers (Childers et al. 2011). Fortunately, phosphorous can be recycled by better farming practices or by using more aquatic plants and other methods to recapture more than the current rate of 50% from human wastes. While phosphorous is a critical macronutrient and prime example, many other fertilizers have similar life cycles and would follow the phosphorous in any reduce, recapture, and reuse applications. Given the water and fertilizer scarcities this century will likely pose to billions of people, we sincerely hope that duckweed-based water treatment systems and many other water and nutrient reclamation technologies will be applied at larger scale to "close the loop" and avoid scarcity.

One of the earliest companies to work with duckweed, Biolex Therapeutics, saw the rapidly growing high protein biomass of *Lemna* as a great expression platform for transgenic proteins. They produced several complex antibodies, including one to target Leukemia, and trademarked the term Plantibodies$^{TM}$, yet sadly went out of business. Since their closure, there have been improvements in the transgenic expression within several duckweed species. There have even been academic papers reporting over 20 transgenic therapeutic proteins in duckweed reaching as high as 7% of total soluble protein (Balaji et al. 2016). Given the lower cost of production and lower risk of transmissible pathogens compared to mammalian cell lines, duckweed may once again provide genetically engineered proteins for medical or other applications.

Catalyst Agri-Innovations Society (CAIS) works with a number of diverse companies in several locations including an on-farm anaerobic digester with nutrient extraction and at a land-based fish farm. All of their work is on efficiency and resource recovery at the food/energy/water nexus in the overall agriculture domain. They currently work with several wastewater treatments like the Trident Processes system for separating manure solids, anaerobic digestion to extract energy, and duckweed or algae to remove solutes. Wastewater from multiple species is anaerobically digested to generate methane and energy, and the digestate moves on to enclosed stacked shelf growth chambers filled with duckweed. After doubling in under 48 h and cleaning the water, duckweed is fermented to separate protein from high-value simple sugars. Christopher Bush, Co-founder of CAIS, has worked with the XPRIZE Foundation, designing competitions including "Feeding the Next Billion." The team also works with the HeroX platform where a sponsor can publicly host a problem and cash prize for the solution, greatly increasing the number of scientists who can see and solve the problem and learn from the winning solution. This type of modern interdisciplinary research center, consulting firm, or incubator that relies on datasets from large sensor arrays and crowd sourcing looks to be increasing in popularity, and we look forward to the variety of applications that will be developed where duckweed will play a role as one of several options to reclaim resources or feed people and livestock more effectively.

Perhaps given their ability to clean wastewater while providing food and fresh air, duckweeds can be seen as not only a crop species, but a life support system. The current water recovery system on the International Space Station relies on complex chemical treatments and reagents while generating wastes, which has NASA interested in developing closed-loop life support systems for long-term missions. Many plants develop poorly in microgravity and produce inedible biomass, so non-gravitropic aquatic plants and specifically duckweeds have been studied for space flight in closed-loop systems, microgravity simulations, and space flights since 1966 (Landolt and Kandeler 1987; Gale et al. 1989; Bluem and Paris 2003). *Lemna aequinoctialis* was even found to have a 32% increase in growth rate in simulated microgravity (Yuan and Xu 2017). Therefore, Space Lab Technologies, LLC is currently collaborating with the University of Colorado at Boulder on a Phase 2 grant from NASA to develop the μG-LilyPond$^{TM}$ growth chamber as part of a life support system (Escobar and Escobar 2017). Part of their project is studying how bursts of high intensity light can stimulate production of carotenoids, vitamin E, and other nutritious secondary metabolites (Demmig-Adams and Adams 2002), and how these bursts within the light regimen can be optimized for energy use, plant yield, and nutritional content. Thanks to their high growth rate, ability to grow in shallow trays, preference for ammonia, and entirely edible nutritious biomass duckweed are currently the prime candidates for the system. Presently, it is designed to provide fresh food and oxygen, with the eventual goal of converting urine to clean water. Based on the previous literature, the goal is to create a 1 $m^3$ system capable of treating the wastewater and $CO_2$ of 4 crew members and providing an edible vegetable yield up to 250 g of dry weight per day (Gale et al. 1989; Landolt and Kandeler 1987). The µG-LilyPond™ system will need to overcome the unique challenges of space missions including size and weight restrictions, controlled growth and harvest in microgravity, water delivery via capillary action, sterility, minimal human maintenance, and rapid recovery from failures (Escobar and Escobar 2017). This intimate reliance on duckweed in a closed-loop system provides both a technical and a symbolic example of how humans and duckweed complement each other, and how we can use the smallest plants to solve the largest challenges.

**Acknowledgments** We'd like to thank Paul Skillicorn from Agriquatics, Professor Zhao Hai from Chengdu University, Darren Eng from Greenbelt Resources, Eduardo Mercovich from MamaGrande, Ron Salpeter from Hinoman, Tsipi and Ben Shoham from Green Onyx, Maurits van de Ven from Plantible Foods, Christopher Bush from CAIS, and Christine Escobar from Space Lab Technologies for sharing data on their companies and research and reviewing this document to ensure its accuracy.

## References

Alexandratos N, Bruinsma J (2012) World agriculture towards 2030/2050: the 2012 revision. Food and Agriculture Organization of the United Nations (FAO), Rome

Alvarado S, Guédez M, Lué-Merú MP, Nelson G, Alvaro A, Jesús AC, Gyula Z (2008) Arsenic removal from waters by bioremediation with the aquatic plants water hyacinth (*Eichhornia crassipes*) and lesser duckweed (*Lemna minor*). Bioresour Technol 99 (17):8436–8440

Appenroth KJ, Crawford DJ, Les DH (2015) After the genome sequencing of duckweed—how to proceed with research on the fastest growing angiosperm? Plant Biol (Stuttg) 17:1–4. https://doi.org/10.1111/Plb.12248

Appenroth K-J, Sree KS, Böhm V, Hammann S, Vetter W, Leiterer M, Jahreis G (2017) Nutritional value of duckweeds (*Lemnaceae*) as human food. Food Chem 217:266–273

Appenroth KJ, Sree KS, Bog M, Ecker J, Boehm V, Lorkowski S, Sommer K, Vetter W, Tolzin-Banasch K, Kirmse R (2018) Nutritional value of the duckweed species of the genus *Wolffia* (Lemnaceae) as human food. Frontiers in chemistry 6:483

Balaji P, Satheeshkumar PK, Venkataraman K, Vijayalakshmi MA (2016) Expression of anti-tumor necrosis factor alpha (TNFalpha) single-chain variable fragment (scFv) in *Spirodela punctata* plants transformed with Agrobacterium tumefaciens. Biotechnol Appl Biochem 63(3):354–361. https://doi.org/10.1002/bab.1373

Baldi BG, Maher BR, Slovin JP, Cohen JD (1991) Stable Isotope Labeling, in Vivo, of d- and l-Tryptophan Pools in *Lemna gibba* and the Low Incorporation of Label into Indole-3-Acetic Acid. Plant Physiol 95 (4):1203–1208. https://doi.org/10.1104/pp.95.4.1203

Bhanthumnavin K, Mcgarry MG (1971) *Wolffia* arrhiza as a possible source of inexpensive protein. Nature 232 (5311):495

Bluem V, Paris F (2003) Possible applications of aquatic bioregenerative life support modules for food production in a Martian base. Adv Space Res 31(1):77–86

Boehm R, Kruse C, Voeste D, Barth S, Schnabl H (2001) A transient transformation system for duckweed (*Wolffia columbiana*) using Agrobacterium-mediated gene transfer. J Appl Bot 75(3–4):107–111

Bog M, Baumbach H, Schween U, Hellwig F, Landolt E, Appenroth KJ (2010) Genetic structure of the genus *Lemna* L. (*Lemnaceae*) as revealed by amplified fragment length polymorphism. Planta 232 (3):609–619. https://doi.org/10.1007/s00425-010-1201-2

Bog M, Schneider P, Hellwig F, Sachse S, Kochieva EZ, Martyrosian E, Landolt E, Appenroth KJ (2013) Genetic characterization and barcoding of taxa in the genus *Wolffia* Horkel ex Schleid. (Lemnaceae) as revealed by two plastidic markers and amplified fragment length polymorphism (AFLP). Planta 237 (1):1–13. https://doi.org/10.1007/s00425-012-1777-9

Borisjuk N, Chu P, Gutierrez R, Zhang H, Acosta K, Friesen N, Sree KS, Garcia C, Appenroth KJ, Lam E (2015) Assessment, validation and deployment strategy of a two-barcode protocol for facile genotyping of duckweed species. Plant Biol (Stuttg) 17:42–49. https://doi.org/10.1111/Plb.12229

Cantó-Pastor A, Mollá-Morales A, Ernst E, Dahl W, Zhai J, Yan Y, Meyers BC, Shanklin J, Martienssen R (2015) Efficient transformation and artificial miRNA gene silencing in *Lemna minor*. Plant Biol (Stuttg) 17:59–65. https://doi.org/10.1111/plb.12215

Cao HX, Wang W, Le HTT, Vu GTH (2016) The power of CRISPR-Cas9-induced genome editing to speed up plant breeding. Int J Genomics 2016:10. https://doi.org/10.1155/2016/5078796

Cao HX, Fourounjian P, Wang W (2018) The importance and potential of duckweeds as a model and crop plant for biomass-based applications and beyond. In:

Hussain CM (ed) Handbook of environmental materials management. Springer International Publishing, Cham, pp 1–16. https://doi.org/10.1007/978-3-319-58538-3_67-1

Chhabra G, Chaudhary D, Sainger M, Jaiwal PK (2011) Genetic transformation of Indian isolate of *Lemna minor* mediated by *Agrobacterium tumefaciens* and recovery of transgenic plants. Physiol Mol Biol Plants 17(2):129–136. https://doi.org/10.1007/s12298-011-0059-5

Childers DL, Corman J, Edwards M, Elser JJ (2011) Sustainability challenges of phosphorus and food: solutions from closing the human phosphorus cycle. Bioscience 61(2):117–124. https://doi.org/10.1525/bio.2011.61.2.6

Cui W, Cheng JJ (2015) Growing duckweed for biofuel production: a review. Plant Biol (Stuttg) 17:16–23. https://doi.org/10.1111/plb.12216

Demmig-Adams B, Adams WW 3rd (2002) Antioxidants in photosynthesis and human nutrition. Science 298 (5601):2149–2153. https://doi.org/10.1126/science.1078002

Duan P, Chang Z, Xu Y, Bai X, Wang F, Zhang L (2013) Hydrothermal processing of duckweed: effect of reaction conditions on product distribution and composition. Bioresour Technol 135:710–719. https://doi.org/10.1016/j.biortech.2012.08.106

Escobar CM, Escobar AC (2017, July) Duckweed: a tiny aquatic plant with enormous potential for bioregenerative life support systems. In: 47th International conference on environmental systems, Charleston, South Carolina, 16–20 July 2017. ICES, p 281

Feng B, Fang Y, Xu Z, Xiang C, Zhou C, Jiang F, Wang T, Zhao H (2017) Development of a new marker system for identification of *Spirodela polyrhiza* and *Landoltia punctata*. International Journal of Genomics 2017:8. https://doi.org/10.1155/2017/5196763

Fu L, Huang M, Han B, Sun X, Sree KS, Appenroth KJ, Zhang J (2017) Flower induction, microscope-aided cross-pollination, and seed production in the duckweed *Lemna gibba* with discovery of a male-sterile clone. Sci Rep 7(1):3047. https://doi.org/10.1038/s41598-017-03240-8

Gale J, Smernoff DT, Macler BA, MacElroy RD (1989) Carbon balance and productivity of *Lemna gibba*, a candidate plant for CELSS. Adv Space Res 9(8):43–52

Goopy J, Murray P (2003) A review on the role of duckweed in nutrient reclamation and as a source of animal feed. Asian-Australas J Anim Sci 16(2):297–305

Gwaze FR, Mwale M (2015) The prospect of Duckweed in pig nutrition: a review. J Agric Sci 7(11):189

Henchion M, Hayes M, Mullen AM, Fenelon M, Tiwari B (2017) Future protein supply and demand: strategies and factors influencing a sustainable equilibrium. Foods 6(7). https://doi.org/10.3390/foods6070053

Hou W, Chen X, Song G, Wang Q, Chang CC (2007) Effects of copper and cadmium on heavy metal polluted waterbody restoration by duckweed (*Lemna minor*). Plant Physiol Biochem 45(1):62–69

International Organization of Standardization (2005) Water quality—Determination of the toxic effect of water constituents and waste water on duckweed (*Lemna minor*)—Duckweed growth inhibition test. ISO Geneva (CH), ISO guideline 20079

Khvatkov P, Chernobrovkina M, Okuneva A, Pushin A, Dolgov S (2015) Transformation of *Wolffia arrhiza* (L.) Horkel ex Wimm. Plant Cell Tissue Organ Cult (PCTOC) 123(2):299–307

Kuehdorf K, Jetschke G, Ballani L, Appenroth KJ (2013) The clonal dependence of turion formation in the duckweed *Spirodela polyrhiza*—an ecogeographical approach. Physiol Plant. https://doi.org/10.1111/ppl.12065

Kutschera U, Niklas KJ (2015) Darwin-Wallace Demons: survival of the fastest in populations of duckweeds and the evolutionary history of an enigmatic group of angiosperms. Plant Biol (Stuttg) 17:24–32. https://doi.org/10.1111/Plb.12171

Lam E, Appenroth KJ, Michael T, Mori K, Fakhoorian T (2014) Duckweed in bloom: the 2nd international conference on duckweed research and applications heralds the return of a plant model for plant biology. Plant Mol Biol 84 (6):737–742. https://doi.org/10.1007/s11103-013-0162-9

Landolt E (1986) The family of Lemnaceae—a monographic study, vol 1. Veroffentlichungen des Geobotanischen Institutes der Eidgenossischen Technischen Hochschule, Stiftung Rubel, Zurich

Landolt E, Kandeler R (1987) Biosystematic investigations in the family of duckweeds (*Lemnaceae*), Vol. 4: The family of Lemnaceae-a monographic study, Vol. 2 (phytochemistry, physiology, application, bibliography). Veroeffentlichungen des Geobotanischen Instituts der ETH, Stiftung Ruebel (Switzerland)

Liu C, Feng Q, Yang J, Qi X (2018) Catalytic production of levulinic acid and ethyl levulinate from uniconazole-induced duckweed (*Lemna minor*). Bioresour Technol 255:50–57. https://doi.org/10.1016/j.biortech.2018.01.087

Martirosyan EV, Ryzhova NN, Skryabin KG, Kochieva EZ (2008) RAPD analysis of genome polymorphism in the family Lemnaceae. Russ J Genet 44(3):360–364. https://doi.org/10.1134/s1022795408030198

Mkandawire M, Dudel EG (2005) Accumulation of arsenic in *Lemna gibba* L. (duckweed) in tailing waters of two abandoned uranium mining sites in Saxony, Germany. Sci Total Environ 336 (1):81–89

Nauheimer L, Metzler D, Renner SS (2012) Global history of the ancient monocot family Araceae inferred with models accounting for past continental positions and previous ranges based on fossils. New Phytol 195 (4):938–950. https://doi.org/10.1111/j.1469-8137.2012.04220.x

OECD/FAO (2017) OECD-FAO agricultural outlook 2017–2026. OECD Publishing, Paris. https://doi.org/10.1787/19991142

OECD (2006) Test No. 221: *Lemna* sp. growth inhibition test. OECD Guidelines for the Testing of Chemicals, Section 2. OECD Publishing, Paris. https://doi.org/10.1787/9789264016194-en

Posner HB (1962) Characteristics of X-ray-induced aberrants of *Lemna perpusilla* 6746. Plant Cell Physiol 3(3):275–284

Rezania S, Taib SM, Din MFM, Dahalan FA, Kamyab H (2016) Comprehensive review on phytotechnology: heavy metals removal by diverse aquatic plants species from wastewater. J Hazard Mater 318:587–599

Rodell M, Famiglietti JS, Wiese DN, Reager JT, Beaudoing HK, Landerer FW, Lo MH (2018) Emerging trends in global freshwater availability. Nature 557 (7707):651–659. https://doi.org/10.1038/s41586-018-0123-1

Schmitz U, Kelm H (2017) First discovery of flowering *Wolffia arrhiza* in Central Europe. Aquat Bot 143:33–35

Shoham T (2018) On a mission to introduce duckweed as the 21st century cash crop: Current active industrial efforts. Duckweed Forum 6(3):92–97

Skillicorn P (2013) The agriquatics system vs activated sludge: a comparative analysis. https://paulskillicorn.wordpress.com/2013/04/17/the-agriquatics-system-vs-activated-sludge-a-comparative-analysis/. Accessed 9 Sept 2019

Skillicorn P, Spira W, Journey W (1993) Duckweed aquaculture: a new aquatic farming system for developing countries. World Bank

Slovin JP, Tobin EM (1982) Synthesis and turnover of the light-harvesting chlorophylla/b-protein in *Lemna gibba* grown with intermittent red light: possible translational control. Planta 154(5):465–472. https://doi.org/10.1007/BF01267815

van der Spiegel M, Noordam MY, van der Fels-Klerx HJ (2013) Safety of novel protein sources (Insects, microalgae, seaweed, duckweed, and rapeseed) and legislative aspects for their application in food and feed production. Compr Rev Food Sci F 12(6):662–678. https://doi.org/10.1111/1541-4337.12032

Sree KS, Bog M, Appenroth KJ (2016) Taxonomy of duckweeds (*Lemnaceae*), potential new crop plants. Emir J Food Agr 28(5):291–302. https://doi.org/10.9755/ejfa.2016-01-038

Sree KS, Maheshwari SC, Boka K, Khurana JP, Keresztes A, Appenroth KJ (2015a) The duckweed *Wolffia* microscopica: a unique aquatic monocot. Flora 210:31–39. https://doi.org/10.1016/j.flora.2014.10.006

Sree KS, Sudakaran S, Appenroth K-J (2015b) How fast can angiosperms grow? Species and clonal diversity of growth rates in the genus *Wolffia* (Lemnaceae). Acta Physiol Plant 37(10):204

Tang J, Li Y, Ma J, Cheng J (2015) Survey of duckweed diversity in Lake Chao and total fatty acid, triacylglycerol, profiles of representative strains. Plant Biology 17(5):1066–1072

Tippery N, Les D, Crawford D (2015) Evaluation of phylogenetic relationships in Lemnaceae using nuclear ribosomal data. Plant Biology 17(s1):50–58

Vunsh R, Li J, Hanania U, Edelman M, Flaishman M, Perl A, Wisniewski JP, Freyssinet G (2007) High expression of transgene protein in Spirodela. Plant Cell Rep 26(9):1511–1519. https://doi.org/10.1007/s00299-007-0361-4

Wang W, Wu Y, Yan Y, Ermakova M, Kerstetter R, Messing J (2010) DNA barcoding of the Lemnaceae, a family of aquatic monocots. BMC Plant Biol 10:205. https://doi.org/10.1186/1471-2229-10-205

Xue HL, Xiao Y, Jin YL, Li XB, Fang Y, Zhao H, Zhao Y, Guan JF (2012) Genetic diversity and geographic differentiation analysis of duckweed using inter-simple sequence repeat markers. Mol Biol Rep 39(1):547–554. https://doi.org/10.1007/s11033-011-0769-3

Yamamoto Y, Rajbhandari N, Lin X, Bergmann B, Nishimura Y, Stomp A-M (2001) Genetic transformation of duckweed *Lemna gibba* and *Lemna minor*. Vitro Cell Develop Biol-Plant 37(3):349–353. https://doi.org/10.1007/s11627-001-0062-6

Yan Y, Candreva J, Shi H, Ernst E, Martienssen R, Schwender J, Shanklin J (2013) Survey of the total fatty acid and triacylglycerol composition and content of 30 duckweed species and cloning of a Δ6-desaturase responsible for the production of γ-linolenic and stearidonic acids in Lemna gibba. BMC Plant Biol 13(1):201

Yuan J, Xu K (2017) Effects of simulated microgravity on the performance of the duckweeds *Lemna* aequinoctialis and *Wolffia* globosa. Aquat Bot 137:65–71. https://doi.org/10.1016/j.aquabot.2016.11.010

Zhang C, Duan P, Xu Y, Wang B, Wang F, Zhang L (2014a) Catalytic upgrading of duckweed biocrude in subcritical water. Bioresour Technol 166:37–44. https://doi.org/10.1016/j.biortech.2014.05.022

Zhang DQ, Jinadasa K, Gersberg RM, Liu Y, Ng WJ, Tan SK (2014b) Application of constructed wetlands for wastewater treatment in developing countries–a review of recent developments (2000–2013). J Environ Manag 141:116–131

Zhang X, Zhao FJ, Huang Q, Williams PN, Sun GX, Zhu YG (2009) Arsenic uptake and speciation in the rootless duckweed *Wolffia* globosa. New Phytol 182 (2):421–428

Zhao Y, Fang Y, Jin Y, Huang J, Bao S, Fu T, He Z, Wang F, Zhao H (2014) Potential of duckweed in the conversion of wastewater nutrients to valuable biomass: a pilot-scale comparison with water hyacinth. Bioresour Technol 163:82–91. https://doi.org/10.1016/j.biortech.2014.04.018

Zhao H, Appenroth K, Landesman L, Salmean AA, Lam E (2012) Duckweed rising at Chengdu: summary of the 1st international conference on Duckweed application and research. Plant Mol Biol 78 (6):627–632. https://doi.org/10.1007/s11103-012-9889-y

Ziegler P, Adelmann K, Zimmer S, Schmidt C, Appenroth KJ (2015) Relative in vitro growth rates of duckweeds (*Lemnaceae*)—the most rapidly growing higher plants. Plant Biol (Stuttg) 17:33–41. https://doi.org/10.1111/plb.12184

# Tiny Plants with Enormous Potential: Phylogeny and Evolution of Duckweeds

2

Nicholas P. Tippery and Donald H. Les

**Abstract**

Duckweeds (family Lemnaceae) comprise 37 angiosperm species, which are distributed among five genera. Although these tiny specimens represent the smallest flowering plants on earth, the group is practically ubiquitous in water bodies worldwide. The paucity of morphological features in duckweeds has made it difficult to elucidate their evolutionary history, or to present a compelling classification of the group, but more recent molecular evidence has facilitated an improved systematic evaluation of these unique plants. The duckweeds are closely related to aroids (family Araceae), with which they share several morphological features. Within Lemnaceae, the two species of *Spirodela* and the monotypic genus *Landoltia* are more distantly related to other duckweeds than are the larger genera *Lemna*, *Wolffia*, and *Wolffiella* to one another. A substantial amount of plastid DNA sequence data has upheld a phylogeny for the family that is mostly consistent, and many of those relationships have been corroborated by the recent addition of nuclear DNA data. Morphologically, the genera lacking roots (*Wolffia* and *Wolffiella*) comprise a single lineage, as do the three largest genera (*Lemna*, *Wolffia*, and *Wolffiella*) that are more reduced in comparison with *Landoltia* and *Spirodela*. The biogeography of Lemnaceae indicates that numerous dispersal events have occurred in relatively recent evolutionary time, and that several species essentially are cosmopolitan. Although not particularly speciose, duckweeds comprise a surprisingly diverse group with much potential for exploring various genetic, biochemical, and metabolic phenomena.

## 2.1 Introduction

Duckweeds (family Lemnaceae) are a common floristic element of freshwater aquatic habitats worldwide, where they often cover large portions of the water surface. Their growth form, in which the entire plant body floats on or is suspended beneath the water surface, is uncommon among aquatic plants, and their extreme morphological reduction makes them truly unique. Among the reduced features are leaves and stems, which are collectively subsumed into a simplified structure known as a "frond," and roots, which are entirely lacking in two genera (*Wolffia* and *Wolffiella*). Vegetative reproduction commonly occurs by the budding of additional fronds that can separate from the parent frond, as well as by producing

N. P. Tippery
Department of Biological Sciences, University of Wisconsin, Whitewater, USA

D. H. Les (✉)
Department of Ecology and Evolutionary Biology, University of Connecticut, Connecticut, USA
e-mail: les@uconn.edu



X. H. Cao et al. (eds.), *The Duckweed Genomes*, Compendium of Plant Genomes,
https://doi.org/10.1007/978-3-030-11045-1_2

smaller, specialized structures that are more tolerant of freezing and desiccation. Categorized as turions (Landolt 1986), these unusual structures actually represent modified whole plants (Sculthorpe 1967) and often are induced under periods of nutrient deficiency or attenuating daylight (Landolt and Kandeler 1987).

Duckweeds also are capable of sexual reproduction, which is accomplished via tiny flowers that self-pollinate, or are pollinated above the water surface by small arthropod vectors such as aphids, flies, mites, and spiders; despite some reports to the contrary, they are not water-pollinated (Landolt 1986, 1992a, b). Most species produce only one tiny seed per ovary (Landolt 1986). Duckweed fronds range in size from 0.5 to 15 mm in length, with the miniscule members of the genus *Wolffia* representing the smallest of all angiosperm species.

Duckweeds evolved from an ancestor that likely resembled modern aroid plants (e.g., *Arisaema*, *Calla*, *Orontium*, *Philodendron*), and they diversified into 37 currently recognized species (Table 2.1; Cabrera et al. 2008; Cusimano et al. 2011; Nauheimer et al. 2012; Sree et al. 2016). The five genera are rather easily distinguished using vegetative features: *Landoltia* (dotted duckmeat) and *Spirodela* (duckmeat) species have relatively large, floating fronds each with multiple roots; *Lemna* (duckweed) fronds are mostly floating (except for the suspended *L. trisulca*) with one root each; *Wolffia* (watermeal) species have small and stout fronds lacking roots; and *Wolffiella* (mudmidget) plants also lack roots and have small, elongated fronds. (In this chapter, we will use the term "duckweed" to refer broadly to plants in the family Lemnaceae, although this common name sometimes is used more specifically to refer only to members of the genus *Lemna*.) Ecologically, duckweeds are an important component of aquatic habitats: They grow densely and shade out large portions of shallow, still water (Sculthorpe 1967; Landolt 1986; Parr et al. 2002; Roijackers et al. 2004; Driever et al. 2005), and are an important food source for waterfowl, fish, and other herbivores (Sculthorpe 1967; Rusoff et al. 1980).

Duckweed plants are distributed in temperate and tropical latitudes worldwide (Landolt 1986). In temperate environments, they may produce turions or rely more heavily upon seeds to survive colder winter temperatures (Crawford et al. 2006). They are able to disperse rather easily as seeds or fronds, because their small size enables them to adhere to the legs and feathers of waterfowl (Jacobs 1947; Hillman 1961; Landolt 1986; Les et al. 2003; Coughlan et al. 2014, 2017; Green 2016). Duckweeds frequently reproduce vegetatively, and flowering, when it occurs, may be influenced by day length (Hillman 1976).

Duckweeds have become recognized in recent years for their potential applications in biofuels, environmental remediation, carbon sequestration, and biochemical production (Bonomo et al. 1997; Stomp 2005; Cheng and Stomp 2009; Ge et al. 2012; Xu et al. 2012; Su et al. 2014; Cui and Cheng 2015). They are capable of rather rapid vegetative multiplication, require minimal nutrients (Lasfar et al. 2007), are readily cultured (e.g., Oron 1994), and are amenable to genetic manipulation (Yamamoto et al. 2001; Cantó-Pastor et al. 2015). Besides their potential applications for human benefit, duckweeds also offer a study system in which to investigate ecological and evolutionary questions (Crawford et al. 2006; Laird and Barks 2018). Although a few species (e.g., *Lemna gibba*, *Spirodela polyrhiza*) have received the most study, the diversity of Lemnaceae could enable more varied research applications in the future.

## 2.2 Taxonomy

Currently, there are 13 accepted species of *Lemna*, 11 species of *Wolffia*, ten of *Wolffiella*, two species of *Spirodela*, and one of *Landoltia*, for a total of 37 duckweed species worldwide (Sree et al. 2016; Table 2.1). The first duckweed species to be recognized by Linnaeus (1753) all were classified within the single genus *Lemna*. Authors subsequently segregated species among *Wolffia* (named originally by Schreber in 1791, but not validly published until Schleiden in

**Table 2.1** Currently accepted Lemnaceae species, following Sree et al. (2016)

| Subfamily | Genus | Section | Species |
|---|---|---|---|
| Lemnoideae Engl. | *Landoltia* Les & Crawford | | *L. punctata* (G.Mey.) Les & D.J. Crawford |
| | *Lemna* L. | *Alatae* Hegelm. | *L. aequinoctialis* Welw. |
| | | | *L. perpusilla* Torr. |
| | | *Biformes* Landolt | *L. tenera* Kurz |
| | | *Lemna* | *L. disperma* Hegelm. |
| | | | *L. gibba* L. |
| | | | *L. japonica* Landolt |
| | | | *L. minor* L. |
| | | | *L. obscura* (Austin) Daubs |
| | | | *L. trisulca* L. |
| | | | *L. turionifera* Landolt |
| | | *Uninerves* Hegelm. | *L. minuta* Kunth |
| | | | *L. valdiviana* Phil. |
| | | | *L. yungensis* Landolt |
| | *Spirodela* Schleid. | | *S. intermedia* W.Koch |
| | | | *S. polyrhiza* (L.) Schleid. |
| Wolffioideae Engl. | *Wolffia* Horkel ex Schleid. | [unassigned] | *W. australiana* (Benth.) Hartog & Plas |
| | | [unassigned] | *W. borealis* (Engelm. ex Hegelm.) Landolt |
| | | *Pigmentatae* Landolt | *W. brasiliensis* Wedd. |
| | | *Pseudorrhizae* Landolt | *W. microscopica* (Griff.) Kurz |
| | | *Wolffia* | *W. angusta* Landolt |
| | | | *W. arrhiza* (L.) Horkel ex Wimm. |
| | | | *W. columbiana* H.Karst. |
| | | | *W. cylindracea* Hegelm. |
| | | | *W. elongata* Landolt |
| | | | *W. globosa* (Roxb.) Hartog & Plas |
| | | | *W. neglecta* Landolt |
| | *Wolffiella* Hegelm. | *Rotundae* Landolt | *W. rotunda* Landolt |
| | | *Stipitatae* Hegelm. | *W. hyalina* (Delile) Monod |
| | | | *W. repanda* (Hegelm.) Monod |
| | | *Wolffiella* | *W. caudata* Landolt |
| | | | *W. denticulata* (Hegelm.) Hegelm. |
| | | | *W. gladiata* (Hegelm.) Hegelm. |
| | | | *W. lingulata* (Hegelm.) Hegelm. |
| | | | *W. neotropica* Landolt |
| | | | *W. oblonga* (Phil.) Hegelm. |
| | | | *W. welwitschii* (Hegelm.) Monod |

1844), *Spirodela* (Schleiden 1839), *Wolffiella* (Hegelmaier 1895), and finally *Landoltia* (Les and Crawford 1999). Landolt (1986) contributed the most comprehensive modern treatment of the genus, recognizing 34 species in four genera (i.e., prior to the establishment of *Landoltia*). Subsequent to his monograph, Landolt named four additional species (*Lemna yungensis*, *Wolffia cylindracea*, *Wolffia neglecta*, *Wolffiella caudata*; Landolt 1992a, b, 1994, 1998). Additionally, he merged *Lemna ecuadoriensis* with *L. obscura* (Landolt 2000), bringing the most recently accepted number of species to 37 (Sree et al. 2016). Numerous synonyms exist for genera and species of Lemnaceae, and these were outlined clearly by Landolt (1986) and then later by Sree et al. (2016). In this treatment, we provisionally exclude *Lemna landoltii* Halder & Venu, a recently described taxon with affinity to *L. perpusilla* but known to grow in only one pond in India (Halder and Venu 2012) and currently uncorroborated by molecular sequence data.

The majority of duckweed species are clearly differentiated morphologically and resolve as monophyletic. However, a series of investigations have used broader taxon sampling, along with plastid gene sequences and amplified fragment length polymorphism (AFLP) data, to show that a small number of species are incompletely differentiated. These methods support the independence of *Landoltia punctata*, *Spirodela intermedia*, and *S. polyrhiza* (Bog et al. 2015). However, in the genus *Lemna*, *L. valdiviana* and *L. yungensis* are intertwined, and *L. gibba* associates variously with *L. japonica* and *L. turionifera*, potentially as a result of interspecific hybridization (Bog et al. 2010). *Wolffia globosa* clusters with *W. neglecta* and *W. borealis*, whereas plastid data place some *W. elongata* accessions with *W. columbiana* (Bog et al. 2013). Study of the genus *Wolffiella* has revealed extensive genetic overlap among two species groups, *W. gladiata* + *W. lingulata* + *W. oblonga* and *W. hyalina* + *W. repanda* + *W. rotunda* (Bog et al. 2018). Tippery et al. (2015) also sampled multiple accessions for some of these species and recovered consistent plastid and nuclear-based topologies, but with varied levels of internal support for specific associations.

With the exception of *Wolffia*, all other duckweed genera are clearly monophyletic and well differentiated using molecular data (Les et al. 2002; Borisjuk et al. 2015; Tippery et al. 2015). In *Wolffia*, the species *W. australiana*, *W. borealis*, *W. brasiliensis*, and *W. microscopica* resolve within a modestly supported *Wolffia* clade on the plastid phylogeny (Les et al. 2002); however, the nuclear ribosomal phylogeny renders these taxa as unresolved, showing uncertain affinity for either *Wolffia* or *Wolffiella* (Tippery et al. 2015). Although molecular data have raised some concerns regarding the monophyly of *Wolffia*, in light of the morphological distinctness of this genus and the lack of a strongly supported phylogenetic alternative, we continue to recommend the generic recognition of *Wolffia* without taxonomic alteration from its current status.

Several taxonomic sections have been established within Lemnaceae genera. Some of these are congruent with phylogenetic relationships, whereas others have been reevaluated or discarded because of more recent evidence. For maximum clarity, we will highlight only those sections that are reasonably well supported by phylogenetic evidence (Table 2.1). Within *Wolffiella*, Hegelmaier (1868) segregated *W. hyalina* and *W. repanda* into section *Stipitatae*, and Landolt (1986) placed *W. rotunda* into section *Rotundae*, with the remaining species comprising section *Wolffiella*. *Lemna* section *Uninerves* (Hegelmaier 1895) contains *L. minuta*, *L. valdiviana*, and *L. yungensis*, while *L. aequinoctialis* and *L. perpusilla* belong to section *Alatae* (Hegelmaier 1895). *Lemna tenera* is the sole species in section *Biformes* (Landolt 1986), and the remaining species occupy section *Lemna*. In genus *Wolffia*, section *Wolffia* contains seven species that are consistently monophyletic in molecular data analyses, whereas the four poorly resolved species have been assigned variously to section *Pseudorrhizae* (*W. microscopica*; Landolt 1986), the paraphyletic section *Pigmentatae* (*W. borealis* and *W. brasiliensis*; Landolt 1986), or are unassigned (*W. australiana*; Les et al. 2002).

The taxonomic disposition of family Lemnaceae has become controversial in recent years, following phylogenetic studies that merged Lemnaceae within a more broadly inclusive Araceae (Cabrera et al. 2008; Cusimano et al. 2011; Nauheimer et al. 2012; Henriquez et al. 2014). Analyses using extensive plastid DNA data indicate that the most recent common ancestor of the core Araceae and the morphologically divergent genera *Gymnostachys*, *Lysichiton*, *Orontium*, and *Symplocarpus* also gave rise to duckweeds. Thus, to preserve Araceae as a monophyletic family, the broader community (including the influential Angiosperm Phylogeny Group; APG IV 2016) has opted to sink duckweeds within Araceae. However, we continue to advocate the continued recognition of Lemnaceae as a distinct family for several reasons.

First, the available phylogenetic evidence supporting the descent of Lemnaceae from within Araceae cannot be considered comprehensive. Previous phylogenetic analyses of Araceae used only plastid sequence data (Cusimano et al. 2011; Nauheimer et al. 2012), without sampling any nuclear loci. Plastid phylogenies for Araceae sensu lato consistently show an alternative topology for duckweeds that places *Lemna* as sibling to *Landoltia* + *Wolffia* + *Wolffiella*, rather than showing the topology obtained in duckweed-specific studies, i.e., *Landoltia* sibling to *Lemna* + *Wolffia* + *Wolffiella* (Les et al. 2002; Tippery et al. 2015). Numerous examples exist of nuclear and plastid phylogenetic discordance across land plants (e.g., Tippery and Les 2011; Bruun-Lund et al. 2017), as well as persistent taxonomic issues that have not been resolved unambiguously using molecular data (e.g., Mathews 2009; Wickett et al. 2014). Moreover, the Araceae s.l. studies recovered exceptionally long-branch lengths for all Lemnaceae genera and highlighted their morphological distinctness in fundamental traits such as pollen structure and chromosome number (Cusimano et al. 2011). Although maximum likelihood and Bayesian phylogenetic methods are less sensitive to long-branch artifacts (Gaut and Lewis 1995; Philippe et al. 2005), there remains evidence that long-branch lineages should be evaluated with caution (Kück et al. 2012).

Secondly, even if the phylogeny depicted by some authors (e.g., Cabrera et al. 2008; Cusimano et al. 2011) correctly depicts evolutionary relationships, a morphologically more satisfying taxonomy could be achieved by recognizing family Orontiaceae R.Br. ex Lindl. (i.e., Gymnostachydoideae Bogner & Nicolson and Orontioideae Mayo, Bogner & P.C.Boyce) as sibling to Lemnaceae + Araceae. Orontiaceae long have been considered distinct from "true" Araceae (sensu Mayo et al. 1997; Cusimano et al. 2011), and they differ in some noteworthy features, such as largely dimerous flowers, unidirectional tepal and stamen development, and orthotropous ovule orientation, although these features also are homoplasic among Araceae s.l. genera (Grayum 1991; Buzgo 2001). The circumscription of family Orontiaceae is not without precedent. Brown (1810) initially published "Orontiaceae" as a section of family "Aroideae" (i.e., Araceae), and later, Lindley (1846) elevated Orontiaceae to the level of "order," at a level equivalent to "order" Araceae. The International Code of Nomenclature (ICN; Turland et al. 2018) Art. 18.2 allows such "orders" to be treated as validly published families, if that was the intent of the author. Given the equivalent presentation of Araceae by Lindley, it appears quite clear that he intended for Orontiaceae to be considered at the rank of family also. Segregating Orontiaceae from Araceae s.l. would allow a distinct lineage to be understood more clearly in terms of its morphological evolution and biogeography. In contrast, sinking the undeniably distinct Lemnaceae within a broader Araceae stands to upset the morphological uniformity of Araceae.

Finally, the amount of nomenclatural confusion would be reduced if the taxa of Orontiaceae were removed from Araceae. The family Orontiaceae contains only nine species in four genera (*Gymnostachys*: 1, *Lysichiton*: 2, *Orontium*: 1, *Symplocarpus*: 5), whereas Lemnaceae comprise 37 species in five genera. The concept of family Lemnaceae extends back nearly two centuries (Gray 1821; IPNI 2018), and over that time, there has been little confusion about what species belong to the family. We argue that family Lemnaceae deserves to continue as a unique angiosperm family, for the sake of nomenclatural stability and morphological clarity.

## 2.3 Phylogenetic Relationships

Because of their extreme morphological reduction, duckweeds traditionally were difficult to place among other angiosperms. Fossils of apparent duckweed relatives have supported a link with the large family Araceae (Kvaček 1995; Stockey et al. 1997, 2007, 2016; Coiffard and Mohr 2018), and various morphological data also support the current understanding of Lemnaceae as closely related to Araceae (summarized by Les et al. 2002). Ongoing molecular phylogenetic studies (Cabrera et al. 2008; Cusimano et al. 2011; Henriquez et al. 2014) have begun to portray a consistent phylogenetic position for Lemnaceae, albeit based solely on plastid DNA data.

Duckweeds and other Araceae belong to the monocot order Arales Dumort., containing Araceae, Lemnaceae, and Orontiaceae (Les and Tippery 2013). Arales are closely related to the diverse order Alismatales Dumort., which comprises predominantly aquatic plants (e.g., *Aponogetonaceae*, *Hydrocharitaceae*, *Potamogetonaceae*), as well as family Tofieldiaceae, a family of uncertain phylogenetic position (Les and Tippery 2013). In an analysis that combined extensive molecular data with numerous fossil calibration points, Magallón et al. (2015) noted that Arales diverged from alismatid families approximately 128.9 Ma (i.e., Arales stem node). In a prior study, Nauheimer et al. (2012) obtained a similar stem node estimate (135 Ma), and also provided a crown node estimate for the diversification of Arales at 121.7 Ma, as well as estimates for the stem (103.6 Ma) and crown (73.4 Ma) ages of Lemnaceae.

We extended the divergence time estimates from the study by Nauheimer et al. (2012) by applying their crown age estimate to a duckweed tree that was generated using plastid and nuclear data (Tippery et al. 2015). We converted the tree to be ultrametric using the *chronos* function of the *ape* package in R, which employs the penalized likelihood method of tree calibration (Sanderson 2002; Kim and Sanderson 2008; Paradis 2013; R Core Group 2018). Using this strategy, we dated the crown node diversification of *Spirodela* to 35.5 Ma, the stem node divergence of *Landoltia* to 56.8 Ma, as well as nodes involving the more species-rich genus *Lemna* (stem 54.4 Ma, crown 41.7). Because the phylogeny does not clearly differentiate the genera *Wolffia* and *Wolffiella*, we only could date the crown diversification ages of *Wolffia* sect. *Wolffia* (8.0 Ma) and genus *Wolffiella* (15.6 Ma) (Fig. 2.1).

Extensive molecular data have been obtained for all duckweed species, and they largely support a single phylogenetic hypothesis. Early studies (Crawford and Landolt 1995; Crawford et al. 1996, 1997, 2005) used allozyme data to evaluate interspecific boundaries, and these upheld the taxonomy used by Landolt (1986). More recently, Bog et al. (2010, 2013, 2015, 2018) also studied multiple individuals of closely related species, using AFLP and plastid data, and identified a number of species groups with potential interbreeding or incomplete divergence (see Taxonomy section above). Comprehensive phylogenetic studies of Lemnaceae have employed flavonoid biochemical data (Les et al. 1997), as well as DNA sequence data from plastid (*rbcL*, *matK/trnK*, *rpl16*, *rps16*; Jordan et al. 1996; Les et al. 2002; Martirosyan et al. 2009) and nuclear (internal transcribed spacer (ITS); Tippery et al. 2015) gene regions. Molecular data helped to justify establishing *Landoltia* as a separate genus (Les and Crawford 1999), and they have provided important support for modern sectional classifications of the larger genera.

The genera *Spirodela* and *Landoltia* are easily differentiated from each other and from other duckweeds using morphological or molecular data (Les et al. 1997, 2002; Les and Crawford 1999; Tippery et al. 2015). Among the larger genera, *Wolffiella* consistently has received strong support in phylogenetic analyses for being monophyletic, and also for the monophyly and

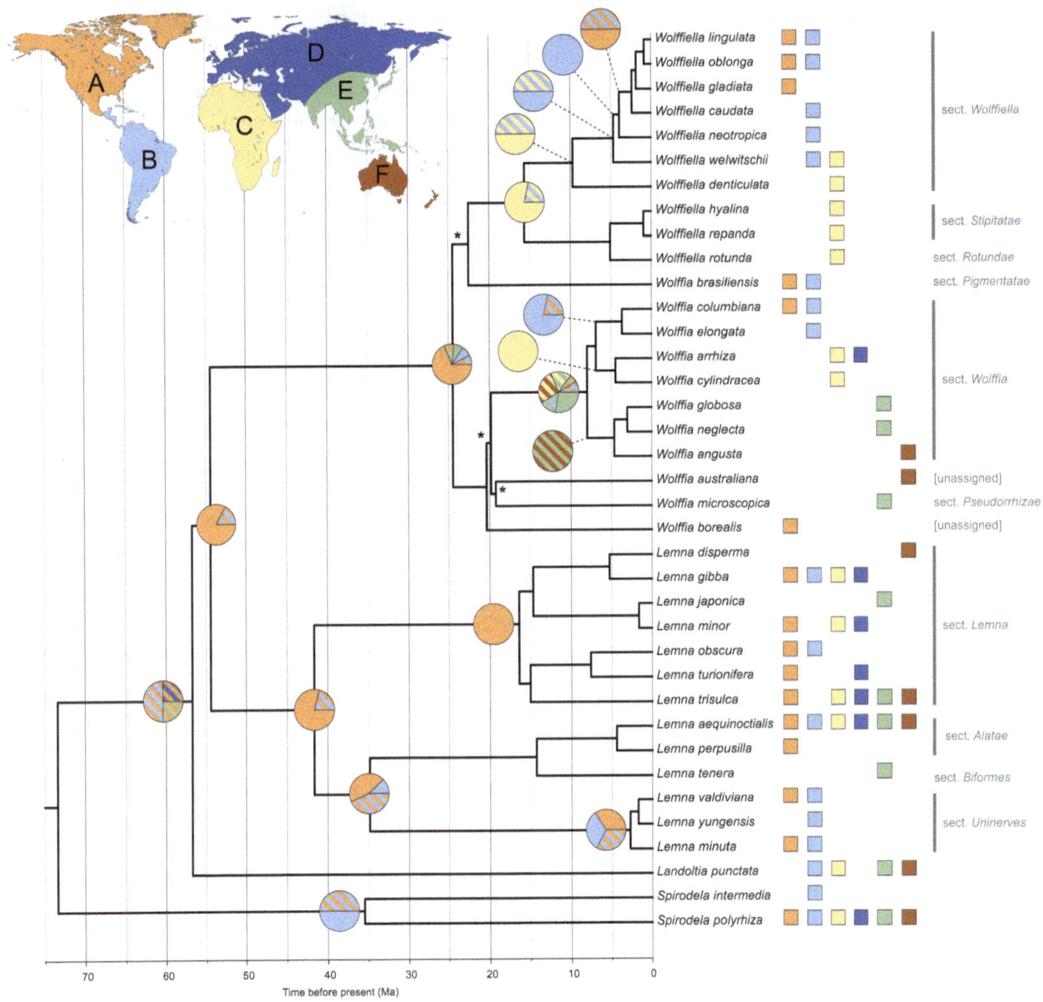


**Fig. 2.1** Phylogeny of Lemnaceae, constructed using molecular sequence data from plastid and nuclear gene regions. The tree reflects relationships that were determined previously (Tippery et al. 2015), with branches made ultrametric to enable estimation of past divergence times. Asterisks indicate nodes that received <0.95 Bayesian posterior probability support; all other nodes depicted received Bayesian posterior probability $\geq 0.99$ (and typically also maximum likelihood bootstrap support $\geq 95\%$). Note that the resolution of *Wolffia brasiliensis* with *Wolffiella* is likely an analytical artifact rather than an indication of its actual phylogenetic association (see text for details). Ancestral area biogeographic reconstructions are shown at taxonomically significant nodes, color-coded according to the areas shown on the inset map. Pie charts show the relative proportion of reconstructions that include a particular geographic area; reconstructions that include two areas are colored with diagonal stripes

distinctness of sections *Rotundae*, *Stipitatae*, and *Wolffiella* (Les et al. 1997, 2002; Tippery et al. 2015). The monophyly of *Lemna* also has been confirmed by molecular data, and sections *Alatae*, *Biformes*, *Lemna*, and *Uninerves* are phylogenetically distinct. Section *Hydrophylla* Dumort., containing only the submersed species *L. trisulca* (Landolt 1986), is nested within section *Lemna* and thus cannot be considered taxonomically acceptable. Whereas the other genera are rather cleanly differentiated in phylogenetic analyses, the monophyly of genus *Wolffia* is less certain. Analysis of plastid data gave modest support to a monophyletic *Wolffia* (Les et al. 2002), but the addition of nuclear data caused *W. brasiliensis* to associate more strongly with *Wolffiella* than with other *Wolffia* species (Tippery et al. 2015). Landolt (1986) included *W. brasiliensis* in section *Pigmentatae* along with *W. borealis*, but these two species do not associate strongly in phylogenetic analyses. *Wolffia australiana* and *W. microscopica*, the latter belonging to section *Pseudorrhizae* (Landolt 1986), likewise have uncertain relationships with other *Wolffia* species (Les et al. 2002; Tippery et al. 2015). The phylogenetic distinctness of these four species is such that each might merit its own monotypic section; however, we do not advocate for any taxonomic revisions at this time. We do retain the sectional classification according to ICN guidelines, whereby named sections must contain their respective type species. Thus, *W. microscopica* belongs to section *Pseudorrhizae*. We here designate *W. brasiliensis* to be the type species of section *Pigmentatae*, because Landolt (1986), when establishing the section, mentioned that Hegelmaier (1868, 1895) originally considered *W. brasiliensis* to be the sole representative of this group. The remaining *Wolffia* species, which belong to section *Wolffia*, are comfortably monophyletic and distinct from other lineages.

Some phylogenetic uncertainty inevitably persists, even after analyzing over 5000 nucleotides of DNA sequence data (Tippery et al. 2015). Next-generation sequencing methods are poised to drastically amplify the amount of phylogenetically informative sequence data in duckweeds (Appenroth et al. 2015). Researchers have obtained whole-genome nucleotide sequence data for *Spirodela polyrhiza* (Wang et al. 2012, 2014) and two species of *Lemna* (*L. gibba* and *L. minor*; lemna.org), as well as plastid genome sequences for *Wolffia australiana* and *Wolffiella lingulata* (Wang and Messing 2011). Such large-scale data may be able to resolve persistent phylogenetic uncertainty among duckweed lineages.

## 2.4 Morphological Evolution

The 37 duckweed species share a large number of distinctive morphological characters, and they clearly evolved from a single common ancestor. Duckweeds are exceptional among flowering plants for having their vegetative and reproductive organs severely reduced in size, and in some cases lost entirely (e.g., the roots of *Wolffia* and *Wolffiella*). A unique anatomical synapomorphy for Lemnaceae among other Arales is the existence of pollen with an ulcerate aperture (Keating 2002; Cusimano et al. 2011). Prior to the advent of molecular data, many authors considered water lettuce (*Pistia*, Araceae) to be the closest relative of duckweeds, because it shares a pleustonic habit (i.e., with the entire plant floating on the water surface), reduced flower number, and similar leaf morphology. *Pistia* indeed is a close relative of duckweeds, but no closer than any other "true" Araceae (sensu Mayo et al. 1997; Cusimano et al. 2011), and the morphological similarity between the two kinds of plants resulted from convergent evolution rather than descent from a morphologically similar ancestor. Instead of having a pleustonic growth form, the ancestor that gave rise to duckweeds would have looked more like a modern aroid, with an erect habit supported by terrestrial roots, basal leaves, and a multi-flowered inflorescence (i.e., features that are widespread in Araceae, Orontiaceae, and Tofieldiaceae).

On morphological grounds, duckweed genera are split broadly between two groups: plants that produce roots (*Landoltia*, *Lemna*, and *Spirodela*) belong to subfamily Lemnoideae, and those lacking roots (*Wolffia* and *Wolffiella*) belong to subfamily Wolffioideae (Landolt 1986). Phylogenetic data support Landolt's (1986) assertion that rooted plants represent the ancestral condition, with rootless plants being more derived. Moreover, the number of roots per frond follows a reduction series as lineages branch off: *Spirodela* (8–21 roots) branches first, followed by *Landoltia* (2–7 roots), then *Lemna* (one root), then the two rootless genera. *Wolffia* species differ from *Wolffiella* by having thick, nearly globose fronds and flowers that are located medially (as opposed to off-center) on the upper frond surface (Landolt 1986).

As the lineage most distantly related to the other Lemnaceae, it is tempting to interpret *Spirodela* as having morphological features that are ancestral for the family, yet it is important to bear in mind that the genus also acquired synapomorphic traits during its evolution. *Spirodela* species have fronds with the largest number of veins and roots, and at least some of their roots do not perforate the associated scale-like leaf. *Landoltia punctata* is a distinct duckweed lineage, with fewer leaves per frond than *Spirodela* species, and roots that all perforate the associated leaf (Landolt 1986).

There are several synapomorphies that help define the clade of *Lemna* + *Wolffia* + *Wolffiella*. Species belonging to these genera have a disporic embryo sac (versus monosporic in *Landoltia* and *Spirodela*), and they also differ from other Lemnaceae by lacking dorsal/ventral scales.

*Lemna* species, in addition to their reduced vein number and root number, differ from other Lemnaceae by lacking root tracheids, guard cell plastids, and anther pigment cells. Their crystal cells consist only of raphides, without the additional druses that are produced by *Landoltia* and *Spirodela* (Landolt 1986). Phylogenetic data support the monophyly of section *Alatae* (*L. aequinoctialis* and *L. perpusilla*), containing two species with short, sharply pointed roots with sheaths that are winged at the base. *Lemna tenera* constitutes the sole species in section *Biformes*, and it differs from most other *Lemna* species in having elongate fronds that commonly grow submersed. Obvious synapomorphies are lacking for the sibling sections *Alatae* and *Biformes*. Species in the monophyletic section *Uninerves* (*L. minuta*, *L. valdiviana*, *L. yungensis*) share the trait of having fronds with only one nerve each (Landolt 1986, 1998). Sections *Alatae*, *Biformes*, and *Uninerves* together are monophyletic. The remaining *Lemna* species belong to section *Lemna*, including the morphologically distinct *L. trisulca*, which molecular data cannot sufficiently distinguish from the rest of the section. A number of morphological characters do distinguish section *Lemna*, however, including relatively long (>3 cm) roots with sheaths that are not winged and tips that are mostly rounded; fronds often are dotted or completely red beneath, and ovules are orthotropous.

*Wolffia* and *Wolffiella* are sibling lineages assigned to subfamily Wolffioideae (Landolt 1986). Besides lacking roots and frond veins, Wolffioideae have a great deal of morphological features in common that distinguish them from the rest of Lemnaceae, including the lack of a floral organ prophyllum and crystal cells, and having a single reproductive pouch (versus two in the rest of the family), unilocular anthers (versus bilocular), ribbed seeds (versus smooth), a single stamen per flower (versus two), apical anther dehiscence (versus transverse), and an ovary inserted at the base of the stamen (versus above the stamens).

Molecular data suggest that species of *Wolffia* do not all comprise a clade, but they share quite a few features, including thick fronds and flowers that emerge from the middle of the frond. Turions are widely present in genus *Wolffia*, with the exception of *W. microscopica*, whereas turions otherwise are found in only a few Lemnaceae species (*S. polyrhiza*, *Lemna aequinoctialis*, and *L. turionifera*). The *Wolffia* species that are most difficult to place phylogenetically also are distinct morphologically. In addition to being the only *Wolffia* species without turions, *W. microscopica* (the sole species in section *Pseudorrhizae*) also features a conical appendage on the lower frond surface. The two species belonging to section *Pigmentatae*, *W. borealis* and *W. brasiliensis*, have pigment cells in their vegetative tissue (visible on dead fronds). *Wolffia australiana*, which Landolt (1986) placed in section *Wolffia* on morphological grounds, clearly is separate from that group phylogenetically (Les et al. 1997; Tippery et al. 2015) and remains unassigned to a section. Section *Wolffia*, containing the largest number of species, resolves consistently as monophyletic with strong support; species in this section uniformly have globular to ovoid fronds lacking pigment cells and conical appendages.

The genus *Wolffiella* also resolves consistently as monophyletic. Plants in the genus have flattened fronds and flowers that are located laterally on the upper frond surface. Sections of *Wolffiella* largely have been upheld by molecular data and are morphologically distinct. Section *Stipitatae* (*W. hyalina* and *W. repanda*) includes species having an elongated appendage that originates from the vegetative budding pouch. Sections *Stipitatae* and *Rotundae*, the latter comprising only *W. rotunda*, together are monophyletic and contain the only *Wolffiella* species that have smaller (<3 mm), orbicular to ovate fronds that float on the water surface. Section *Wolffiella* also is monophyletic and contains all other species in the genus, in contrast to sections *Rotundae* and *Stipitatae*; species in section *Wolffiella* have slender, elongated fronds (>3 mm) that grow at least partially submersed.

## 2.5 Biogeography

Several duckweed species have broad distributions that extend over multiple continents (e.g., *S. polyrhiza*, *Landoltia punctata*, *Lemna aequinoctialis*, *L. minor*, *L. trisulca*), whereas others have narrower ranges (e.g., *Lemna tenera*, *Wolffia elongata*, *Wolffiella denticulata*; Fig. 2.2 a–e). Les et al. (2003) evaluated pairs of closely related species and concluded that duckweeds, as well as numerous other aquatic plant groups, effected transoceanic dispersal in the relatively recent past. Duckweeds are capable of external biotic dispersal (i.e., exozoochory), which they accomplish by adhering to the surfaces of ducks and other waterfowl that are known to travel large distances (Jacobs 1947; Hillman 1961; Kimball et al. 2003; Coughlan et al. 2014). Moreover, their habit of growing over large areas of water surface makes it relatively easier for them to attach to a biotic vector, and their extremely small size and propensity for vegetative reproduction increase the likelihood that populations will establish in new territory (Coughlan et al. 2017; Green 2016).

In a phylogenetic context, it becomes apparent just how many times duckweeds have established in novel locations relative to their immediate ancestors. We conducted a biogeographic analysis in RASP (reconstruct ancestral state in phylogenies) ver. 3.2 (Yu et al. 2015), with the S-DIVA reconstruction (Yu et al. 2010), using geographic regions that commonly are not occupied by the same duckweed species (i.e., Americas north of Yucatán; Americas south of Yucatán; Africa; Eurasia west of India and north of Japan/Korea; eastern Eurasia including India, Japan, Korea, and Malaysia; and Australia and New Zealand; Fig. 2.1), to reconstruct the colonization history of duckweed species. The reconstruction was constrained to allow a maximum of two areas to be occupied simultaneously. Dispersal events were plotted onto an ultrametric tree that was calibrated using the duckweed crown age of 73.4 Ma. In their biogeographic analysis of Araceae, Nauheimer et al. (2012) recognized the dispersal ability of aquatic taxa and analyzed duckweeds as having a "water associated" range with high dispersal ability. They reconstructed the common ancestor of all Araceae to be North American, whereas the sibling lineage of duckweeds was reconstructed to have a Eurasian ancestor. Fossil duckweed relatives recovered from late Cretaceous formations in North America (Kvaček 1995; Stockey et al. 1997, 2007, 2016) may represent extinct lineages that predated the clade of extant duckweeds, and it is noteworthy that contemporary fossils putatively related to duckweeds also have been recovered from Africa (Coiffard and Mohr 2018).

In our analysis, the phylogenetic positions of *S. polyrhiza* and *Landoltia punctata*, two species with worldwide distributions (Fig. 2.2a), along with the disparate distribution ranges of related Araceae lineages, resulted in broad uncertainty about the geographic range of the duckweed ancestor (Fig. 2.1). The common ancestor of *Lemna* species (41.7 Ma [crown age]—54.4 Ma [stem age]) was reconstructed with highest probability to be in North America. Within genus *Lemna*, section *Uninerves* was reconstructed to have an ancestor in the Americas (2.7–34.8 Ma), where all three species are native (Fig. 2.2b). The ancestor of sections *Alatae* and *Biformes* likely

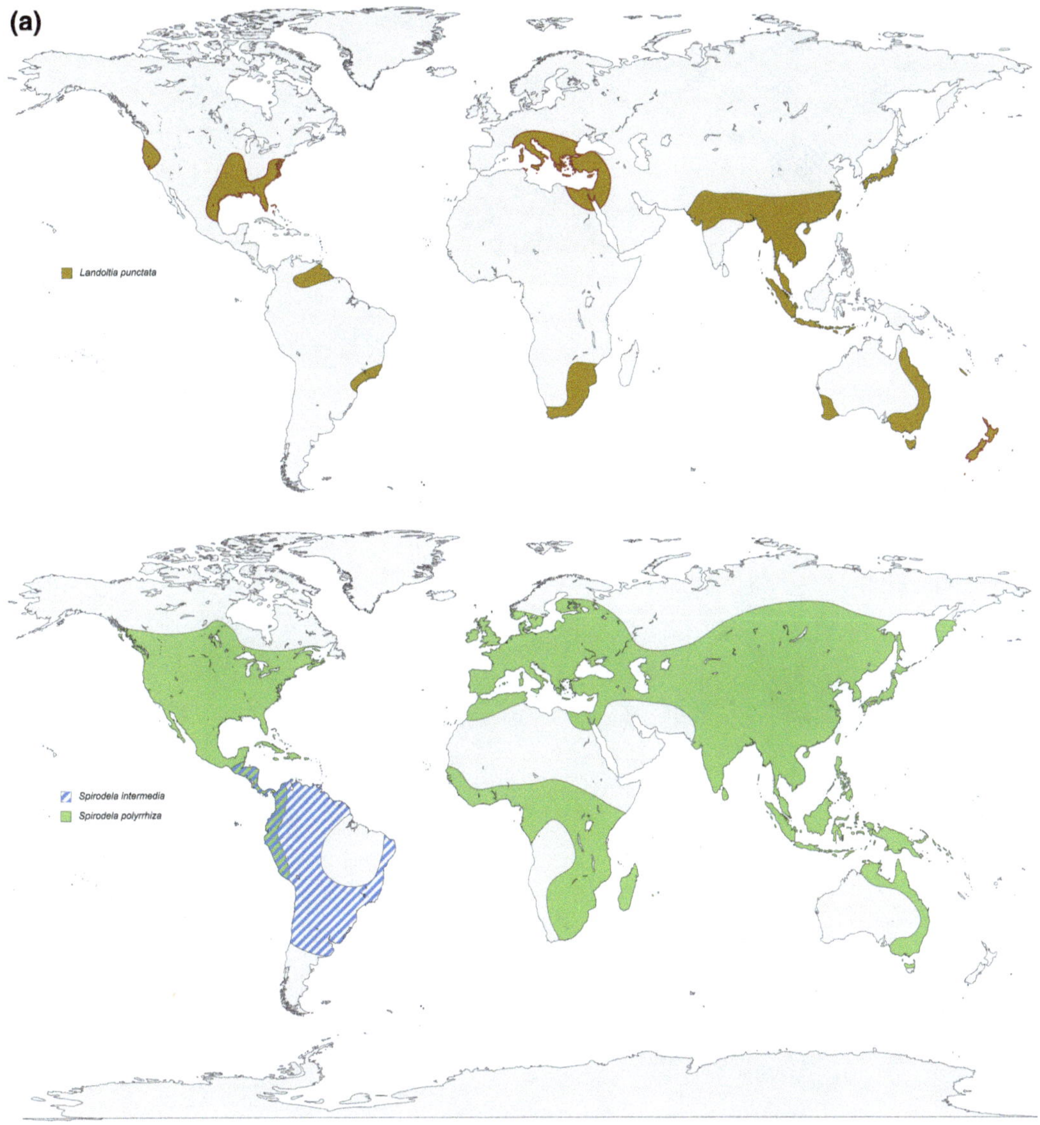


**Fig. 2.2** Global distributions of duckweed species, redrawn from Landolt (1986). Maps show groups of closely related species: **a** *Landoltia* and *Spirodela*, **b** *Lemna* sects. *Alatae*, *Biformes*, and *Uninerves*, **c** *Lemna* sect. *Lemna*, **d** *Wolffia*, and **e** *Wolffiella*

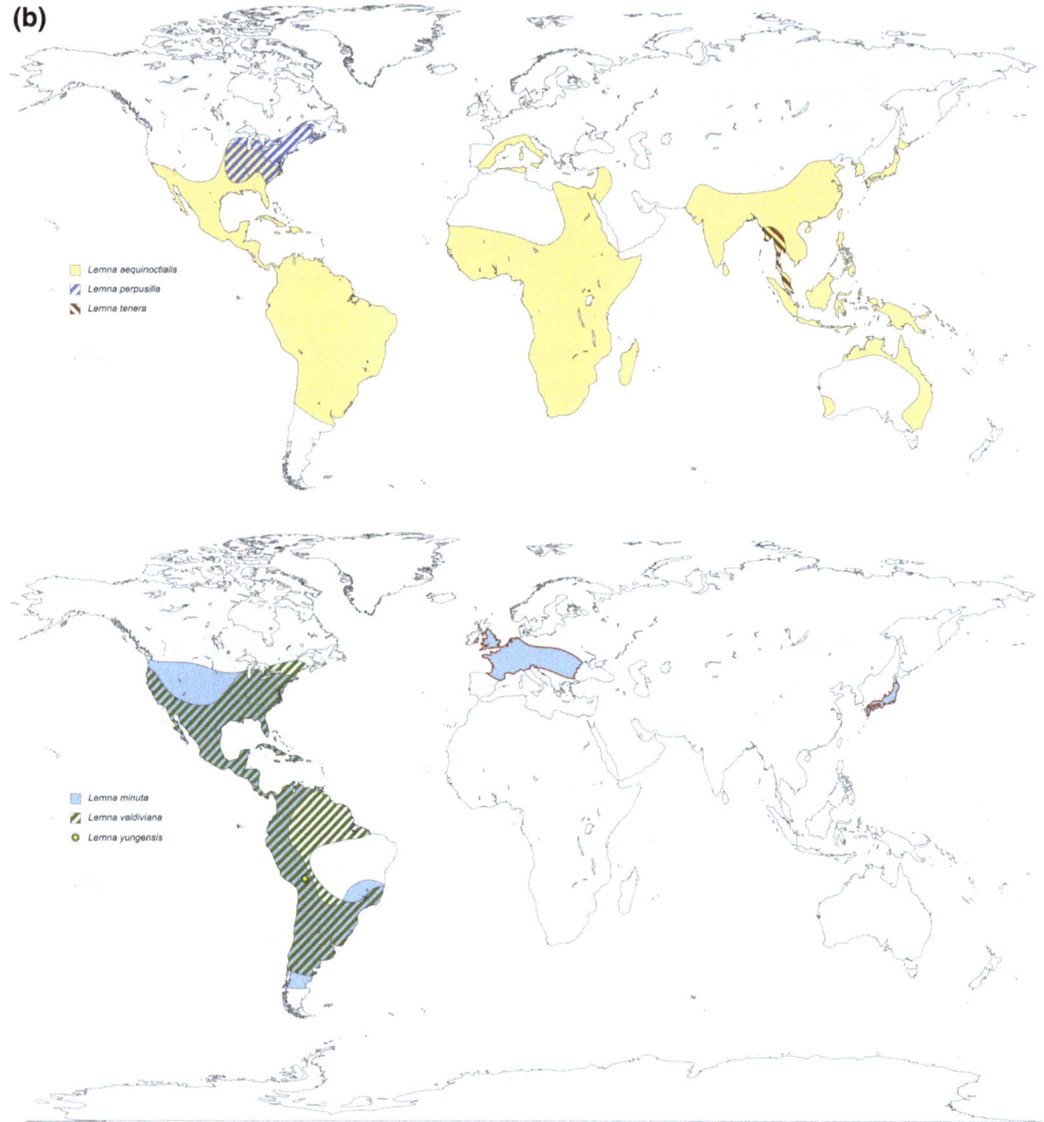


**Fig. 2.2** (continued)

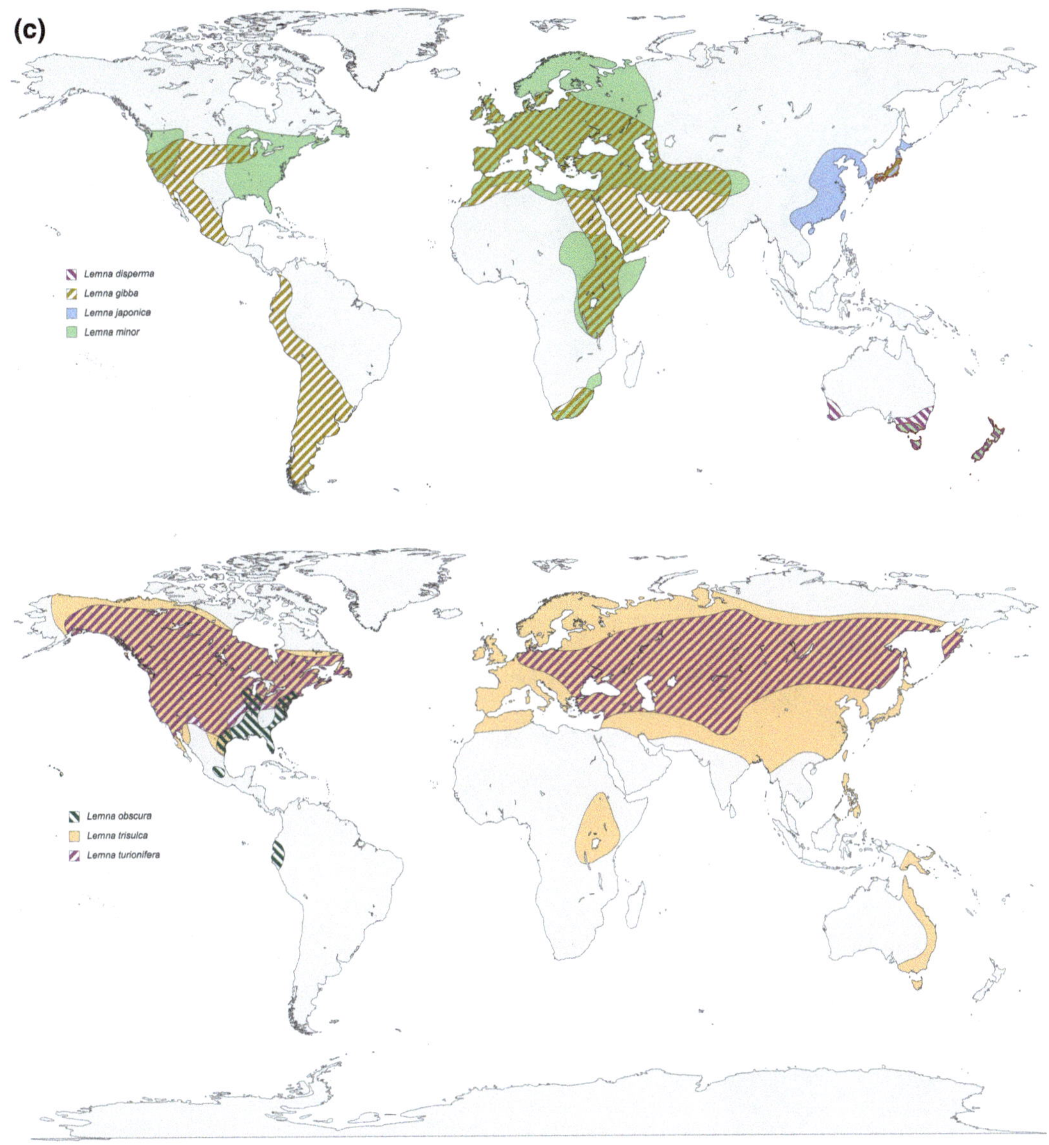


**Fig. 2.2** (continued)

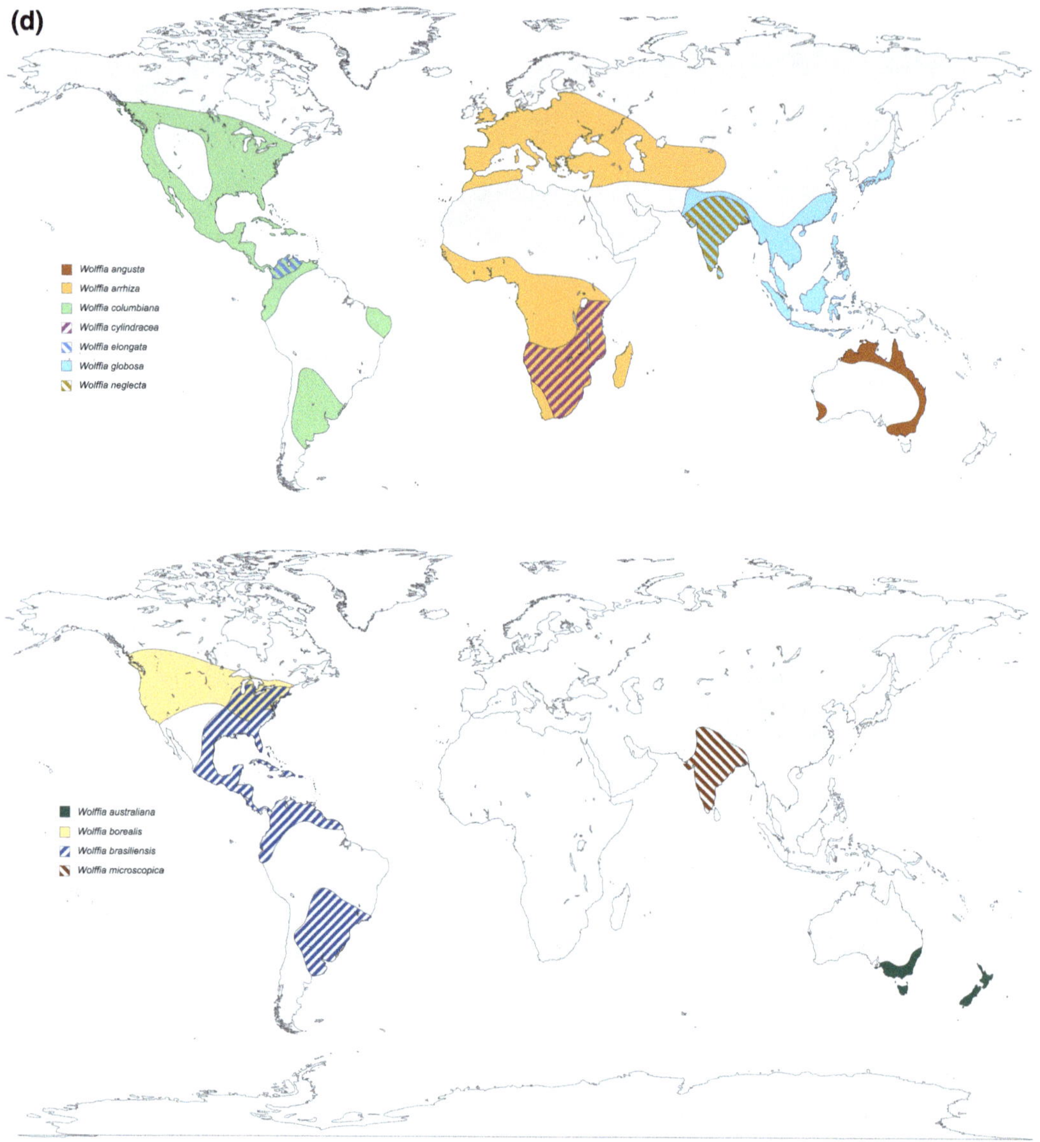


**Fig. 2.2** (continued)

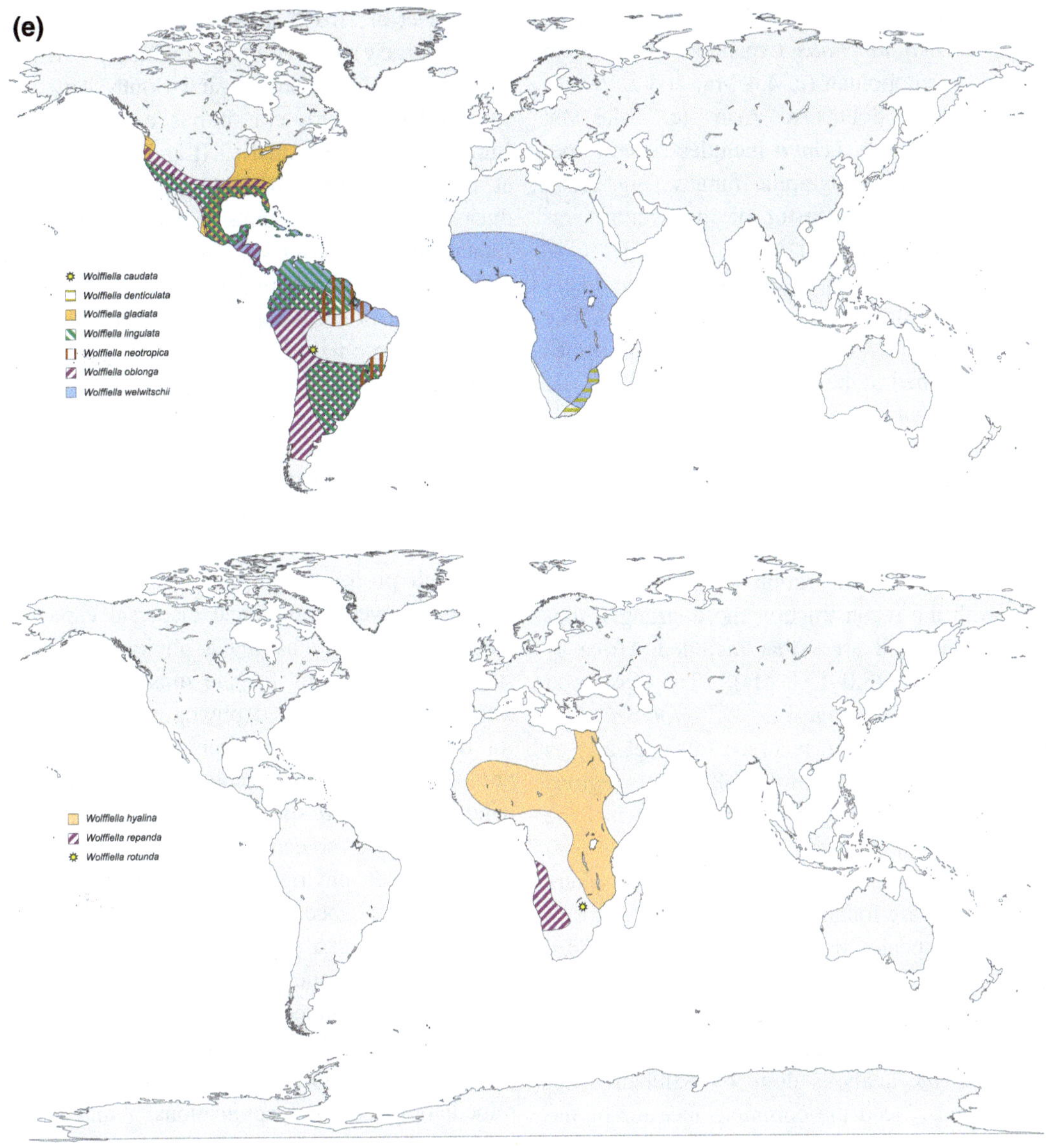


**Fig. 2.2** (continued)

grew in North America (14.3–34.8 Ma), from which *L. aequinoctialis* expanded its range to become cosmopolitan (c. 4.4 Ma) and *L. tenera* dispersed to Southeast Asia (c. 14.3 Ma; Fig. 2.2b). Section *Lemna* includes several species with broad geographic ranges (Fig. 2.2c), and the common ancestor of the section was reconstructed to inhabit North America (16.4–41.7 Ma). The ancestor of Wolffioideae likely grew in the Americas also (24.4–54.4 Ma). The phylogenetically ambiguous *Wolffia* species either remained in the Americas (*W. borealis*, *W. brasiliensis*) or dispersed independently to Australia (*W. australiana*) or eastern Asia (*W. microscopica*) (all dispersals of these species were reconstructed to have occurred roughly 19.2–22.5 Ma). The geographic range for the common ancestor of section *Wolffia* was uncertain, with the reconstruction more strongly supporting ancestral areas that included Africa or eastern Asia (8.0–19.8 Ma). The common ancestor of *W. angusta*, *W. globosa*, and *W. neglecta*, however, resolved rather clearly as east Asian or Australian (4.6–8.0 Ma), regions where these species grow today. Likewise, the sibling species *W. arrhiza* and *W. cylindracea* had an ancestor in Africa (4.5–6.9 Ma) where both species are found today, and the ancestor of the sibling species *W. columbiana* and *W. elongata* likely dispersed to South America (3.7–6.9 Ma) where these species grow today. Our biogeographic reconstruction for *Wolffiella* supported the analyses done by Kimball et al. (2003) and placed the common ancestor of the genus in Africa (15.6–22.5 Ma), where *W. denticulata*, some populations of *W. welwitschii*, and species in sections *Rotundae* and *Stipitatae* grow today. Starting about 3.9–9.7 Ma and possibly coincident with the range expansion of *W. welwitschii*, the ancestor of *W. caudata*, *W. gladiata*, *W. lingulata*, *W. neotropica*, and *W. oblonga* dispersed into the Americas.

Anthropogenic dispersal events have been responsible for the recent introductions of some species. *Lemna minuta*, native to the Americas, is invasive in Europe and Japan (Landolt 2000; Ceschin et al. 2017). *Landoltia punctata* has a rather large native range, including Africa, Southeast Asia, Australia, and South America, and in the modern era, it has expanded into Europe and North America (Landolt 1986; Les et al. 1997; Jacono 2018). Other anthropogenic duckweed dispersal events include *L. gibba* to Japan and *L. minor* to Australasia (Landolt 1986). Given their abundance and capacity for dispersal, duckweeds represent a likely group of plants for colonizing new territory, potentially aided by humans.

## 2.6 Conclusions

Many aspects that make Lemnaceae distinct, from their profound morphological evolution to their extensive geographical dispersal capacity, are enlightened by having a phylogenetic perspective. Duckweeds for the most part have a stable and well-resolved phylogeny that allows for meticulous reconstruction of ancestral morphological states and biogeography. Only a handful of *Wolffia* species remain ambiguously resolved on the molecular phylogeny, and future genomic-level phylogenetic studies may soon resolve these species satisfactorily. Genomic sequence data also stand to offer new perspectives into the unique physiology of duckweeds, and molecular biotechnology may be able to deliver duckweed strains that are maximally productive for biofuel production, wastewater remediation, or other applications. Amid all of the potential new benefits from studying duckweeds, it will be important to retain a phylogenetic perspective, for example, to explore enzyme variants or similar physiological pathways in related species. Individual duckweed species each contain a wealth of biological potential, and taken together, the Lemnaceae are an extensive resource for biotechnology.

**Acknowledgements** We are grateful to Daniel J. Crawford for reviewing the novel biogeographical research presented here and to Paul Fourounjian, Wenqin Wang, and Hieu Cao for editorial assistance.

## References

Apg, IV (The Angiosperm Phylogeny Group) (2016) An update of the Angiosperm Phylogeny Group classification for the orders and families of flowering plants: APG IV. Bot J Linn Soc 181:1–20. https://doi.org/10.1111/boj.12385

Appenroth KJ, Crawford DJ, Les DH (2015) After the genome sequencing of duckweed—how to proceed with research on the fastest growing angiosperm? Plant Biol 17(s1):1–4. https://doi.org/10.1111/plb.12248

Bog M, Baumbach H, Schween U et al (2010) Genetic structure of the genus *Lemna* L. (Lemnaceae) as revealed by amplified fragment length polymorphism. Planta 232:609–619. https://doi.org/10.1007/s00425-010-1201-2

Bog M, Schneider P, Hellwig F et al (2013) Genetic characterization and barcoding of taxa in the genus Wolffia Horkel ex Schleid. (Lemnaceae) as revealed by two plastidic markers and amplified fragment length polymorphism (AFLP). Planta 237:1–13. https://doi.org/10.1007/s00425-012-1777-9

Bog M, Lautenschlager U, Landrock MF et al (2015) Genetic characterization and barcoding of taxa in the genera *Landoltia* and *Spirodela* (Lemnaceae) by three plastidic markers and amplified fragment length polymorphism (AFLP). Hydrobiologia 749:169–182. https://doi.org/10.1007/s10750-014-2163-3

Bog M, Landrock MF, Drefahl D, Sree KS, Appenroth KJ (2018) Fingerprinting by amplified fragment length polymorphism (AFLP) and barcoding by three plastidic markers in the genus *Wolffiella* Hegelm. Plant Syst Evol 304:373–386. https://doi.org/10.1007/s00606-017-1482-z

Bonomo L, Pastorelli G, Zambon N (1997) Advantages and limitations of duckweed-based wastewater treatment systems. Water Sci Technol 35:239–246. https://doi.org/10.1016/S0273-1223(97)00074-7

Borisjuk N, Chu P, Gutierrez R et al (2015) Assessment, validation and deployment strategy of a two-barcode protocol for facile genotyping of duckweed species. Plant Biol 17(s1):42–49. https://doi.org/10.1111/plb.12229

Bruun-Lund S, Clement WL, Kjellberg F et al (2017) First plastid phylogenomic study reveals potential cyto-nuclear discordance in the evolutionary history of *Ficus* L. (Moraceae). Mol Phyl Evol 109:93–104. https://doi.org/10.1016/j.ympev.2016.12.031

Buzgo M (2001) Flower structure and development of Araceae compared with alismatids and Acoraceae. Bot J Linn Soc 136:393–425. https://doi.org/10.1111/j.1095-8339.2001.tb00582.x

Brown R (1810) Prodromus florae Novae Hollandiae et Insulae Van-Diemen. Taylor et Socii, London

Cabrera LI, Salazar GA, Chase MW et al (2008) Phylogenetic relationships of aroids and duckweeds (Araceae) inferred from coding and noncoding plastid DNA. Am J Bot 95:1153–1165. https://doi.org/10.3732/ajb.0800073

Cantó-Pastor A, Mollá-Morales A, Ernst E et al (2015) Efficient transformation and artificial miRNA gene silencing in *Lemna minor*. Plant Biol 17(s1):59–65. https://doi.org/10.1111/plb.12215

Ceschin S, Abati S, Leacche I et al (2017) Ecological comparison between duckweeds in central Italy: The invasive *Lemna minuta* vs the native *L. minor*. Plant Biosyst 152:674–683. https://doi.org/10.1080/11263504.2017.1317671

Cheng JJ, Stomp AM (2009) Growing duckweed to recover nutrients from wastewaters and for production of fuel ethanol and animal feed. Clean-Soil Air Water 37:17–26. https://doi.org/10.1002/clen.200800210

Coiffard C, Mohr BA (2018) Cretaceous tropical Alismatales in Africa: diversity, climate and evolution. Bot J Linn Soc 188:117–131. https://doi.org/10.1093/botlinnean/boy045

Coughlan NE, Kelly TC, Jansen MAK (2014) Mallard duck (*Anas platyrhynchos*)-mediated dispersal of Lemnaceae: a contributing factor in the spread of invasive *Lemna minuta*? Plant Biol 17(s1):108–114. https://doi.org/10.1111/plb.12182

Coughlan NE, Kelly TC, Jansen MAK (2017) “Step by step”: high frequency short-distance epizoochorous dispersal of aquatic macrophytes. Biol Invasions 19:625–634. https://doi.org/10.1007/s10530-016-1293-0

Crawford DJ, Landolt E (1995) Allozyme divergence among species of *Wolffia* (Lemnaceae). Plant Syst Evol 197:59–69. https://doi.org/10.1007/BF00984632

Crawford DJ, Landolt E, Les DH (1996) An allozyme study of two sibling species of *Lemna* (Lemnaceae) with comments on their morphology, ecology and distribution. Bull Torrey Bot Soc 123:1–6. https://doi.org/10.2307/2996300

Crawford DJ, Landolt E, Les DH et al (1997) Allozyme variation and the taxonomy of *Wolffiella*. Aquat Bot 58:43–54. https://doi.org/10.1016/S0304-3770(97)00012-0

Crawford DJ, Landolt E, Les DH et al (2005) Allozyme variation within and divergence between *Lemna gibba* and *L. disperma*: systematic and biogeographic implications. Aquat Bot 83:119–128. https://doi.org/10.1016/j.aquabot.2005.06.001

Crawford DJ, Landolt E, Les DH et al (2006) Speciation in duckweeds (Lemnaceae): phylogenetic and ecological inferences. Aliso 22:231–242. https://doi.org/10.5642/aliso.20062201.19

Cui W, Cheng JJ (2015) Growing duckweed for biofuel production: a review. Plant Biol 17(s1):16–23. https://doi.org/10.1111/plb.12216

Cusimano N, Bogner J, Mayo SJ et al (2011) Relationships within the Araceae: comparison of morphological patterns with molecular phylogenies. Am J Bot 98:654–668. https://doi.org/10.3732/ajb.1000158

Driever SM, van Nes EH, Roijackers RMM (2005) Growth limitation of *Lemna minor* due to high plant

density. Aquat Bot 81:245–251. https://doi.org/10.1016/j.aquabot.2004.12.002
Gaut B, Lewis PO (1995) Success of maximum likelihood phylogeny inference in the four-taxon case. Mol Biol Evol 12:152–162. https://doi.org/10.1093/oxfordjournals.molbev.a040183
Ge X, Zhang N, Phillips GC et al (2012) Growing *Lemna minor* in agricultural wastewater and converting the duckweed biomass to ethanol. Bioresource Technol 124:485–488. https://doi.org/10.1016/j.biortech.2012.08.050
Gray SF (1821) A natural arrangement of British plants, vol 2. Baldwin, Cradock, and Joy, London
Grayum MH (1991) Systematic embryology of the Aracae. Bot Rev 57:167–203. https://doi.org/10.1007/BF02858562
Green AJ (2016) The importance of waterbirds as an overlooked pathway of invasion for alien species. Divers Distrib 22:239–247. https://doi.org/10.1111/ddi.12392
Halder S, Venu P (2012) *Lemna landoltii* sp. nov. (Lemnaceae) from India. Taiwania 58:12–14. https://doi.org/10.6165/tai.2013.58.12
Hegelmaier F (1868) Die Lemnaceen: eine monographische Untersuchung. Engelmann, Leipzig
Hegelmaier F (1895) Systematische übersicht der Lemnaceen. Bot Jahrb Syst 21:268–305
Henriquez CL, Arias T, Pires JC et al (2014) Phylogenomics of the plant family Araceae. Mol Phyl Evol 75:91–102. https://doi.org/10.1016/j.ympev.2014.02.017
Hillman WS (1961) The Lemnaceae, or duckweeds. Bot Rev 27:221–287. https://doi.org/10.1007/BF02860083
Hillman WS (1976) Calibrating duckweeds: light, clocks, metabolism, flowering. Science 193:453–458. https://doi.org/10.1126/science.193.4252.453
IPNI (The International Plant Names Index) (2018). Published on the Internet. http://www.ipni.org. accessed 22 May 2018
Jacobs DL (1947) An ecological life-history of *Spirodela polyrhiza* (greater duckweed) with emphasis on the turion phase. Ecol Monogr 17:437–469. https://doi.org/10.2307/1948596
Jacono CC (2018) *Landoltia punctata* (G. Mey.) Les & D. J. Crawford: U.S. Geological Survey, Nonindigenous Aquatic Species Database, Gainesville, FL, https://nas.er.usgs.gov/queries/FactSheet.aspx?SpeciesID=1116. Revision Date: 6/13/2002, Access Date: 5/23/2018
Jordan WC, Courtney MW, Neigel JE (1996) Low levels of intraspecific genetic variation at a rapidly evolving chloroplast DNA locus in North American duckweeds (Lemnaceae). Am J Bot 83:430–43. https://doi.org/10.1002/j.1537-2197.1996.tb12724.x
Keating RC (2002) Acoraceae and Araceae. In: Gregory M, Cutler DF (eds) Anatomy of the monocotyledons, vol 9. Oxford University Press, Oxford p, pp 1–327
Kim J, Sanderson MJ (2008) Penalized likelihood phylogenetic inference: bridging the parsimony-likelihood gap. Syst Biol 57:665–674. https://doi.org/10.1080/10635150802422274
Kimball RT, Crawford DJ, Les DH et al (2003) Out of Africa: molecular phylogenetics and biogeography of *Wolffiella* (Lemnaceae). Biol J Linn Soc 79:565–576. https://doi.org/10.1046/j.1095-8312.2003.00210.x
Kück P, Mayer C, Wägele JW et al (2012) Long branch effects distort maximum likelihood phylogenies in simulations despite selection of the correct model. PLoS ONE 7:e36593. https://doi.org/10.1371/journal.pone.0036593
Kvaček Z (1995) *Limnobiophyllum* Krassilov—a fossil link between the Araceae and the Lemnaceae. Aquat Bot 50:49–61. https://doi.org/10.1016/0304-3770(94)00442-O
Laird RA, Barks PM (2018) Skimming the surface: duckweed as a model system in ecology and evolution. Am J Bot 105:(not yet paginated). https://doi.org/10.1002/ajb2.1194
Landolt E (1986) The family of Lemnaceae—a monographic study. Ver Geobot Inst ETH Stift Rübel 71:1–563
Landolt E (1992a) *Wolfiella caudata*, a new Lemnaceae species from the Bolivian Amazon region. Ber Geobot Inst ETH Stift Rübel 58:121–123
Landolt E (1992b) The flowers of *Wolffia australiana* (Lemnaceae). Ber Geobot Inst ETH Stift Rübel 58:132–137
Landolt E (1994) Taxonomy and ecology of the section *Wolffia* of the genus *Wolffia* (Lemnaceae). Ber Geobot Inst ETH Stift Rübel 60:137–151
Landolt E (1998) *Lemna yungensis*, a new duckweed species from rocks of the Andean Yungas in Bolivia. Bull Geobot Inst ETH 64:15–21
Landolt E (2000) Contribution on the Lemnaceae of Ecuador. Fragm Flor Geobot 45:221–237
Landolt E, Kandeler R (1987) The family of Lemnaceae—a monographic study, vol. 2. Ver Geobot Inst ETH Stift Rübel 2:211–234
Lasfar S, Monette F, Millette L et al (2007) Intrinsic growth rate: a new approach to evaluate the effects of temperature, photoperiod and phosphorus–nitrogen concentrations on duckweed growth under controlled eutrophication. Water Res 41:2333–2340. https://doi.org/10.1016/j.watres.2007.01.059
Les DH, Crawford DJ (1999) *Landoltia* (Lemnaceae), a new genus of duckweeds. Novon 9:530–533. https://doi.org/10.2307/3392157
Les DH, Tippery NP (2013) In time and with water… the systematics of alismatid monocotyledons. In: Wilkin P, Mayo SJ (eds) Early events in monocot evolution. Cambridge University Press, Cambridge, pp 118–164
Les DH, Landolt E, Crawford DJ (1997) Systematics of the Lemnaceae (duckweeds): Inferences from micromolecular and morphological data. Plant Syst Evol 204:161–177. https://doi.org/10.1007/BF00989203
Les DH, Crawford DJ, Landolt E et al (2002) Phylogeny and systematics of Lemnaceae, the duckweed family.

Syst Bot 27:221–240. https://doi.org/10.1043/0363-6445-27.2.221

Les DH, Crawford DJ, Kimball RT et al (2003) Biogeography of discontinuously distributed hydrophytes: a molecular appraisal of intercontinental disjunctions. Int J Plant Sci 164:917–932. https://doi.org/10.1086/378650

Lindley J (1846) The vegetable kingdom, ed 2. Bradbury and Evans, London

Linnaeus C (1753) Species plantarum, vol 2. Laurentius Salvius, Stockholm

Magallón S, Gómez-Acevedo S, Sánchez-Reyes LL et al (2015) A metacalibrated time-tree documents the early rise of flowering plant phylogenetic diversity. New Phyt 207:437–453. https://doi.org/10.1111/nph.13264

Martirosyan EV, Ryzhova NN, Kochieva EZ et al (2009) Analysis of chloroplast rps16 intron sequences in Lemnaceae. Mol Biol 43:32–38. https://doi.org/10.1134/S002689330

Mathews S (2009) Phylogenetic relationships among seed plants: persistent questions and the limits of molecular data. Am J Bot 96:228–236. https://doi.org/10.3732/ajb.0800178

Mayo SJ, Bogner J, Boyce PC (1997) The genera of Araceae. The Trustees, Royal Botanic Gardens, Kew

Nauheimer L, Metzler D, Renner SS (2012) Global history of the ancient monocot family Araceae inferred with models accounting for past continental positions and previous ranges based on fossils. New Phytol 195:938–950. https://doi.org/10.1111/j.1469-8137.2012.04220.x

Oron G (1994) Duckweed culture for wastewater renovation and biomass production. Agr Water Manage 26:27–40. https://doi.org/10.1016/0378-3774(94)90022-1

Paradis E (2013) Molecular dating of phylogenies by likelihood methods: a comparison of models and a new information criterion. Mol Phyl Evol 67:436–444. https://doi.org/10.1016/j.ympev.2013.02.008

Parr LB, Perkins RG, Mason CF (2002) Reduction in photosynthetic efficiency of *Cladophora glomerata*, induced by overlying canopies of *Lemna* spp. Water Res 36:1735–1742. https://doi.org/10.1016/S0043-1354(01)00395-5

Philippe H, Zhou Y, Brinkmann H et al (2005) Heterotachy and long-branch attraction in phylogenetics. BMC Evol Biol 5:50. https://doi.org/10.1186/1471-2148-5-50

R Core Group (2018) R: A language and environment for statistical computing. R Foundation for Statistical Computing, Vienna. https://www.r-project.org

Roijackers R, Szabó S, Scheffer M (2004) Experimental analysis of the competition between algae and duckweed. Archiv für Hydrobiologie 160:401–412. https://doi.org/10.1127/0003-9136/2004/0160-0401

Rusoff LL, Blakeney EW Jr, Culley DD Jr (1980) Duckweeds (Lemnaceae family): a potential source of protein and amino acids. J Agric Food Chem 28:848–850. https://doi.org/10.1021/jf60230a040

Sanderson MJ (2002) Estimating absolute rates of molecular evolution and divergence times: a penalized likelihood approach. Mol Biol Evol 19:101–109. https://doi.org/10.1093/oxfordjournals.molbev.a003974

Schleiden MJ (1839) Prodromus monographiae Lemnacearum. Linnaea 13:385–392

Schreber JCD (1791) Genera plantarum, vol 2. Varrentrapp & Wenner, Frankfurt

Sculthorpe CD (1967) The biology of aquatic vascular plants. St. Martin's Press, New York

Sree KS, Bog M, Appenroth KJ (2016) Taxonomy of duckweeds (Lemnaceae), potential new crop plants. Emir J Food Agr 28:291–302. https://doi.org/10.9755/ejfa.2016-01-038

Stockey RA, Hoffman GL, Rothwell GW (1997) The fossil monocot *Limnobiophyllum scutatum*: resolving the phylogeny of Lemnaceae. Am J Bot 84:355–368. https://doi.org/10.2307/2446009

Stockey RA, Rothwell GW, Johnson KR (2007) Cobbania corrugata gen. et comb. nov. (Araceae): a floating aquatic monocot from the Upper Cretaceous of western North America. Am J Bot 94:609–624. https://doi.org/10.3732/ajb.94.4.609

Stockey RA, Rothwell GW, Johnson KR (2016) Evaluating relationships among floating aquatic monocots: a new species of *Cobbiana* (Araceae) from the Upper Maastrichtian of South Dakota. Int J Plant Sci 177:706–725. https://doi.org/10.1086/688285

Stomp AM (2005) The duckweeds: a valuable plant for biomanufacturing. Biotechnol Ann Rev 11:69–99. https://doi.org/10.1016/S1387-2656(05)11002-3

Su H, Zhao Y, Jiang J et al (2014) Use of duckweed (*Landoltia punctata*) as a fermentation substrate for the production of higher alcohols as biofuels. Energ Fuel 28:3206–3216. https://doi.org/10.1021/ef500335h

Tippery NP, Les DH (2011) Phylogenetic relationships and morphological evolution in *Nymphoides* (Menyanthaceae). Syst Bot 36:1101–1113. https://doi.org/10.1600/036364411X605092

Tippery NP, Les DH, Crawford DJ (2015) Evaluation of phylogenetic relationships in Lemnaceae using nuclear ribosomal data. Plant Biol 17(s1):50–58. https://doi.org/10.1111/plb.12203

Turland NJ, Wiersema JH, Barrie FR et al (eds) (2018) International Code of Nomenclature for algae, fungi, and plants (Shenzhen Code) adopted by the Nineteenth International Botanical Congress Shenzhen, China, July 2017. Regnum Vegetabile 159. Koeltz Botanical Books, Glashütten. https://doi.org/10.12705/Code.2018

Wang W, Messing J (2011) High-throughput sequencing of three Lemnoideae (duckweeds) chloroplast genomes from total DNA. PLoS ONE 6:e24670. https://doi.org/10.1371/journal.pone.0024670

Wang W, Wu Y, Messing J (2012) The mitochondrial genome of an aquatic plant. Spirodela polyrhiza. PLoS One 7:e46747. https://doi.org/10.1371/journal.pone.0046747

Wang W, Haberer G, Gundlach H et al (2014) The *Spirodela polyrhiza* genome reveals insights into its neotenous reduction fast growth and aquatic lifestyle. Nat Commun 5:3311. https://doi.org/10.1038/ncomms4311

Wickett NJ, Mirarab S, Nguyen N et al (2014) Phylotranscriptomic analysis of the origin and early diversification of land plants. Proc Nat Acad Sci 111: E4859–E4868. https://doi.org/10.1073/pnas.1323926111

Xu J, Zhao H, Stomp AM et al (2012) The production of duckweed as a source of biofuels. Biofuels 3:589–601. https://doi.org/10.4155/bfs.12.31

Yamamoto YT, Rajbhandari N, Lin X et al (2001) Genetic transformation of duckweed *Lemna gibba* and *Lemna minor*. Vitro Cell Dev-Pl 37:349–353. https://doi.org/10.1007/s11627-001-0062-6

Yu Y, Harris AJ, He XJ (2010) S-DIVA (statistical dispersal-vicariance analysis): a tool for inferring biogeographic histories. Mol Phyl Evol 56:848–850. https://doi.org/10.1016/j.ympev.2010.04.011

Yu Y, Harris AJ, Blair C et al (2015) RASP (reconstruct ancestral state in phylogenies): a tool for historical biogeography. Mol Phyl Evol 87:46–49. https://doi.org/10.1016/j.ympev.2015.03.008

# Worldwide Genetic Resources of Duckweed: Stock Collections

3

K. S. Sree and K.-J. Appenroth

**Abstract**

Duckweeds are aquatic monocotyledonous plants that include the smallest and the fastest growing angiosperms known till date. The family Lemnaceae presently includes 36 accepted species of duckweeds categorized into five genera. Although having the capacity to flower, these unique aquatic plants usually propagate by vegetative means producing clones. It has already been known through various reports that duckweed plants although belonging to the same species collected from different geographical locations have variations not only in their genetic properties but also in their physiological properties, which is important for many of the practical applications of this plant family. Hence, collection and maintenance of different clones of the same species from different locations are important. The international duckweed clonal collections from around the world help the duckweed community by maintaining the clones, characterizing their species identity and providing the required genetic resources to the users for research and application purposes.

## 3.1 Introduction

Duckweeds, because of their miniature size and unique structure, had gained attention of the researchers as early as eighteenth (von Linnaeus 1753) and nineteenth centuries (Schleiden 1839; Griffith 1851; Hegelmaier 1868). This was a time when some of the species of duckweeds were just discovered and some were still hidden. The researchers of this time had worked in great details on morphology and anatomy of duckweeds of different species collected worldwide, with the then-available tools. In this tradition, the two-volume monographic study of Lemnaceae by Landolt (1986), and Landolt and Kandeler (1987) pronounced the use of duckweeds as model plants for physiological investigations during the twentieth century.

Duckweeds are free-floating, aquatic monocotyledonous flowering plants that inhabit the lentic ecosystems like ponds, pools, lakes and so on. These plants have the distinction to include the smallest angiosperm and the fastest growing angiosperm known till date on earth (Sree et al. 2015b; Ziegler et al. 2015). The anatomy of duckweeds is highly miniaturized and reduced, with the smallest one majorly constituting a ball of parenchymatous cells with a meristematic zone

K. S. Sree (✉)
Department of Environmental Science, Central University of Kerala, Periye 671316, India
e-mail: ksowsree9@cukerala.ac.in; ksowsree@gmail.com

K.-J. Appenroth
Matthias Schleiden Institute—Plant Physiology, Friedrich Schiller University of Jena, 07743 Jena, Germany
e-mail: Klaus.Appenroth@uni-jena.de



X. H. Cao et al. (eds.), *The Duckweed Genomes*, Compendium of Plant Genomes,
https://doi.org/10.1007/978-3-030-11045-1_3

in its vegetative phase (Landolt 1980, 1986; Sree et al. 2015c). Although duckweeds are angiosperms, their dominant reproductive strategy is via vegetative propagation. A daughter frond buds from the vegetative pouch of a mother frond, thus forming a colony of plants having the same genetic structure. The plants of such a colony and their progeny (by vegetative propagation) are called to be a clone of a particular species. So, the plants with the same species identity, collected from different geographical locations, which might have been adapted to grow in different climatic conditions, can be cultured as different clones (Bog et al. 2018). These clones constitute the rich genetic resource of duckweeds.

Apart from the enthusiasm of the researchers to study some of the basic questions of plant biology using duckweeds as model plants, these plants are also a treasure for the entrepreneurs. Duckweeds, in view of their physiological and biotechnological properties, are used in a wide array of practical applications (Appenroth et al. 2015). The high nutritious value of duckweeds makes it a good food and feed supplement for humans (Appenroth et al. 2017, 2018), fish, cattle, poultry and so on (Landolt and Kandeler 1987). Depending on the growth conditions, duckweeds can be modulated to accumulate high amounts of starch, which finds use in the bioenergy field (Cui and Cheng 2015; Ma et al. 2018). Other applications of duckweed include biomass production, wastewater treatment (Ziegler et al. 2018), biotechnological uses in the pharmaceutical industry and so on (Appenroth et al. 2015).

Through numerous research studies conducted worldwide, it is evident that the physiological properties of duckweed are not dependent on the species but on the clones of duckweed collected from different geographical locations and so are the relevant practical applications. Some of the physiological properties which have been tested include growth rate (Sree et al. 2015a; Ziegler et al. 2015), turion formation (Kuehdorf et al. 2013), starch accumulation capacity (Sree et al. 2015c; Ma et al. 2018) and protein content (Appenroth et al. 2018). Hence, over the years, it became extremely important to study several clones of the same species collected from different geographical locations, in order to understand better the biodiversity, genetics and the inherent potential of duckweeds. For a researcher to travel to different places in order to collect different clones and decipher their species identity cannot only be expensive but is also time consuming and requires expertise in the field. In this respect, the duckweed clonal stock collections that are repositories of duckweed genetic resources help other researchers to have access to clones of different duckweed species.

## 3.2 Duckweed: Genera and Species

As of now, in the family Lemnaceae, 36 different species of duckweed have been identified and taxonomically categorized into five genera (Sree et al. 2016; Bog et al. 2020a, b) (Figs. 3.1 and 3.2). Based on their morphology, they have been categorized into two subfamilies: Lemnoideae and Wolffioideae. The Lemnoideae possess roots and have two lateral pouches from where new fronds are vegetatively produced. The Wolffioideae, on the other hand, have only one such pouch and are devoid of roots (Landolt 1986; Les et al. 2002). The complete botanical names and recent changes in nomenclature of all the currently accepted species were given by Sree et al. (2016) and Bog et al. (2020a, b). Very recently, the total number of duckweed species was reduced from 37 to 36 (Bog et al. 2020a, b).

## 3.3 Duckweed Stock Collection: The Pioneer

The late Elias Landolt evidently realized the importance of duckweed clones grown under different climatic conditions, and during his stay as a young post-doc at the Stanford University, CA, from 1953 to 1955, he had started collecting duckweed plants from widespread geographical locations, which grew into the Landolt Duckweed Collection at Zurich, Switzerland, with numerous duckweed clones collected from around the globe. E. Landolt added these collected duckweed plants to an already existing non-duckweed plant

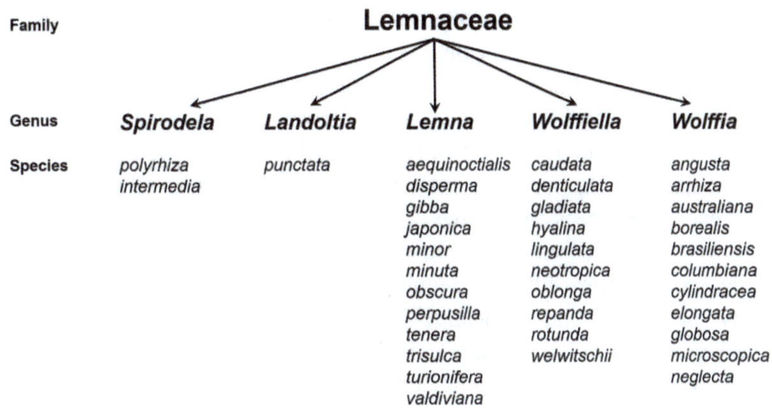


**Fig. 3.1** Schematic representation of the 36 duckweed species categorized into five genera of the family Lemnaceae

collection at Stanford which had a four-digit ID numbering system, in that time already occupied up to ID of ca. 6000 plus. This is the reason why originally all ID numbers for duckweed clones in his collection were above 6000. This ID numbering system was adopted and continued by the duckweed community (Zhao et al. 2012). Accordingly, in order to share the knowledge resources and to correlate the results obtained in different laboratories across the world, it was decided that all duckweed axenic clones used for scientific projects be registered at the Rutgers Duckweed Stock Cooperative at the Rutgers University, NJ, USA, to obtain a uniform ID for the duckweed clone.

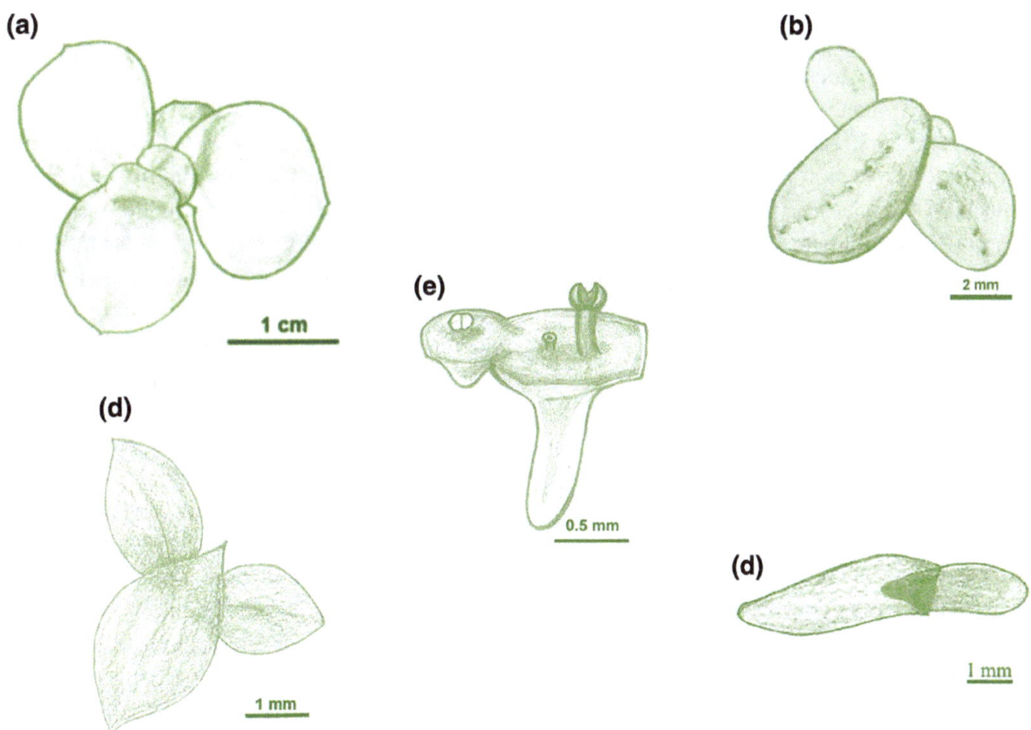


**Fig. 3.2** Drawings of one of the species of duckweed from each of the five genera. **a**. *Spirodela polyrhiza*, **b**. *Landoltia punctata*, **c**. *Lemna tenera*, **d**. *Wolffiella oblonga and* **e**. *Wolffia microscopica. Drawings by KSS*

E. Landolt not only collected duckweed clones but also taxonomically identified and classified them, which bestowed his duckweed collection the "gold standard" for duckweed research (Lam and Appenroth 2013). From the very beginning, he was not only in search for new duckweed species (see Sree et al. 2016) but also was interested in collecting clones of the same species from different geographical locations, indicating that evidently, he was very much interested in biodiversity and adaptation of plants to specific environmental conditions.

In the 90 s of the twentieth century, E. Landolt passed a part of the duckweed collection (ca. 1000 clones) to Prof. Anne-Marie Stomp from the North Carolina State University, USA. In 1997, Prof. Stomp had transferred this collection to Biolex Therapeutics, a company that was interested in making use of duckweeds (Cross 2015), which later transferred a part of the duckweed collection to Rutgers State University, NJ (Prof. Dr. Eric Lam) forming the initial stock of the Rutgers Duckweed Stock Cooperative with presently more than 1000 clones of duckweed.

The collection in Zurich comprised of much more than 1000 duckweed clones. After the retirement of E. Landolt in 1996 (Urbanska et al. 2013), gradually it became difficult to maintain the full stock collection in the cultivation rooms of the ETH Zurich. The number of clones was therefore reduced in the collection. Finally, in 2009 it was transferred to private rooms and was named as Landolt Duckweed Collection in Zurich (Laemmler 2018).

## 3.4 Requirements of a Duckweed Stock Collection

Often each stock collection optimizes the standard conditions and requirements according to the available resources and conducts of that laboratory. The following is established at the Duckweed stock collection at Friedrich Schiller University of Jena being headed by one of the co-authors, KJA.

The difference between "normal" cultivation of duckweed, i.e. mainly under optimal growth conditions for experimental purpose, and stock cultivation for long-term availability of duckweed clones in the laboratory exists in the point that the growth of the plants in stock cultivation is slowed down by regulating different factors. Maintaining a stock collection not only involves financial resources but is also time and manpower consuming. The slower growth of the plants in the stock cultivation optimizes the use of these resources. Duckweeds in stock cultivation need to be transferred to fresh nutrient medium from time to time failing which the plants will die once the nutrient reservoirs are exhausted. Slower growth of plants in a stock cultivation means less number of subcultures in a year, which also helps to avoid introduction of microbial contaminations to the plant cultures.

The medium for stock cultivation should be rich of nutrients in order to prevent quick nutrient shortage. Several media are suitable, e.g. N-medium with increased phosphate concentration to 1 mM or Steinberg's medium with increased Fe-EDTA concentration to 25 μM (Appenroth 2015).

The growth rate is decreased by regulating several of the controlled conditions in the laboratory. Use of solidified medium, either by addition of agar (e.g. 0.9%) or Gelrite (0,45%), is one of the methods. The light intensity is reduced from the standard 100 μmole $m^{-2}$ $s^{-1}$ continuous white light to 30 μmole $m^{-2}$ $s^{-1}$ white light often with a photoperiod. All duckweed clones from around the world, even those from tropical regions, can survive 17 °C environmental temperature, which further reduces the plant growth during stock cultivation. Clones of *Spirodela polyrhiza* and *S. intermedia* are often grown on liquid medium in the stock cultivation because their relatively large frond size forms different vertical layers of plants over the solidified medium and after a certain time, the fronds of the upper layer start to die, as they do not any more have access to the nutrient medium. There exists an alternative method for the long-term live storage of the species *S. polyrhiza*. This species forms turions or starch-filled submerged resting bodies, which can be easily separated from the floating fronds. These turions under favourable conditions can germinate and form active floating fronds. The turions from each of the *S. polyrhiza* clones can be easily harvested and stored in darkness for several years. However, we do not know a single laboratory where this is realized.

All plants in stock collections have to be kept under axenic ("sterile") conditions because contamination by bacteria or fungi may inhibit favourable growth conditions of the plants leading to death of the plants after some time. Especially dangerous are the contaminations by green algae as they compete with the duckweeds for resources. These contaminations can be removed by surface sterilization of the plants with a commercially available household bleach like Eau de Javel or DanKlorix (Sree and Appenroth 2017). The availability of axenic cultures allows studies on duckweeds alone without having any interference from other organisms (Appenroth et al. 2016). All media need to be sterilized (autoclaved), and inoculation should be carried out in laminar flow boxes to avoid any contaminations during the process.

In order to examine and ensure axenic conditions, usually glucose (25 mM or less) is added to the solidified media. Any microbial contamination can then be easily recognized. However, this has two disadvantages. 1. If there is a contamination, plants are usually quickly killed in the presence of sugars as the microbial population explodes. 2. Sugars accelerate senescence and shorten the time between two subculturing events. To avoid these disadvantages, the duckweed clones can be kept in both the presence and absence of glucose, thereby taking advantage of both cultivation methods.

Cultivation vessels for stock cultivation are very different in different collections as this depends on the available space and manpower. They range from 100-ml Erlenmeyer flasks with cotton wool stoppers (Fig. 3.3) to Petri dishes of different diameter (from 35 to 90 mm) closed with parafilm, to standard glass test tubes 13 × 100 mm with cotton wool stoppers.

## 3.5 Species Identification of the Clones

For a duckweed clone collected from nature to be successfully integrated into a stock collection, accurate identification of the species is a prerequisite. Species identification using morphological markers would be the first choice (Landolt 1980). Based on morphology, it is easy to identify the duckweed genus to which the clone belongs. In some cases, it is also easy to identify the species, e.g. *Lemna trisulca* or *S. polyrhiza*. However, in many cases even experts of duckweed morphology have the biggest problems to assign species identity based on morphological markers. This is partially caused by the miniaturized phenotype of the duckweeds. In some cases, the ease of identity might also depend on the growth conditions. For instance, *L. gibba* can be easily recognized when it possesses the typical gibbous structure. Unfortunately, this is often not the case in natural environment and then not many duckweed researchers can distinguish *L. gibba* from e.g. *L. minor*. In most cases, for determining the species identity, integration of molecular barcoding is unavoidable. One of the standardized methods includes sequencing of plastidic fragments that can be used as molecular markers. It turned out that for different genera, different sets of molecular markers have to be used (Bog et al. 2013, 2015, 2018). Unfortunately, even then the results are not always clear-cut (Borisjuk et al. 2015) and have to be integrated together with the morphological markers to eliminate any uncertainty. Several molecular taxonomic methods are presently in development but are either still very expensive (e.g. genotyping by sequencing; Bog et al. 2020b) or need further investigations in order to expand it to the whole plant family (e.g. polymorphic NB-ARC-related genes; Chu et al. 2018).

**Fig. 3.3** A view of the duckweed stock collection at the Matthias Schleiden Institute at Friedrich Schiller University of Jena

## 3.6 Duckweed Stock Collections Around the Globe

There are several stock collections in the world that house defined duckweed species or clones within the limits of their resources:

- Rutgers Duckweed Stock Cooperative, Rutgers State University, New Jersey, USA—Prof. Dr. Eric Lam.
- Matthias Schleiden Institute—Plant Physiology, University of Jena, Jena, Germany—Dr. Klaus-J. Appenroth. It also houses a herbarium of all the 36 species of duckweeds.
- Landolt Duckweed Collection Zurich, Zurich, Switzerland—Walter Lämmler.
- Chinese Academy of Science, Chengdu Institute of Biology, Chengdu, China—Prof. Hai Zhao.
- Institute of Tropical Bioscience and Biotechnology, Hainan Bioenergy Centre, CATAS, Haikou, Hainan, China—Prof. Dr. Jiaming Zhang—Especially numerous clones from Hainan and China.
- Chinese Academy of Science, Qingdao Institute of Bioenergy & Bioprocess Technology, Qingdao, China—Dr. Yubin Ma—Especially numerous clones from China.
- Dr. K. Sowjanya Sree's Lab, Department of Environmental Science, Central University of Kerala, Periye, India—Presently having a wide collection of clones of different duckweed species collected from India. It also houses a herbarium of all the 36 species of duckweeds.

The duckweed collection at the Matthias Schleiden Institute at Friedrich Schiller University of Jena, Germany, headed by one of the co-authors, KJA, has more than 500 clones of duckweeds. Initially, a large portion of the collection came from the Landolt Duckweed Collection from ETH, Zurich. Many clones were also investigated by molecular markers (Bog et al. 2010, 2013, 2015, 2018; Borisjuk et al. 2015). This clonal collection of the 36 species of duckweeds, over the years, is constantly being used for several comparative physiological studies, investigations on biodiversity of duckweed species, laboratory and field-testing of several practical applications of duckweeds by researchers and entrepreneurs and, also for the development of molecular markers for determining species identity.

## References


Appenroth KJ (2015) Media for in vitro-cultivation of duckweed. Duckweed Forum 3:180–186. http://www.ruduckweed.org/uploads/1/0/8/9/10896289/iscdra_issue11-2015-11_final.pdf

Appenroth KJ, Sree KS, Böhm V, Hammann S, Vetter W, Leiterer M, Jahreis G (2017) Nutritional value of duckweeds (Lemnaceae) as human food. Food Chem 217:266–273

Appenroth KJ, Sree KS, Bog M, Ecker J, Seeliger C, Böhm V, Lorkowski S, Sommer K, Vetter W, Tolzin-Banasch K, Kirmse R, Leiterer M, Dawczynski C, Liebisch G, Jahreis G (2018) Nutritional value of the duckweed species of the genus Wolffia (Lemnaceae) as human food. Front Chem 6(483)

Appenroth KJ, Sree KS, Fakhoorian T, Lam E (2015) Resurgence of duckweed research and applications: report from the 3rd International Duckweed Conference. Plant Mol Biol 89:647–654

Appenroth KJ, Ziegler P, Sree KS (2016) Duckweed as a model organism for investigating plant-microbe interactions in an aquatic environment and its applications. Endocytobiosis Cell Res 27:94–106

Bog M, Baumbach H, Schween U, Hellwig F, Landolt E, Appenroth K-J (2010) Genetic structure of the genus *Lemna* L. (Lemnaceae) as revealed by amplified fragment length polymorphism. Planta 232:609–619

Bog M, Landrock MF, Drefahl D, Sree KS, Appenroth KJ (2018) Fingerprinting by amplified fragment length polymorphism (AFLP) and barcoding by three plastidic markers in the genus *Wolffiella* Hegelm. Plant Syst Evol 304:373–386

Bog M, Lautenschlager U, Landrock MF, Landolt E, Fuchs J, Sree KS, Oberprieler C, Appenroth KJ (2015) Genetic characterization and barcoding of taxa in the genera *Landoltia* and *Spirodela* (Lemnaceae) by three plastidic markers and amplified fragment length polymorphism (AFLP). Hydrobiologia 749:169–182

Bog M, Schneider P, Hellwig F, Sachse S, Kochieva EZ, Martyrosian E, Landolt E, Appenroth KJ (2013) Genetic characterization and barcoding of taxa in the genus *Wolffia* Horkel ex Schleid. (Lemnaceae) as revealed by two plastidic markers and amplified fragment length polymorphism (AFLP). Planta 237:1–139

Bog M, Sree KS, Fuchs J, Hoang PTN, Schubert I, Kuever J, Rabenstein A, Paolacci S, Jansen MAK, Appenroth KJ (2020a) A taxonomic revision of the section Uninerves within the genus *Lemna*. Taxon, in press

Bog M, Xu S, Himmelbach A, Brandt R, Wagner F, Appenroth, KJ, Sree KS (2020b) Genotyping-By-Sequencing for species delimitation in the section Uninerves Hegelm. (genus *Lemna* L.). In: Cao XH, Fourounjian P, Wang W (ed) Compendium of plant genomes: the duckweed genomes 1st Edn. Springer. ISBN. 978-3-030-11045-1

Borisjuk N, Chu P, Gutierrez R, Zhang H, Acosta K, Friesen N, Sree KS, Garcia C, Appenroth KJ, Lam E (2015) Assessment, validation and deployment strategy of a two barcode protocol for facile genotyping of duckweed species. Plant Biology 17(Suppl. 1):42–49

Chu P, Wilson GM, Michael TP, Vaiciunas J, Honig J, Lam E (2018) Sequence-guided approach to genotyping plant clones and species using polymorphic NB-ARC-related genes. Plant Mol Biol 98:219–231

Cross JW (2015) the rise and fall of a duckweed biotechnology firm: what can it tell us? Duckweed Forum 3:83–88. (http://www.ruduckweed.org/uploads/1/0/8/9/10896289/iscdra_issue09-03_03-2015-07-2.pdf)

Cui W, Cheng JJ (2015) Growing duckweed for biofuel production: a review. Plant Biol (Suppl. 1):16–23

Griffith W (1851) Notulae ad plantas asiaticas. Bishop College Press, Calcutta, India, 3:216–229

Hegelmaier F (1868) Die Lemnaceen: Eine monographische Untersuchung. Engelmann, Leipzig, Germany, p 169

Kuehdorf K, Jetschke G, Ballani L, Appenroth KJ (2013) The clonal dependence of turion formation in the duckweed *Spirodela polyrhiza*—an ecogeographical approach. Physiol Plant 150:46–54

Laemmler W (2018) Landolt Duckweed Collection Zurich. Duckweed Forum 6:98–99 (http://www.ruduckweed.org/uploads/1/0/8/9/10896289/duckweed_forum_issue_22.pdf)

Lam E, Appenroth KJ (2013) The second international conference on duckweed research and applications. In Dedication to Elias Landolt (1926–2013). http://duckweed2013.rutgers.edu/duckweed-2013-conf-program.pdf

Landolt E (1980) Key to the determination of taxa within the family of Lemnaceae. Veröffentlichungen des Geobotanischen Institutes der ETH. Stiftung Ruebel (Zurich) 70:13–21

Landolt E (1986) The family of lemnaceae—a monographic study. vol 1. Biosystematic Investigations in the Family of Duckweeds (Lemnaceae). Zurich: Veröffentlichungen des Geobotanischen Institutes der ETH, Stiftung Rübel

Landolt E. Kandeler R (1987) The family of lemnaceae—a monographic study. vol 2. Biosystematic Investigations in the Family of Duckweeds (Lemnaceae). Zurich: Veröffentlichungen des Geobotanischen Institutes der ETH, Stiftung Rübel

Les DH, Crawford DJ, Landolt E, Gabel JD, Kimball RT (2002) Phylogeny and systematics of Lemnaceae, the duckweed family. Syst Bot 27:221–240

Linnaeus, Carl von (1753) Species Plantarum 2:970

Ma YB, Zhu M, Yu CJ, Wang Y, Liu Y, Li ML, Sun YD, Zhao JS, Zhou GK (2018) Large-scale screening and characterisation of *Lemna aequinoctialis* and *Spirodela polyrhiza* strains for starch production. Plant Biology 20:357–364

Schleiden MJ (1839) Prodromus Monographiae Lemnacearum oder Conspectus generum atque specierum. Linnaea 13:385–392

Sree KS, Adelmann K, Garcia C, Lam E, Appenroth KJ (2015a) Natural variance in salt tolerance and induction of starch accumulation in duckweeds. Planta 241:1395–1404

Sree KS, Appenroth KJ (2017) Cultivation, sterilization and stock cultivation of duckweeds; an update. Duckweed Forum 5:42–44. http://www.ruduckweed.org/uploads/1/0/8/9/10896289/iscdra-duckweedforum_issue17-2017-04.pdf

Sree KS, Bog M, Appenroth KJ (2016) Taxonomy of duckweeds (Lemnaceae), potential new crop plants. Emirate J Food Agric 28:291–302

Sree KS, Sadakaran S, Appenroth KJ (2015b) How fast can angiosperms grow? Species and clonal diversity of growth rates in the genus *Wolffia* (Lemnaceae). Acta Physiologia Plantarum 37:204

Sree KS, Maheshwari SC, Boka K, Khurana JP, Keresztes A, Appenroth KJ (2015c) *Wolffia microscopica* (Griff.) Kurz: A unique member of Lemnaceae (Duckweed). Flora 210:31–39

Urbanska KM, Crawford DJ, Appenroth KJ, Les DH (2013) Obituary Elias Landolt (July 21, 1926–April 1, 2013). Aquat Bot 11:A1–A2

Zhao H, Appenroth K, Landesman L, Salmean AA, Lam E (2012) Duckweed rising at Chengdu: summary of the 1st International conference on duckweed application and research. Plant Mol Biology 78:627–632

Ziegler P, Adelmann K, Zimmer S, Schmidt C, Appenroth KJ (2015) Relative in vitro growth rates of duckweeds (Lemnaceae)—the most rapidly growing higher plants. Plant Biology 17(Suppl. 1):33–41

Ziegler P, Sree KS, Appenroth KJ (2018) Duckweed biomarkers for identifying toxic water contaminants? Environ Sci Pollut Res. https://doi.org/10.1007/s11356-018-3427-7

# 4 Cytogenetics, Epigenetics and Karyotype Evolution of Duckweeds

Xuan Hieu Cao and Giang T. H. Vu

**Abstract**

Duckweed is an interesting model for studying cytogenetics, epigenetics and karyotype evolution. Belonging to the monocotyledonous arum or aroid family Araceae, these aquatic plants present an approximately 12-fold range of different genome sizes, from 158 Mbp (*Spirodela* sps.) to 1881 Mbp (*Wolffia arrhiza*) and variable chromosome numbers. In addition to reduced gene repertoires that found in so far all published duckweed genomes of *Spirodela*, *Lemna* and *Wolffia* species, several peculiar genome and epigenome features (e.g. the lowest copy number of genes coding for rRNA, extreme levels of global DNA methylation, and atypical patterns of heterochromatic and euchromatic territories) indicate a unique and interesting history of duckweed genome evolution, organization and adaptation to plants with simplified body architecture and extremely fast growth rate. Together with the high-throughput long-read, long-range information and optical mapping technologies, available cytogenetic resources, including an efficient and robust protocol of multicolour fluorescence in situ hybridization (mcFISH) and mitotic chromosome preparation, and a *Spirodela* genome-integrated bacterial-artificial-chromosome (BAC) map with ancestral chromosome linkages allow further comprehensive comparative genomic and cytogenetic analysis between duckweed, its close relatives, and other monocotyledonous plants of interest.

## 4.1 Introduction

Duckweeds are very small, free-floating aquatic plants that most people can easily and morphologically distinguish from species of any other flowering plants, even closely related aquatic plants, due to their simplified body structure. Morphological traits and molecular data both support a placement of duckweeds as an early and monophyletic branch of the arum or aroid family *Araceae* (Cabrera et al. 2008; Cusimano et al. 2011; Nauheimer et al. 2012; Henriquez et al. 2014). Duckweeds were first described as a family, the *Lemnaceae*, by S. F. Gray in 1821 (Gray 1821). In contrast, Engler (1876) in his early works on aroids classified the duckweeds as a subfamily of *Araceae* under the name *Lemnoideae*. However, later he considered

X. H. Cao (✉)
Institute of Biology/Plant Physiology, Martin-Luther University of Halle-Wittenberg, Weinbergweg 10, 06120 Halle, Germany
e-mail: xuan.cao@biologie.uni-halle.de

G. T. H. Vu
Leibniz Institute of Plant Genetics and Crop Plant Research (IPK), Corrensstrasse 3, 06166 Gatersleben, Germany


X. H. Cao et al. (eds.), *The Duckweed Genomes*, Compendium of Plant Genomes,
https://doi.org/10.1007/978-3-030-11045-1_4

duckweeds as a separate family (Engler 1889). In the nineteenth and twentieth centuries, several monographs on the duckweeds were published (Hegelmaier 1868, 1895; Daubs 1962; Landolt 1986; Landolt and Kandeler 1987), making duckweeds as one of the earliest and most thoroughly investigated model of flowering plants at that time. Until now, scholars still debate whether the duckweeds should be recognized at the familial (*Lemnaceae*) or subfamilial (*Lemnoideae*) level (Appenroth et al. 2015; Sree et al. 2016).

The present duckweeds comprise five well-defined genera (*Spirodela*, *Landoltia*, *Lemna*, *Wolffia* and *Wolffiella*) and altogether 37 species (Sree et al. 2016). Except for *Wolffiella* that is restricted to America and Africa, species of other duckweed genera are cosmopolitan in distribution from tropical to moderate climates, excluding polar regions (Landolt 1986). Of all the continents, the central diversity of duckweeds seems to be in America. Furthermore, *Spirodela*, the genus representing the most primitive characteristics of today's duckweeds, has its centre of distribution in South America. Fossil collections provide evidence that ancestors of the duckweeds (with the genus *Limnobiophyllum* Krassilov) were widely distributed in North America and Eurasia for approximately 50 million years (from the Late Cretaceous until the Miocene). To date, phylogenetic and phylogenomic studies suggest an early offshoot of duckweeds in the family *Araceae* starting about 104 Ma ago in the Early Cretaceous when the breakup of Pangea was in its final stage (Nauheimer et al. 2012; Henriquez et al. 2014). However, it remains unclear why *Limnobiophyllum* died out during the Miocene.

The morphology of duckweeds is highly abbreviated and has evolved by extreme neoteny. The poorly differentiated leaf-like body (called frond or thallus) of *Spirodela*, *Landoltia* and *Lemna* species contains the distal part with the venation. This structure is homologous to the veined leaf blade, whereas the pouches (or cavities) at the basal sections of the fronds represent the sheath of the petiole of *Araceae* plants. The duckweed shoot is highly reduced to a vegetative point (tiny dot) in the middle of the pouches where new fronds, root(s) and also inflorescence are produced. The body plans of *Wolffiella* and *Wolffia* species are even more reduced, without roots and veins in fronds. Generative and vegetative reproductions in these duckweeds are spatially separated with a floral cavity on the upper side of the frond and a budding pouch where the new frond emerges. The shape of *Wolffia* fronds is more or less globular to ellipsoid or ovoid boat-shaped, while *Wolffiella* fronds are long and flat. Putting together, there is a trend of reduction in body complexity, leading from the most ancestors *Spirodela*, through *Landoltia*, *Lemna* and *Wolffiella* to *Wolffia* (cf. Landolt 1986). It is important to note that most of the duckweeds propagate mainly or exclusively via vegetative proliferation (Landolt and Kandeler 1987).

During evolution, the body size of duckweeds decreased in parallel with a successive reduction of morphological structures, resulting in the smallest flowering plants in the genus *Wolffia*. Frond dimensions (length/width/thickness) of *W. globose* and *W. angusta* are 0.8/0.4/0.7 mm and 0.8/0.4/1.0 mm, respectively (Landolt 1998b). Interestingly, the cell size of *Wolffia* is larger than the more primitive duckweeds (e.g. Spirodela, Lemna, Landolt and Kandeler 1987). In another aspect, the cell number per individual plant or frond in *Wolffia* species is much smaller than in Spirodela (Landolt 1986), and eventually in most angiosperms plants, providing an excellent model system for studying genome and karyotype evolution in the relation to the body size, cell size and nuclear DNA content (Cao et al. 2015, 2016).

## 4.2 Genome Size, Chromosome Number and Karyotype Evolution

A passive adaptation, which is already originated at an early stage of monocotyledons evolution more than 100 My ago, to the highly specialized way of life in the water contributed to the low degree of differentiation in duckweeds. It likely comprises one or more reduction series from

*Spirodela*, the most primitive genus with higher differentiated plant body over *Landoltia*, *Lemna* and *Wolffiella* to *Wolffia*, the most derived genus with the lowest degree of differentiation. The question about the relationship between nuclear DNA content (aka genome size) and the level of body differentiation, that was known a C-value paradox (Cavalier-Smith 1985) or later referred to as the C-value enigma (Gregory 2005), attracted Geber to determine cytophotometrically the DNA content of five different duckweed species (Geber 1989). He found a remarkable small amount of DNA in *Spirodela polyrhiza* (0.15 pg/1C) and a surprising increase of DNA content (over *Landoltia punctata* (aka. *Spirodela punctata*), *Lemna minor* and *Wolffiella oblonga* to *Wolffia arrhiza*) with successive reduction of morphological structures within studied duckweeds.

### 4.2.1 Genome Size Evolution

An extensive survey of duckweed genome sizes with 115 accessions (clones) of 23 duckweed species representing all five genera was performed by using flow cytometry confirmed the continuous increase of DNA content in parallel with a morphological reduction and body size (Wang et al. 2011). Among duckweed species, a nearly 13-fold range from 150 Mbp in *Spirodela polyrhiza* to 1881 Mbp in *Wolffia arrhiza* was observed, indicating duckweeds as an interesting model for studying genome size evolution (Fig. 3.1). Interestingly, the five genera show different degrees of interspecific genome size variation from very small in *Spirodela* (150–165 Mbp) and *Landoltia* (372–427 Mbp) to 1.6- or two-fold variation in *Wolffiella* (623–973 Mbp) or *Lemna* (323–760 Mbp) and up to 5.3-fold in the genus *Wolffia* (357–1881 Mbp). In addition, large intraspecific DNA content variances were found among *Lemna* species (e.g. *Lemna minor*, *L. aequinoctialis*, *L. trisulca* and *L. japonica*), while other genera did not show such intraspecific variation. Furthermore, there was no significant overall correlation of genome size values with environmental conditions determined by latitude, longitude and altitude (Wang et al. 2011).

### 4.2.2 Chromosome Number and Karyotype Evolution

There are only few studies of chromosome numbers in duckweeds. The likely first chromosome count of duckweeds was documented in 1933 with a note "the notable feature of the chromosomes of the whole group is their extremely small size" (Blackburn 1933). The smallest chromosomes in *Spirodela polyrhiza* were measured only 0.1 × 0.18 µm (cf. Geber 1989), suggesting the smallest recorded chromosomes in flowering plants at that time. The largest duckweed chromosomes are in *Wolffia* species that have a mean length of about 1.7 µm (Urbanska-Worytkiewicz 1980). In general, duckweed chromosomes were a very difficult material for handling in cytogenetic studies and it has been often the case that different chromosome counts were reported for the same accessions (clones) and species, due to different and likely not-yet-optimal preparation techniques.

The most intensive investigation on duckweed chromosome numbers was accomplished in a 15-year work of Urbanska-Worytkiewicz (1980) in which the author studied about 1500 accessions (clones) of total 30 duckweed species. Surprisingly, reported chromosome counts range from $2n = 20$–126, but do not correlate well with genome size variation (Urbanska-Worytkiewicz 1980; Wang et al. 2011; Cao et al. 2016). Using the squash technique of mitotic dividing cells and chromosome staining by lacto-propionic-orcein, Urbanska-Worytkiewicz proposed the basic chromosome number of duckweed as $x = 10$. The most common karyotype within duckweed species is a diploid number of $2n = 40$. In addition, the euploid numbers ($2n = 20$, 30, 50, 60, 70, 80) and the aneuploid numbers ($2n = 36$, 42) have been recorded. Furthermore, three levels of cytological variations including intra-individual (e.g. aneusomaty, mixoploidy),

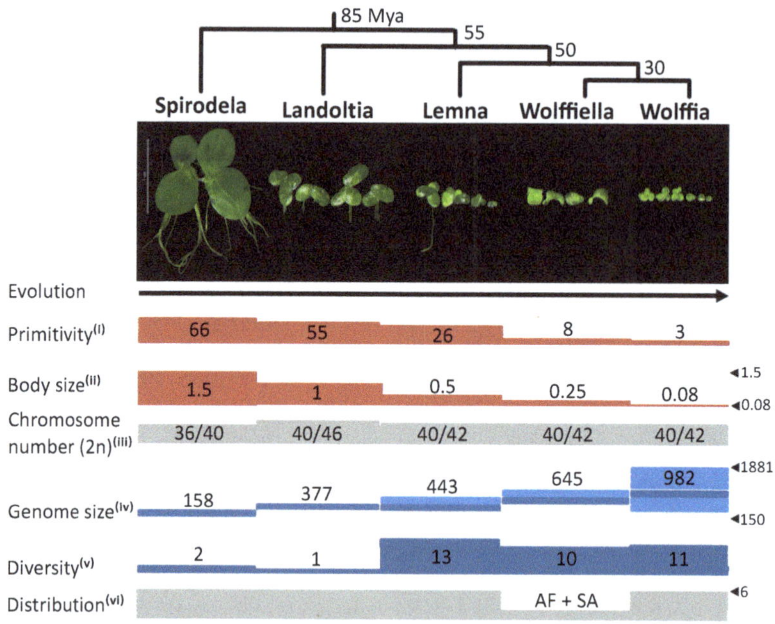


**Fig. 4.1** **Duckweed as an interesting model for studying genome and karyotype evolution**. Representative species for five duckweed genera (*S. polyrhiza* clone 7498, *La. punctata* clone 7260, *L. minor* clone 8623, *Wolffiella lingulata* clone 7655 and *Wolffia arrhiza* clone 8872) demonstrate body size and reduction of morphological of duckweeds (Cao et al. 2015). (i) Degree of primitivity (from the most primitive 71 to the most derived 3) was according to the index of 26 characteristics (Landolt 1986); (ii) the minimum size of frond width in mm (summarized from Landolt 1986, 1998a, 1992, 1994); (iii) the most common chromosome counts (summarized from Urbanska-Worytkiewicz 1980; Geber 1989; Landolt 1986; Wang et al. 2011; Cao et al. 2016) (iv) the mean value of genome sizes among genera (summarized from Wang et al. 2011; Bog et al. 2015); (v) diversity by the number of species among genera; and (vi) geography distributions (summarized from Landolt 1986; Landolt 1998a, 1992, 1994) by the number of continents (AF, Africe; SA, South America). More details can be found in Table 4.1

intra-populational (e.g. aneuploidy, polyploidy) and cytological variation (or technical limitations) were observed. For instance, among 187 studied accessions of *Spirodela polyrhiza*, $2n = 40$ is the most prevalent number (in 175 accessions), whereas $2n = 30$, 50 was revealed as cytological variation in 11 and 1 accession, respectively. Intra-individual variation was observed in three accessions as aneusomatic (cells with $2n = 34$ among $2n = 40$ cells), mixoploid (cells with $2n = 40$ and 80) and the presumably both ($2n = 62$ cells in the $2n = 50$ accession). Intra-populational variation was found in plants from North America consisting of $2n = 40$ and $2n = 38$. A further cytological work using air-drying technique with DAPI staining (Geber 1989) showed more uniform results (e.g. $2n = 40$ in 6 accessions of *S. polyrhiza* with one accession of $2n = 80$). According to his investigations, the basic number for *Lemna*, *Wolffiela*, and *Wolffia* is $2n = 42$. Due to technical difficulties, chromosome counts of duckweed species have remained for two decades with conflict and missing information (Table 3.1). Consequently, there was no further study on molecular cytogenetics or chromosomes of duckweed during this period of time.

Thanks to the development of probes for fluorescent in situ hybridizations and the advent of next-generation sequencing technologies (NGS), modern molecular cytogenetics in duckweeds subsequently got a boost. The protocol for

**Table 4.1** Summary of duckweed cytogenetic features in relation to evolution, development and diversity

| Species | Degree of primitivity[b] | Chromosome number (2n)[c] | Genome size (Mbp/1C)[d] | Frond width (mm)[e] | Continental distribution[f] |
|---|---|---|---|---|---|
| Spirodela | | | | | |
| *S. intermedia* | 69–71 | 30/36[(20)] | 154–162 | 2–9 | SA |
| *S. polyrhiza* | 66–68 | 40[(30, 38, 50, 80)] | 150–165 | 1.5–8 | WW–AN |
| Landoltia | | | | | |
| *L. punctata* | 55–60 | 40/46[(50)] | 372–427 | 1–5 | AS (WW–AN) |
| Lemna | | | | | |
| *L. aequinoctialis* | 30–32 | 40/42[(20, 50, 60, 70, 80, 84)] | 424–760 | 0.8–4.5 | WW–AN |
| *L. perpusilla* | 31–33 | 40/42 | | 0.8–3 | NA |
| *L. tenera* | 27–29 | | | 1.2–3 | AS |
| *L. disperma* | 43–45 | 40/44 | | 0.6–3 | OC |
| *L. gibba* | 46–49 | 40/42[(44, 50, 60, 64, 70, 80, 84)] | 440–486 | 0.8–6 | WW–AN, OC |
| *L. japonica* | 39–41 | 40/63[(50)] | 426–600 | 0.6–4 | AS |
| *L. minor* | 39–41 | 40/42[(20, 30, 50, 63, 126)] | 356–604 | 0.6–7 | WW–AN, (OC) |
| *L. obscura* | 38–40 | 40/42[(50)] | 487 | 0.7–3 | NA, SA |
| *L. trisulca* | 34–36 | 40/42[(20, 44, 60, 63, 80)] | 446–709 | 1–5 | WW–AN, OC, SA |
| *L. turionifera* | 38–40 | 40/42[(50, 80)] | | 0.8–3.5 | NA, AS |
| *L. minuta* | 26–28 | 40/42[(36)] | | 0.5–2.5 | SA (EU, AS) |
| *L. valdiviana* | 26–28 | 40/42 | 323 | 0.6–3 | NA, SA |
| *L. yungensis*[a] | 26–28? | | | 1.5–4 | SA |
| Wolffiella | | | | | |
| *W. caudata* | 8–13? | | | 2–5 | SA |
| *W. denticulata* | 8–9 | 40[(20)] | | 0.3–0.8 | AF |
| *W. gladiata* | 8–9 | 40/42 | 623 | 0.25–0.8 | NA |
| *W. lingulata* | 9–11 | 40/42[(20, 42, 50)] | 629–655 | 0.8–5 | SA |
| *W. neotropica* | 11–13 | 40 | | 1–5.5 | SA |
| *W. oblonga* | 8–10 | 40/42[(70)] | | 0.4–2.5 | SA, AF |
| *W. welwitschii* | 9–11 | 40 | | 2.5–5 | SA, AF |
| *W. rotunda* | 10 | | | 1–3 | AF |
| *W. hyalina* | 9–10 | 40 | 894–973 | 0.8–2 | AF (AS) |
| *W. repanda* | 8–9 | | | 0.4–1.2 | AF |
| Wolffia | | | | | |
| *W. australiana* | 6 | 40[(20)] | 357–375 | 0.3–0.8 | OC |
| *W. borealis* | 8–9 | 40[(20, 22, 30)] | 889 | 0.5–1 | NA |
| *W. brasiliensis* | 8–9 | 40[(20, 42, 50, 60, 80)] | 776 | 0.7–1.5 | SA, NA |
| *W. microscopica* | 6–7 | 40[(35, 42, 70, 80)] | | 0.3–0.8 | SA |
| *W. angusta* | 4 | 40 | | 0.2–0.4 | OC, AS |

(continued)

**Table 4.1** (continued)

| Species | Degree of primitivity[b] | Chromosome number (2n)[c] | Genome size (Mbp/1C)[d] | Frond width (mm)[e] | Continental distribution[f] |
|---|---|---|---|---|---|
| *W. arrhiza* | 4–6 | $40/42^{(30, 50, 60, 62, 63, 70, 80)}$ | 1881 | 0.4–1.2 | EU, AF, AS, SA (NA) |
| *W. columbiana* | 3–4 | $40/42^{(30, 50, 70)}$ | 874 | 0.5–1.2 | NA, SA |
| *W. cylindracea* | ca 3? | | 1076 | 0.3–0.7 | AF |
| *W. elongata* | ca 4 | | | 0.3–0.6 | SA |
| *W. globosa* | 3 | $40^{(16, 23, 30, 46, 50, 60)}$ | 1295 | 0.3–0.6 | AS, AF (NA?) |
| *W. neglecta* | ca 4? | | | 0.4–0.6 | AS |

[a]This species is subjected to revision (see Chap. 11)
[b]Degree of primitivity (from the most primitive 71 to the most derived 3) according to the index of 26 characteristics (modified from Landolt 1986)
[c]The most common chromosome counts, and found karyotype variations, in superscript and brackets (summarized from Urbanska-Worytkiewicz 1980; Geber 1989; Landolt 1986; Wang et al. 2011; Cao et al. 2016)
[d]The range of duckweed genome sizes (summarized from Wang et al. 2011; Bog et al. 2015)
[e]The range of frond width in mm (summarized from Landolt 1986, 1998a, 1992, 1994)
[f]Geography distributions (summarized from Landolt 1986, 1998a, 1992, 1994) in worldwide (WW) or in continents (*AF* Africe; *NA* North America; *OC* Oceania; *AN* Antarctica; *AS* Asia; *EU* Europe; *SA* South America). Brackets or superscript with prefix minus, indicates new introduction of duckweed species; or the absence of duckweed in respective continents, respectively

chromosome preparation of duckweed species has been recently optimized and advanced (Cao et al. 2016). With the availability of a *Spirodela polyrhiza* BAC library and a genome-integrated minimum tiling path (Wang et al. 2014), fundamental resources for molecular cytogenetic studies with multicolour fluorescence in situ hybridization (mcFISH) of duckweed have been established. That includes a Spirodela cytogenetic map containing 96 BAC markers with an average distance of 0.89 Mbp and a cocktail of 41 BACs in three colours for simultaneous identification of all chromosome pairs. The mcFISH system has been demonstrated as an independent and unique way to validate, correct and integrate the NGS assemblies and genomic optical maps by offering long-range linkage information over the whole chromosome. Importantly, the seven ancestral chromosome blocks which were emerged from two rounds of whole-genome duplications approximately 90 My ago were reconstructed, enabling future studies on the chromosome homoeology and karyotype evolution of duckweed species by comparative chromosome painting for instance.

## 4.3 Genome Features, Organization and Adaptation

The famous couplet of Dr. Kihara (1947) "The history of the earth is recorded in the layers of its crust. The history of all organisms is inscribed in the chromosomes". It has increasingly been become evident. Evolutionary path to flowering plants began more than 400 million years ago with the marine green algae ancestor, which evolved to cope with terrestrial habitats and to produce flowers and seeds (de Vries and Archibald 2018). Then duckweeds performed a remarkable accomplishment when they adapted to a freshwater lifestyle (about 100 Mya in the Cretaceous, Nauheimer et al. 2012) and became able to compete with other aquatic plants, which managed also similar extreme habitat shifts. The recent and imminent release of duckweed genomes has and will lead to a progressively greater understanding of genome features and organization in this plant family. At the time of writing, there are three duckweed genome sequences publicly available from two geographical accessions of the Greater

Duckweed (*Spirodela polyrhiza*, Michael et al. 2017; Wang et al. 2014) and one accession of the Common Duckweed (*Lemna minor*, Van Hoeck et al. 2015).

Spirodela genomes have a size of 158 Mbp and organized into 20 chromosomes ($2n = 40$). Interestingly, Spirodela has the fewest protein-coding genes of sequenced angiosperms at 18,507 for accession 9509 and at 19,623 for accession 7498. With a three-fold larger genome (481 Mbp) but also with 20 chromosome pairs, *Lemna minor* accession 5500 contains a similar number of protein-coding genes (18,744 high confidence genes out of total 22,382 predicted genes, Van Hoeck et al. 2015). These numbers correspond to 30% less than that of *Arabidopsis thaliana* and 50% less than that of monocotyledonous rice *Oryza sativa*. It is worthy to note that Spirodela has undergone two successive rounds (ca 95 Mya, likely coincident with the split between duckweeds and the remaining monocot family Araceae) of whole-genome duplications, but yet has maintained a small genome size and a small number of protein-coding genes. Comparing Spirodela predicted proteins with those of plant reference genomes (e.g. *A. thaliana*, tomato (*Solanum lycopersicum*), banana (*Musa acuminata)* and rice (*Oryza sativa spp. indica)*), there are 1745 Spirodela-specific genes (8.9% of the total predicted protein-coding genes), that are enriched for various defence-related processes (Wang et al. 2014). In addition, *Spirodela* gene families showed a significantly reduced gene number and preferential removal of duplicated genes. Furthermore, the loss of several plant conserved gene clusters, including genes involved in water transport, biosynthesis of phenylpropanoid and lignin as well as cell wall organization is consistent with the miniature plant body architecture and specialized ecological adaptation of *Spirodela*. However, in order to optimize extremely fast growth, a few specific gene families involved in ammonium assimilation and light harvesting are exceptionally amplified. Lemna proteome is mostly (66%) shared with the *Spirodela* proteome (Van Hoeck et al. 2015), including a lineage-specific enrichment of proteins involved in adaptation to various climate conditions, in removal of surplus nutrients from wastewater, and in providing nutritional value and high biomass productivity.

The increasing wealth of complete whole-genome sequences highlights the critical role of transposable elements (TEs) in plant genome evolution. These mobile genetic elements function as a driver of drastic changes in genome size and as an important source of new variants in coding and regulatory sequences. Transposable elements (TEs) constitute 16 to 25% of the *Spirodela* genomes 7498 and 9509, respectively, while *A. thaliana* genomes has roughly similar nuclear genome size (135–157 Mbp, Bennett et al. 2003) and similar TE content (15–24%, The Arabidopsis Genome Initiative 2000; Hu et al. 2011). Given that repetitive sequences comprise 62% of the *L. minor* genome assembly, repeat content explains 94% of the genome size difference between sequenced *Spirodela* and Lemna genomes (Van Hoeck et al. 2015). A detailed look on sequences of long terminal repeats (LTRs) in the *Spirodela* genomes suggests that the intact LTRs are old ($\sim$4 Mya) and a high proportion of LTRs contains one terminal repeat without its pair or internal sequence (solo LTRs). Moreover, *Spirodela* genomes are streamlined with less than 100 copies of ribosomal DNA, corresponding to 15% of that in *Arabidopsis* genome (Michael et al. 2017).

The whole-genome sequence of seagrass *Zostera marina* or eelgrass (the only other sequenced alismatid with the 202 Mb genome and 20,450 protein-coding genes) which diverged from the common ancestor (Alismatales, Araceae) with duckweeds between 135 and 107 million years ago (Olsen et al. 2016) provides new insights on the alismatid lineage with the stage for the terrestrial-freshwater and the subsequent freshwater-marine transitions. Since the split from *Z. marina*, *Spirodela* has gained 292 protein domains (Pfam database), 40 of which appear to be unique. Interestingly, 519 protein domains have been lost, while 467 protein domains have contracted in the *Spirodela* proteome. These numbers are more than twice of the corresponding numbers of *Z. marina* (146 and 162 protein domains, respectively). In addition, transposable elements account for 63% of the *Zostera* assembly and tend to accumulate in stretches of repeat elements which separating gene-dense islands. Those observations indicate typical genome shrinkage in *S. polyrhiza* by eliminating non-essential protein-coding genes as well as repetitive sequences (e.g. LTRs and ribosomal DNA) probably due to deletion-towards-biases in DNA repair mechanisms (Schubert and Vu 2016).

Duckweeds mainly proliferate by vegetative budding of new fronds from the meristematic zones, in a macroscopic manner analogous to asexual propagation in yeast. With such clonal propagation, duckweeds rapidly achieve massive population sizes in nature as several millions of individuals (i.e. ramets) can be found in a single pond. In ecological aspect, using this reproductive strategy, duckweeds have apparently well adapted to large fluctuations that may rapidly occur in aquatic habitats. The recurrent rejuvenation of duckweed populations enhances the phenotypical plasticity, allowing duckweeds to respond more flexibly towards their changing environment. Asexuality (i.e. lack of efficient recombination, independent reassortment and segregation) is supposed to allow faster accumulation of chromosomal mutations than in sexual or more frequently sexual species (commonly known as the Meselson effect, Weir et al. 2016). It could be a possible explanation for controversial observations (Urbanska-Worytkiewicz 1980) of complex cytological variations of duckweeds at multiple levels (i.e. intra-individual, intra-population and inter-population variations with aneuploid, mixoploid and (endo)polyploids). However, how and to what extent this reproductive strategy can have differential consequences on the plasticity and evolution of duckweed genomes remain an open question and a challenge for future work. Recent population genomics investigations on geographically widespread clones of *S. polyrhiza* revealed a very low genetic diversity which is likely due to a very low mutation rate (i.e. one to two orders of magnitude lower than Arabidopsis, Xu et al. 2019; Ho et al. 2019). Interestingly, diversity at nonsynonymous sites relative to synonymous sites is shown to be high, possibly suggesting a relaxed selection on many genes due to the simplified form and lifestyle of duckweeds. As more duckweed genomes are reported from several ongoing whole-genome sequencing projects of duckweeds, including *Spirodela intermedia*, *La. punctata*, *L. minor*, *L. gibba* and *Wolffia australiana*, these will pave the way for the detailed understanding of such exciting genome features, organization and genome adaptation.

## 4.4 Epigenetic and Epigenomic Aspects

Genomic DNA in eukaryotic cells is condensed and packaged with histones to form a DNA-protein complex known as chromatin at various levels of folded structures. In interphase nuclei, chromatin structures exhibit spatio-temporal dynamics both in the context of their local condensation and nuclear positioning, presenting as the highly compact heterochromatin and the less condensed euchromatin patterns. Chromatin modifications (such as histone modifications and DNA methylation) are epigenetic marks that had the potential to modify the degree of chromatin condensation, and thus are the key determinants of gene activities, cell fate and genome stability. The advent of high-throughput methods for genome-wide profiling of epigenetic marks such as bisulphite sequencing (BSseq) for DNA methylation and chromatin immuno-precipitation DNA-sequencing (ChIPseq) for histone modifications has sparked interest in studying the epigenetic modifications of chromatin in duckweeds, which to date has received surprisingly little attention.

In conventional cytogenetic approaches, chromatin staining, for example, with DNA fluorochromes helps to visualize heterochromatin and euchromatin as strongly stained and weakly stained regions, respectively, under the microscope. The former usually corresponds to transcriptionally inactive, repetitive or gene-poor genome regions, whereas the later corresponds to transcriptionally active parts of the genome. Günter Geber, in his thesis (1989), could recognize heterogeneity in duckweed nuclei by chromatin staining with different DNA base-specific fluorochromes. He observed chromocenter structures, masses of heterochromatin, by guanine-specific chromomycin A3 (CMA) staining but not by thymine-specific 4′,6-diamidino-2-phenylindole (DAPI), suggesting the presence of GC-rich repetitive heterochromatin fractions in duckweed genomes. The more recent investigation on nuclei of representative duckweed species, including *S. polyrhiza*, *La. punctata*, *L. minor*, *W. lingulata* and *W. arrhiza* (Cao et al. 2015) with different sizes of genome, found AT-rich chromocenters only in nuclei of *W. arrhiza* (1C = 1881 Mbp). Interestingly, the immunostaining, the identification of DNA methylation (5-mC) and histone modifications (H3K9me2 and H3K27me1) by means of fluorescence-labelled antibodies, showed dispersed signals of those heterochromatin marks throughout the duckweed nuclei. Such chromatin organization, especially in small genome duckweed species, does resemble the global distribution pattern of *A. thaliana* nuclei particularly from tissue culture or from young seedlings while heterochromatin immunosignals of *A. thaliana* typical nuclei are restricted to constitutive and pronounced heterochromatin chromocenters. The recent study on genome-wide cytosine methylation of *S. polyrhiza* from bisulphite sequencing revealed an extremely low global DNA methylation level (as low as 9% of the genome, Michael et al. 2017) as compared with *A. thaliana* (32%), rice (39%) or *Brachypodium distachyon* (54%). Given that despite an astonishingly low genetic diversity duckweeds demonstrate a high level of phenotypic plasticity enabling thriving in a worldwide range of environmental conditions, this ability to adapt might be underlined by amazingly efficient epigenetic mechanisms including also in TE silencing machinery (for no recent TE activity), in DNA repair (for extremely low mutation rate) and in transgenerational epigenetic inheritance (for the fastest growing rate in flowering plants). In future, detailed epigenomics and epigenetic modifications will help researchers to delve into exactly which (epi)genetic elements or epigenetic pathways facilitate such high effectiveness and efficiency in duckweed systems.

**Acknowledgements** We thank Charlotte Ost (University of Halle-Wittenberg) for reading the article and for critical discussions.

## References

Appenroth KJ, Crawford DJ, Les DH (2015) After the genome sequencing of duckweed—how to proceed with research on the fastest growing angiosperm? Plant Biol (Stuttg) 17:1–4. https://doi.org/10.1111/Plb.12248

Bennett MD, Leitch IJ, Price HJ, Johnston JS (2003) Comparisons with Caenorhabditis (approximately 100 Mb) and Drosophila (approximately 175 Mb) using flow cytometry show genome size in Arabidopsis to be approximately 157 Mb and thus approximately 25% larger than the *Arabidopsis* genome initiative estimate of approximately 125 Mb. Ann Bot 91(5):547–557

Blackburn KB (1933) Notes on the chromosomes of the duckweeds (Lemnaceae) introducing the question of chromosome size. In: Proceedings of University Durham Philosophy Society. pp 84–90

Bog M, Lautenschlager U, Landrock M, Landolt E, Fuchs J, Sowjanya Sree K, Oberprieler C, Appenroth K-J (2015) Genetic characterization and barcoding of taxa in the genera *Landoltia* and *Spirodela* (Lemnaceae) by three plastidic markers and amplified fragment length polymorphism (AFLP). Hydrobiologia 1–14. https://doi.org/10.1007/s10750-014-2163-3

Cabrera LI, Salazar GA, Chase MW, Mayo SJ, Bogner J, Dávila P (2008) Phylogenetic relationships of aroids and duckweeds (Araceae) inferred from coding and noncoding plastid DNA. Am J Bot 95(9):1153–1165. https://doi.org/10.3732/ajb.0800073

Cao HX, Vu GT, Wang W, Appenroth KJ, Messing J, Schubert I (2016) The map-based genome sequence of *Spirodela polyrhiza* aligned with its chromosomes, a reference for karyotype evolution. New Phytol 209(1):354–363. https://doi.org/10.1111/nph.13592

Cao HX, Vu GTH, Wang W, Messing J, Schubert I (2015) Chromatin organisation in duckweed interphase nuclei in relation to the nuclear DNA content. Plant Biol (Stuttg) 17:120–124. https://doi.org/10.1111/plb.12194

Cavalier-Smith T (1985) The evolution of genome size. Wiley Ltd, Chichester, UK

Cusimano N, Bogner J, Mayo SJ, Boyce PC, Wong SY, Hesse M, Hetterscheid WLA, Keating RC, French JC (2011) Relationships within the araceae: comparison of morphological patterns with molecular phylogenies. Am J Bot 98(4):654–668. https://doi.org/10.3732/ajb.1000158

Daubs EH (1962) A monograph of Lemnaceae. University of Illinois at Urbana-Champaign
de Vries J, Archibald JM (2018) Plant evolution: landmarks on the path to terrestrial life. New Phytol. https://doi.org/10.1111/nph.14975
Engler A (1876) Vergleichende Untersuchungen über die morphologischen Verhältnisse der Araceae, vol 39. vol 3. Druck von E. Blochmann & Sohn für die Akademie in Commission bei Wilh. Engelmann in Leipzig
Engler A (1889) Lemnaceae In: Engler A, Prantl K (eds) Die Natürlichen Pflanzen Familien. pp 154–164
Geber G (1989) Zur Karyosystematik der Lemnaceae. University of Vienna, Vienna, Austria
Gray SF (1821) A natural arrangement of British plants
Gregory TR (2005) The evolution of the genome. Academic Press, Burlington. https://doi.org/10.1016/B978-0-12-301463-4.X5000-1
Hegelmaier F (1868) Die Lemnaceen: eine monographische Untersuchung. Engelmann
Hegelmaier F (1895) Systematische übersicht der Lemnaceen
Henriquez CL, Arias T, Pires JC, Croat TB, Schaal BA (2014) Phylogenomics of the plant family Araceae. Mol Phylogenet Evol 75:91–102. https://doi.org/10.1016/j.ympev.2014.02.017
Ho EKH, Bartkowska M, Wright SI, Agrawal AF (2019) Population genomics of the facultatively asexual duckweed *Spirodela polyrhiza*. New Phytol. https://doi.org/10.1111/nph.16056
Hu TT, Pattyn P, Bakker EG, Cao J, Cheng JF, Clark RM, Fahlgren N, Fawcett JA, Grimwood J, Gundlach H, Haberer G, Hollister JD, Ossowski S, Ottilar RP, Salamov AA, Schneeberger K, Spannagl M, Wang X, Yang L, Nasrallah ME, Bergelson J, Carrington JC, Gaut BS, Schmutz J, Mayer KF, Van de Peer Y, Grigoriev IV, Nordborg M, Weigel D, Guo YL (2011) The *Arabidopsis lyrata* genome sequence and the basis of rapid genome size change. Nat Genet 43 (5):476–481. https://doi.org/10.1038/ng.807
Kihara H (1947) Ancestors of common wheat. Tokyo: Sogensha Tokyo (in Japanese) Cited by Kimura M, Ohta T On some principles governing molecular evolution. Proc Natl Acad Sci USA 71:2848–2852
Landolt E (1986) The family of Lemnaceae—a monographic study, vol 1. Veroffentlichungen des Geobotanischen Institutes der Eidgenossischen Technischen Hochschule, Stiftung Rubel, Zurich
Landolt E (1992) Wolffiella caudata, a new Lemnaceae species from the Bolivian Amazon region. Berichte des Geobotanischen Institutes der Eidgenössischen Technischen Hochschule Stiftung Ruebel 58:121–123
Landolt E (1994) Taxonomy and ecology of the section Wolffia of the genus Wolffia (Lemnaceae). Ber Geobot Inst ETH, Stiftung Rübel, Zürich 60:137–151
Landolt E (1998a) Lemna yungensis, a new duckweed species from rocks of the Andean Yungas in Bolivia. Bull Geobot Inst ETH 64:15–21
Landolt E (1998b) Lemnaceae. In: Flowering plants monocotyledons. Springer, pp 264–270
Landolt E, Kandeler R (1987) Biosystematic investigations in the family of duckweeds (Lemnaceae), vol 4: The family of Lemnaceae-a monographic study, vol 2 (phytochemistry, physiology, application, bibliography). Veroeffentlichungen des Geobotanischen Instituts der ETH, Stiftung Ruebel (Switzerland)
Michael TP, Bryant D, Gutierrez R, Borisjuk N, Chu P, Zhang H, Xia J, Zhou J, Peng H, El Baidouri M, Ten Hallers B, Hastie AR, Liang T, Acosta K, Gilbert S, McEntee C, Jackson SA, Mockler TC, Zhang W, Lam E (2017) Comprehensive definition of genome features in *Spirodela polyrhiza* by high-depth physical mapping and short-read DNA sequencing strategies. Plant J 89(3):617–635. https://doi.org/10.1111/tpj.13400
Nauheimer L, Metzler D, Renner SS (2012) Global history of the ancient monocot family Araceae inferred with models accounting for past continental positions and previous ranges based on fossils. New Phytol 195 (4):938–950. https://doi.org/10.1111/j.1469-8137.2012.04220.x
Olsen JL, Rouze P, Verhelst B, Lin YC, Bayer T, Collen J, Dattolo E, De Paoli E, Dittami S, Maumus F, Michel G, Kersting A, Lauritano C, Lohaus R, Topel M, Tonon T, Vanneste K, Amirebrahimi M, Brakel J, Bostrom C, Chovatia M, Grimwood J, Jenkins JW, Jueterbock A, Mraz A, Stam WT, Tice H, Bornberg-Bauer E, Green PJ, Pearson GA, Procaccini G, Duarte CM, Schmutz J, Reusch TB, Van de Peer Y (2016) The genome of the seagrass Zostera marina reveals angiosperm adaptation to the sea. Nature. https://doi.org/10.1038/nature16548
Schubert I, Vu GT (2016) Genome stability and evolution: Attempting a holistic view. Trends Plant Sci 21 (9):749–757. https://doi.org/10.1016/j.tplants.2016.06.003
Sree KS, Bog M, Appenroth KJ (2016) Taxonomy of duckweeds (Lemnaceae), potential new crop plants. Emir J Food Agr 28(5):291–302. https://doi.org/10.9755/ejfa.2016-01-038
The Arabidopsis Genome Initiative (2000) Analysis of the genome sequence of the flowering plant *Arabidopsis thaliana*. Nature 408(6814):796–815. https://doi.org/10.1038/35048692
Urbanska-Worytkiewicz K (1980) Cytological variation within the family of Lemnaceae. In: Landolt E (ed) Biosystematic investigations in the family duckweeds (Lemnaceae), vol 70. Veroff. Geobot. Inst. ETH, Stiftung Rubel, Zürich, pp 30–101
Van Hoeck A, Horemans N, Monsieurs P, Cao HX, Vandenhove H, Blust R (2015) The first draft genome of the aquatic model plant Lemna minor opens the route for future stress physiology research and biotechnological applications. Biotechnol Biofuels 8 (1):1–13. https://doi.org/10.1186/s13068-015-0381-1
Wang W, Haberer G, Gundlach H, Glasser C, Nussbaumer T, Luo MC, Lomsadze A, Borodovsky M, Kerstetter RA, Shanklin J, Byrant DW, Mockler TC, Appenroth KJ, Grimwood J, Jenkins J, Chow J, Choi C, Adam C, Cao XH, Fuchs J, Schubert I,

Rokhsar D, Schmutz J, Michael TP, Mayer KF, Messing J (2014) The *Spirodela polyrhiza* genome reveals insights into its neotenous reduction fast growth and aquatic lifestyle. Nat Commun 5:3311. https://doi.org/10.1038/ncomms4311

Wang WQ, Kerstetter R, Michael TP (2011) Evolution of genome size in duckweeds (Lemnaceae). J Botany 2011:1300099. https://doi.org/10.1155/2011/570319

Weir W, Capewell P, Foth B, Clucas C, Pountain A, Steketee P, Veitch N, Koffi M, De Meeus T, Kabore J, Camara M, Cooper A, Tait A, Jamonneau V, Bucheton B, Berriman M, MacLeod A (2016) Population genomics reveals the origin and asexual evolution of human infective trypanosomes. Elife 5:e11473. https://doi.org/10.7554/eLife.11473.006

Xu S, Stapley J, Gablenz S, Boyer J, Appenroth KJ, Sree KS, Gershenzon J, Widmer A, Huber M (2019) Low genetic variation is associated with low mutation rate in the giant duckweed. Nat Commun 10(1):1243. https://doi.org/10.1038/s41467-019-09235-5

# Genetic Diversity and DNA Barcoding in the Duckweed Family

5

Jiaming Zhang and Azizullah Azizullah

**Abstract**
Lemnaceae is a globally distributed aquatic plant family. This family is composed of 37 species with distinct ecological characteristics and distribution patterns. The intra-specific diversity in this family is relatively low compared to other angiosperms, in which the proposed primordial duckweed, *Spirodela polyrhiza,* has the lowest intra-specific diversity. Due to simplified morphology, species identification in this family is difficult without the help of DNA barcoding.

## 5.1 Species Diversity and Distribution in the Duckweed Family

The duckweed family is comprised of five genera and 37 species (Appenroth et al. 2013). They are distributed in different areas of the planet with some species distributed more globally, while others are local to a small area (Table 5.1). For global distribution maps of the species, please also see Chap. 15, phylogenetic relationships.

The genus Spirodela contains two species, including *Spirodela polyrhiza* and *S. intermedia.* The widely studied *S. polyrhiza* is the more globally distributed species, located in the tropical, subtropical, and temperate zones. It is found in a wide range of habitats between 62°N and 35°S in Asia, Europe, North America, South America, Africa, and Australia. It has been collected in high latitude regions such as Irkutsk, Siberia (53°N), and Moscow (56°N) in Russia; it has even been collected in Pukila, Finland, close to the Arctic Circle. Interestingly, another species in the genus Spirodela, *S. intermedia,* is local to the tropical region in South America; however, it was once collected at a botanical garden in Delhi, India.

*Landoltia punctata* is the only species in the genus Landoltia and is widely distributed in the tropical and subtropical regions. The latitude of its habitats is between 40°N and 40°S in South America, North America, Asia, Oceania, and Europe. However, it is not known whether it is distributed in Africa.

The genus Lemna contains 13 species and is collected all around the six human habituated continents (Table 5.1). The species in this genus are the most divergent in ecological characteristics. *Lemna minor* and *L. gibba* are probably most widely distributed species in this genus. They are located in the subtropical and temperate zones and have been collected from as far south

J. Zhang (✉)
Institute of Tropical Bioscience and Biotechnology, Chinese Academy of Tropical Agricultural Sciences, Xueyuan Road 4, Haikou 571101, China
e-mail: zhangjiaming@itbb.org.cn

A. Azizullah
Department of Botany, Kohat University of Science and Technology, 26000 Kohat, Pakistan


X. H. Cao et al. (eds.), *The Duckweed Genomes*, Compendium of Plant Genomes,
https://doi.org/10.1007/978-3-030-11045-1_5

**Table 5.1** Distribution of duckweed species

| Genus | Species | Continent | Latitude |
|---|---|---|---|
| Spirodela | Polyrhiza | Asia, Europe, North America, South America, Africa, and Oceania | 62°N–35°S |
| | Intermedia | South America, Asia | 18°N–35°S |
| Landoltia | Punctata | South America, North America, Asia, Oceania, and Europe | 40°N–40°S |
| Lemna | Minor | Asia, Europe, Africa, North America, and Oceania, | 60°N–15°N<br>15°S–40°S |
| | Gibba | Asia, Europe, North and South America, and Africa | 63°N–10°N<br>12°S–34°S |
| | Aequinoctialis | Asia, North America, South America, Africa, and Europe | 35°N–35°S |
| | Minuta | Europe, North America, South America, and Asia | 52°N–13°N<br>13°S–34°S |
| | Valdiviana | South America, North America | 35°N–35°S |
| | Japonica | East Asia | 45°N–26°N |
| | Tenera | Northern Australia, Southeast Asia | 18°N–18°S |
| | Yungensis | South America | 10°S–26°S |
| | Turionifera | Asia, Europe, North America | 62°N–35°N |
| | Obscura | North America, South America | 35°N–5°N |
| | Perpusilla | North America | 40°N–30°N |
| | Trisulca | Asia, Europe, North America, Africa | 60°N–10°S |
| | Disperma | Oceania | 25°S–40°S |
| Wolffiella | Lingulata | North America, South America | 35°N–35°S |
| | Oblonga | North America, South America | 35°N–35°S |
| | Gladiata | North America | 40°N–25°N |
| | Neotropica | South America | 15°N–15°N |
| | Welwitschii | Africa, South America | 15°N–30°S |
| | Senticulata | Africa | 15°S–30°S |
| | Caudata | South America | 5°N–25°S |
| | Repanda | Africa | 15°S–30°S |
| | Hyalina | Asia, Africa | 15°N–15°S |
| | Rotunda | Africa | 15°S–30°S |
| Wolffia | Arrhiza | Asia, Europe, South America, Africa | 50°N–35°S |
| | Cylindracea | Africa | 10°S–35°S |
| | Columbiana | North America, South America | 45°N–10°S |
| | Elongata | Asia, South America | 35°N–5°N |
| | Neglecta | Asia | 30°N–20°N |
| | Angusta | Oceania, Southeast Asia | 10°N–35°S |
| | Globosa | Asia, North America, South America | 35°N–5°N |
| | Microscopica | Asia | 30°N–15°N |
| | Australiana | Oceania | 30°S–45°S |
| | Borealis | North America | 45°N–30°N |
| | Brasiliensis | South America, North America | 45°N–30°S |

as New Zealand (45°S) and as far north as Finland, Norway (60°N), and Saskatoon, Canada. Their habitats spread over six continents including Asia, Europe, Africa, Oceania, North America, and South America. Thanks to this cosmopolitan distribution, the International Organization for Standardization (ISO) and the Organization for Economic Co-operation and Development (OECD) both use *L. minor* and *L .gibba* in water quality and chemical safety tests. *L. minuta* has a similar distribution pattern; however, it is spread across a comparatively smaller area. These three Lemna species are not collected in typical tropical zones.

In contrast to *L. minor* and *L. gibba*, *L. aequinoctialis*, *L. valdiviana*, and *L. obscura* are mostly habituated in the tropical and subtropical areas between 35°N and 27°S. *L. valdiviana* and *L. obscura* are mainly distributed in South America and North America, while *L. aequinoctialis* is distributed more widely in Asia, North America, South America, Africa, and Europe. *L. aequinoctialis* has occasionally been collected in high latitude in Europe, such as France (45°N). *L. tenera* is a typical tropical species and mainly located in Northern Australia and Southeast Asia. *L. trisulca* is mainly distributed in the subtropical region of the northern hemisphere; however, it is also collected in the tropical region (Uganda and Kenya).

*L. turionifera* and *L. perpusilla* are mainly distributed in the temperate zone of the northern hemisphere in Asia and North America, while *L. yungensis* and *L. disperma* are distributed in small regions in South America and Oceania, respectively. *L. japonica* is mainly located in East Asia such as China and Japan. Some accessions have been collected in Finland, Germany, and USA, but their identities have not been confirmed by DNA barcodes.

There are ten species in the genus Wolffiella. Most species in this genus have narrow distribution. They are mostly distributed in North and South America and Africa in a latitude continuous pattern from the southern hemisphere to the northern hemisphere. *W. lingulata*, *W. oblonga*, and *W. gladiate* are collected from North America, while *W. lingulata*, *W. oblonga*, *W. neotropica*, *W. welwitschii*, and *W. caudate* are collected from South America. *W. welwitschii*, *W. denticulate*, *W. repanda*, *W. hyalina*, and *W. rotunda* have been collected from Africa.

There are 11 species in the genus Wolffia; they are mostly distributed in Asia, Europe, North America, South America, Oceania, and Africa. Similar to the genus Wolffiella, most species in this genus distributed in a latitude continuous pattern. *W. arrhiza* is the most widely distributed species in this genus. It has been collected from 50°N to 35°S in Asia, Europe, Africa, and South America. *W. columbiana* and *W. brasiliensis* are widely distributed in the tropical and subtropical zones of North and South America. *W. elongate*, *W. neglecta*, *W. globosa*, *W. microscopica*, and *W. borealis* are mainly located in the northern hemisphere, while *W. australiana* and *W. cylindracea* are mostly located in the southern hemisphere. *W. angusta* is collected from both Australia and Southeast Asia.

## 5.2 Population and Genetic Diversity

Duckweeds often exist in bi- or tri-species communities, although occasionally they can be found in single or more complex communities (Fig. 5.1). The presence or absence of a species in a water body is largely determined by occasion and has no significant relation to nitrate and phosphate concentration and pH of the water body (Xu et al. 2015). The genetic diversity in populations of seven species in genus Lemna and Spirodela from China and Vietnam was once analyzed by ISSR-PCR. The genetic distance of Lemna populations varied from 0.127 to 0.784 and from 0.138 to 0.902 for Spirodela (Xue et al. 2012). Clustering analysis indicated that the geographic differentiation of collected sites correlated closely with the genetic differentiation of duckweeds, suggesting that geographic differentiation had great influence on genetic diversity of duckweeds.

**Fig. 5.1** Typical duckweed populations

*S. polyrhiza* is the most widely distributed species and is suggested to be the most primitive duckweed family member (Les et al. 2002) and is therefore expected to have higher intra-specific genetic diversity among the family. However, its genetic diversity is very low as shown in the samples collected in Hainan Province of China (Xu et al. 2015), as well as in the samples of Northeast China (Zhang et al. 2018) using two chloroplast DNA marker sequences. Genetic diversity analysis of global samples from the genera Spirodela and Landoltia using three plastidic sequences (*rpL16*, *rpS16*, *atpF-atpH*) and AFLP fingerprinting also revealed low genetic diversity in Spirodela (Bog et al. 2013). The comparison of the two sequenced *S. polyrhiza* strains from the USA and Germany also reveals few polymorphisms (Wang et al. 2014; Michael et al. 2017) (Chap. 2). Recent studies on duckweed accessions from China reveal that *S. polyrhiza* has the lowest diversity, followed by *Lan. punctata* and *L. aequinoctialis*. Meanwhile, *W. globosa* has the highest diversity, which is not in parallel with the evolutionary history of the duckweed family. However, there is not enough data available to illustrate the underlying mechanism.

## 5.3 Species Identification of Duckweed Based on DNA Barcoding

Traditionally, taxonomists use taxonomic keys based on morphological characters of vegetative and floral parts to identify plant species. For identification in this traditional way, one usually needs to have both vegetative and floral parts, and the missing of anyone can cause difficulties and mistakes in identification. Due to its minute size and rare flowering, duckweed is very difficult to properly identify at the species level using morphological characters (Les et al. 1997).

DNA barcodes are used as molecular identifiers for living species in the same way as machine-readable black and white barcodes applied in the retail industry for identification of commercial products. The application of DNA barcodes needs the establishment of a comprehensive library that links organisms and their barcodes, just as is done in the commercial barcodes (Ali et al. 2016). A strong barcode needs to have appropriate threshold between inter- and intra-specific genetic distances, i.e., the difference in sequences between species should be large enough to place them apart while the sequence differences within species should be small enough to bring them together in the same group (Meyer and Paulay 2005).

The concept of DNA barcodes was originally proposed for animals, and the mitochondrial coxidase subunit I (COI) gene was accepted as an authentic barcode to accurately identify animal species (Hubert et al. 2003). But COI could not be successfully applied in plants due to very little variation in mitochondrial DNA sequences in plants (Kress et al. 2005). This led to testing of a number of non-mitochondrial genes and regions in plants, i.e., nuclear and chloroplast DNA, as potential candidates for reliable DNA barcodes. Based on a number of different studies, CBOL proposed seven plastid-markers including four plastid-encoded genes (*rpoB*, *rpoC1*, *rbcL,* and *matK*) and three noncoding spacers (*atpF-atpH*, *psbK-psbI,* and *trnH-psbA*) to be used as barcodes in species identification of plants (CBOL Group 2009).

Several studies have been published on the identification and classification of duckweeds on the basis of genotyping as nicely reviewed by Appenroth et al. (2013). Attempts to use DNA in barcoding and phylogenetic studies of Lemnaceae can be traced back to the last decade of the previous century when Jordan et al. (1996) compared three species of this family (*L. minor*, *L. valdiviana*, and *Lan. punctata*) on the basis of the *rpL16* region. Other researchers studied phylogenetic relations in different species of duckweed using *rpS16* gene intron sequences (Martyrosian et al. 2009) and the *trnL-trnF* intergenic spacer sequence (Rothwell et al. 2004). Few other studies have made such attempts as reviewed comprehensively by Appenroth et al. (2013). Then in a large-scale study, Bog et al. (2013) sequenced the plastidic *rpS16* and *rpL16* regions for 54 and 55 clones, respectively, from the 11 species of the genus Wolffia. The library of these barcodes, coupled with the results from AFLP, could clearly identify some species including *W. australiana*, *W. columbiana*, and *W. brasiliensis*. However, some species, for example, *W. globosa,* were mixed with *W. borealis and W. neglecta.*

The CBOL proposed plastid-markers' barcodes for plant identification (plastid-encoded genes (*rpoB*, *rpoC1*, *rbcL,* and *matK*), and three noncoding spacers (*atpF-atpH*, *psbK-psbI,* and *trnH-psbA*)) were tested for the first time by Wang et al. (2010) for their potential use as barcodes in species discrimination of the family Lemnaceae (Wang et al. 2010). They used 97 accessions from 31 species representing all five genera. The encoded genes, i.e., *rpoB*, *rpoC1*, *rbcL,* and *matK,* were conserved and could not prove good marker for this purpose, but the noncoding spacers, i.e., *atpF-atpH*, *psbK-psbI,* and *trnH-psbA,* showed more variability than the coding genes. They concluded that the *atpF-atpH* spacer can be used as a universal DNA barcoding marker for species identification in duckweeds. However, as pointed out by Borisjuk et al. (2015), the study of Wang et al. (2010) was not a complete barcode analysis for three main reasons: (1) Only 31 duckweed species were represented while six species were missing, (2) only a single clone was used for several species, and (3) sequence data of only 20 sampled species (out of 31 species) were used for barcode analysis. Borisjuk et al. (2015) attempted to complete the survey work of Wang et al. (2010) by including the missing data for the six species and more clones to provide a more complete database of the two barcodes (*atpF-atpH* and *psbK-psbI*) for the family of duckweed. By utilizing the sequences obtained in their own study and over 300 sequences downloaded from NCBI database for these two barcodes, they constructed a barcodes library for the family Lemnaceae that could successfully identify 30 of the 37 known species of duckweed (Borisjuk et al. 2015). Four species including *L. valdiviana*, *L. minuta*, *W. globosa,* and *W. lingulata* could not be identified by any of these two barcodes (*atpF-atpH* and *psbK-psbI*), but the remaining three species had some ambiguities (Borisjuk et al. 2015).

The plastid DNA barcodes studied in the identification of duckweeds so far give promising results in inter-species identification but may not be fruitful in intra-species discrimination of ecotypes. For example, Feng et al. (2017)

attempted to validate three DNA barcoding markers (*atpF-atpH*, *psbK-psbI*, and *trnH-psbA*) for inter-species and intra-species discrimination in *S. polyrhiza* and *Lan. punctata* using 48 ecotypes of *Lan. punctata* and 49 ecotypes of *S. polyrhiza*. They concluded that these markers were effective in species identification but could not be employed in ecotypes identification. However, the authors claimed that their newly designed SSR markers (SC09/10, SC19/20, and SC35/36) could be used as universal markers for both inter-species and intra-species level identification of *S. polyrhiza* and *Lan. punctata* (Feng et al. 2017).

The interest in the use of DNA barcoding for duckweeds identification has been extended to the small-scale field level. Baker (2018) and his students used *rbcL* and *atpF-atpH* as barcodes to identify duckweed species in a cove in Lake Saint Clair at Michigan, USA. The *rbcL* data did not show enough sequence divergence to identify Lemna samples to the species level, but the results from the BLASTn searches of the *atpF-atpH* barcodes in this study revealed the presence of four species of duckweed including *L. minor*, *S. polyrhiza*, *W. columbiana*, and *L. obscura* (Baker 2018).

Recent advancement in the techniques of DNA extraction, purification, and sequencing coupled with a drop in cost in recent years made DNA barcoding a preferred method for plant identification (Ali et al. 2016; Wang et al. 2010). Especially due to their small size and reduced morphology, DNA barcoding is effectively aiding in the proper identification of duckweed species (Sree et al. 2016). Identification of duckweed species on the basis of DNA barcoding may become essential in the future. As stated by Baker (2018), the Rutgers Duckweed Stock Cooperative does not accept germplasm donation of duckweeds that are not identified by DNA barcoding. Huge progress has been made in barcoding of duckweed species in the last few years, and efforts are continued by researchers trying different regions of plastid and nuclear genomes as potential barcodes.

## References


Ali MA, Pandey AK, Gyulai G, Park SH, Lee C, Kim SY, Lee J (2016) Introduction to plant DNA barcoding. In: Ali MA, Gabor G, Al-Hemaid F (eds) Plant DNA barcoding and phylogenetics. pp 1–14

Appenroth K-J, Borisjuk N, Lam E (2013) Telling duckweed apart: genotyping technologies for the lemnaceae. Chinese J Appl Environ Biol 19(1):1–10. https://doi.org/10.3724/SP.J.1145.2013.00001

Baker SS (2018) Using DNA barcoding to identify duckweed species as part of an undergraduate ecology course. In: Environmental chemistry: undergraduate and graduate classroom, laboratory, and local community learning experiences, vol 1267. American Chemical Society. https://doi.org/10.1021/bk-2018-1276.ch1005

Bog M, Schneider P, Hellwig F, Sachse S, Kochieva EZ, Martyrosian E, Landolt E, Appenroth KJ (2013) Genetic characterization and barcoding of taxa in the genus *Wolffia* Horkel ex Schleid. (Lemnaceae) as revealed by two plastidic markers and amplified fragment length polymorphism (AFLP). Planta 237 (1):1–13

Borisjuk N, Chu P, Gutierrez R, Zhang H, Acosta K, Friesen N, Sree KS, Garcia C, Appenroth KJ, Lam E (2015) Assessment, validation and deployment strategy of a two-barcode protocol for facile genotyping of duckweed species. Plant Biology 17:42–49

Feng B, Fang Y, Xu Z, Xiang C, Zhou C, Jiang F, Wang T, Zhao H (2017) Development of a new marker system for identification of *Spirodela polyrhiza* and *Landoltia punctata*. Int J Genomics 2017:2017

Group CPW (2009) A DNA barcode for land plants. Proc Nat Acad Sci USA 106:12794–12797

Hubert N, Hanner R, Holm E, Mandrak NE, Taylor E, Burridge M, Watkinson D, Dumont P, Curry A, Bentzen P, Zhang J, April J, Bernatchez L (2003) Identifying Canadian freshwater fishes through DNA barcodes. PLoS ONE 3(6):e2490

Kress WJ, Wurdack KJ, Zimmer EA, Weigt LA, Janzen DH (2005) Use of DNA barcodes to identify flowering plants. Proc Natl Acad Sci USA 102:8369–8374

Les DH, Crawford DJ, Landolt E, Gabel JD, Kimball RT (2002) Phylogeny and systematics of Lemnaceae, the duckweed family. Syst Bot 27(2):221–240

Les DH, Landolt E, Crawford DJ (1997) Systematics of the Lemnaceae (duckweeds): inferences from micromolecular and morphological data. P1 Syst Evol 204:161–177

Martyrosian EV, Ryzhova NN, Kochieva EZ, Skryabin KG (2009) Analysis of chloroplast rpS16 intron sequences in Lemnaceae. Mol Biol 43:32–38

Meyer CP, Paulay G (2005) DNA barcoding: error rates based on comprehensive sampling. Plos Bioloy 3(12): e422

Michael TP, Bryant D, Gutierrez R et al (2017) Comprehensive definition of genome features in *Spirodela polyrhiza* by high-depth physical mapping and short-read DNA sequencing strategies. Plant J 89:617–635. https://doi.org/10.1111/tpj.13400

Rothwell GW, Van Atta MR, Ballard HE Jr, Stockey RA (2004) Molecular phylogenetic relationships among Lemnaceae and Araceae using the chloroplast trnL-trnF intergenic spacer. Mol Phylogenet Evol 30 (2):378–385

Sree KS, Bog M, Appenroth K-J (2016) Taxonomy of duckweeds (Lemnaceae), potential new crop plants. Emirates J Food Agric 28(5):291–302. https://doi.org/10.9755/ejfa.2016-01-038

Wang W, Haberer G, Gundlach H et al (2014) The *Spirodela polyrhiza* genome reveals insights into its neotenous reduction fast growth and aquatic lifestyle. Nat Commun 5:1–13. https://doi.org/10.1038/ncomms4311

Wang W, Wu Y, Yan Y, Ermakova M, Kerstetter R, Messing J (2010) DNA barcoding of the Lemnaceae, a family of aquatic monocots. BMC Plant Biol 10:205

Xu Y, Ma S, Huang M, Peng M, Bog M, Sree KS, Appenroth K, Zhang J (2015) Species distribution, genetic diversity and barcoding in the duckweed family (Lemnaceae). Hydrobiologia 743(1):75–87

Xue H, Xiao Y, Jin Y, Li X, Fang Y, Zhao H, Zhao Y, Guan J (2012) Genetic diversity and geographic differentiation analysis of duckweed using inter-simple sequence repeat markers. Mol Biol Rep 39(1):547–554

Zhang Z, Han B, Sun X, Fu L, Ye S, Zhang J (2018) Genetic diversity of duckweed in a typical temperate and monsoon region. Mol Plant Breed 16(2):656–664

# 6 Strategies and Tools for Sequencing Duckweeds

Xiaoli Xiang and Changsheng Li

**Abstract**

Duckweeds belong to the smallest flowering plants that undergo fast vegetative growth in an aquatic environment. Due to their special plant characteristics, they are commonly used in wastewater treatment, biofuel, and animal feed. Sequencing duckweed genomes will promote their development in molecular biology and functional genomics, thereby facilitating its application in feed, energy, and environmental protection. In addition to extremely fast growth speed, the genome sizes of duckweeds are varied from 150 to 1881 Mb with a roughly 13-fold change. But with the rapid development of sequencing technology and dramatic decrease of sequencing cost, sequencing different kinds of duckweed genomes has become feasible. Here, we review the strategies and tools for sequencing and assembling duckweeds genomes. We introduce the platforms of the next-generation sequencing (NGS) (Illumina paired-end sequencing with short reads) and the third-generation sequencing (TGS) (PacBio and Nanopore sequencing with long reads) that are broadly applied in plant genomics. We also overview the recent widely used scaffolding technologies including Bionano, Hi-C, and 10X Genomics. Tools for de novo assembling duckweeds genomes are determined by the sequencing platforms that give short reads or long reads. The programs of SOAPdenovo and ALLPATHS-LG are sufficient to assemble Illumina short reads; whereas, the assemblers of FALCON, CANU, MECAT, and HGAP are broadly used in assembling plant genomes sequenced by the platforms of PacBio or Nanopore. The hybrid assembly tool such as MaSuRCA is required for the integration of short and long reads. We expect that the strategies and tools will accelerate the duckweed genomics and promote their industrial applications.

## 6.1 Strategies for Sequencing Duckweeds

### 6.1.1 Illumina Paired-End Sequencing for Low Complexity Duckweed Genomes

Next-generation sequencing technologies (NGS), also known as high-throughput sequencing technologies, can generate unprecedented amounts of data, greatly facilitating research in genomics and

X. Xiang
Institute of Biotechnology and Nuclear Technology, Sichuan Academy of Agricultural Sciences, Chengdu, China

C. Li (✉)
Institute of Plant Physiology & Ecology, Shanghai Institutes for Biological Sciences, Chinese Academy of Sciences, Beijing, China
e-mail: lcslyh@yahoo.com


X. H. Cao et al. (eds.), *The Duckweed Genomes*, Compendium of Plant Genomes,
https://doi.org/10.1007/978-3-030-11045-1_6

transcriptomics. The high throughput and low cost of NGS sequencing technologies also advance the sequencing of many crops with a genome size of less than 500 Mb, including cucumber (Huang et al. 2009), watermelon (Guo et al. 2013), wild strawberry (*Fragaria vesca*) (Kang et al. 2013), date palm (Al-Dous et al. 2011), and papaya (Ming et al. 2008), providing invaluable genomic resources for the breeding of vegetables and fruit trees. The increase of data volume and the improvement of accuracy make NGS a great advantage in sequencing most duckweed genomes, especially the genomes with small genome size and low complexity. Before the initiation of duckweed genome sequencing (Wang et al. 2011), a series of genome sizes of duckweed species were determined by flow cytometry (FCM). The species with small genome sizes have been selected and sequenced by the next-generation sequencing technologies. *Spirodela polyrhiza* 7498 ($2n = 40$; estimated 158 Mb) was sequenced by Sanger and 454 Roche platforms. The length of contig N50 arrived to 17.8 kb. With the integration of physical map and BAC end sequencing, the scaffolds were generated with the N50 of 7.6 Mb (Wang et al. 2014). *Spirodela polyrhiza* 9509 ($2n = 40$; estimated 158 Mb) was sequenced by Illumina HiSeq 2000 platform. A total of two paired-end (PE) and three mate-pair (MP) sequencing libraries were constructed from its genomic DNA, with insert sizes of 180 bp, 500 bp, 2 kb, 5 kb, and 20 kb. The 180 bp and 500 bp libraries were used for de novo assembling into contigs, and the 2 kb, 5 kb, and 20-kb libraries were targeted for constructing scaffolds, resulting in a size of 145.8 Mb assembled genome with a scaffold N50 of 4.3 Mb and a contig N50 of 19 kb (Michael et al. 2017). The *Lemna minor* 5500 ($2n = 40$; estimated 481 Mb), as the Lemna ancestor genome, was sequenced by Illumina platform. Two paired-end libraries were created including a HiSeq library ($2 \times 100$ bp) and a MiSeq library with longer short reads ($2 \times 300$ bp). The draft genome was assembled into 472.1 Mb containing 46,047 scaffolds with an N50 length of 23.8 kb and a contig N50 length of 20.9 kb (2015). Compared with *Spirodela polyrhiza*, the assembly of *Lemna minor* 5500 was more fragmented and contained more gaps. One reason was that mate-pair (MP) libraries with large DNA insertions were missing, which could help bridge contigs into scaffolds. Another was that 61.5% of the genome size (481 Mb) was repetitive sequences in *Lemna minor* 5500 that was challenging to be determined. In contrast, there was only 15.79% of repeat elements in the *Spirodela polyrhiza* genome (158 Mb). The repeat content could explain 94.5% of the genome size difference between *Lemna minor* and *Spirodela polyrhiza* (Van Hoeck et al. 2015).

Unlike mammalian genomes, plant genomes are abundant of transposons, leading to a huge variation in genome size. The smallest genome, *Genlisea tuberosa*, is only 61 Mb in size that basically maintains the essential genes. The wheat genome has a large genome size of 17 Gb, of which 90% is a repeat sequence (Michael and VanBuren 2015). The loblolly pine genome even reached 22 Gb that is the largest sequenced genome until recently (Zimin et al. 2014). However, the short read from Illumina sequencing ($\sim 150$ bp) limits its applications since the sequence reads cannot fully span the repeat regions and pose a serious problem for genome assembly. The similarity of repeat reads breaks the contiguity of genome into fragmented contigs, resulting in incomplete genome assembly. The unassembled sequences may confound the biological significance due to the missing of a complete gene, partial regulatory elements, impaired centromeres, and telomeres (Li et al. 2018). Duckweeds genome sizes vary enormously, ranging from 150 Mb in *Spirodela* to 1881 Mb in Wolffia, a total 13-fold change (Wang et al. 2011). The difference of their genome sizes is mainly caused by repetitive sequences, especially long terminal repeats (LTR) which have a length of 4–16 kb and huge amount of copies (Kumar and Bennetzen 1999; Phillippy 2017). Only the long reads that expand over the transposons can make their uniqueness so that they can determine where the repeats belong to. Given the availability of third-generation sequencing (TGS), scientists have an extraordinary opportunity to crack the complex duckweed genomes.

### 6.1.2 PacBio and Oxford Nanopore Sequencing Providing an Opportunity to Crack the Complex Duckweed Genomes

PacBio single-molecule real-time (SMRT) sequencing is the most popular TGS technology in the market. Different with NGS, PacBio sequencing simulates the natural processes of DNA replication, enables real-time sequencing of DNA molecules through zero-mode waveguide pores (ZMWs) and phosphorylated nucleotides, and does not require a pause between read steps, and each step of template amplification will generate a light pulse that can be identified as a different labeled nucleotide (Schadt et al. 2010). The sequencing technology can generate extremely long reads with an average read length of more than 15 Kb. Some reads can reach up to 100 Kb that is comparable to a BAC clone (Pacific 2018a). However, the main concern is the high random sequencing errors in the long noisy reads. The latest sequencing platform for PacBio is Sequel System, which is featured with one million ZMWs per SMRT cell instead of 150,000 in RS II, and therefore, produces seven times more reads per SMRT cell than RS II (Pacific 2018b). The increased throughput and the decreased cost make PacBio SMRT sequencing technology available to any individual laboratory. Given long-read lengths and GC-free preference, PacBio reads allow assemblers to span repeat regions. It has been used in de novo assembling multiple plant genomes and dramatically improved the genome reconstruction (VanBuren et al. 2015; Jiao et al. 2017b; Lan et al. 2017).

The recently published review has summarized the broad applications of PacBio SMRT sequencing technology in genome sequencing, as well as the comparisons with the next-generation sequencing (Li et al. 2018). It was found that PacBio long reads can assemble unprecedented contiguity genomes and more complete high repetitive regions, such as LTR retrotransposon, centromeres, and telomeric repeats (Li et al. 2018). The latest study reported that using single-molecule real-time sequencing and a meta-assembly approach obtained one of the most comprehensive plant genome of *Rosa chinensis* that had a contig N50 of 24 Mb. Therefore, PacBio SMRT sequencing has shown a great promise to solve the complex duckweed genomes. But, there are still no released duckweed genomes sequenced by PacBio reads. However, there was a pioneer work done for *Lemna minor* 8627. The additional input of long-read sequencing significantly increased the contiguity of genome assembly that the contig N50 was extended from 65 to 222 Kb (Ernst 2016).

Oxford Nanopore is another long-read sequencing platform and has similar characteristics as PacBio long reads, which can produce reads up to hundreds of kilobases but with a high error rate. Nanopore sequencing determines DNA sequences by the ionic current changes when DNA strands pass through the tiny Nanopores in the flow cell (Li et al. 2018). Thus, it can produce ultra-long-read lengths the same as the DNA molecule lengths. The recent study showed that Nanopore sequencing can produce sequence reads up to one Mb (Willing et al. 2015), which will effectively solve the duckweeds genomes with the highly repetitive region, like *Wolffiella* and *Wolffia*. Compared to PacBio SMRT, Oxford Nanopore sequencing technology can produce ultra-long reads which are more productive to assemble high-continuity genomes, even though it shows overall lower data quality (Weirather et al. 2017; Jain et al. 2018). However, upon now there are few reports about Nanopore genome sequencing plant genome.

### 6.1.3 Bionano, 10X Genomics, and Hi-C for Scaffolding Duckweed Genomes

Due to the redundant sequence and complexity of the plant genome, it is almost impossible to assemble the genome only by sequencing reads. After obtaining contigs from sequence assembly, they are often ordered and oriented into scaffolds by using large fragment libraries like BAC, MP, or Fosmid to improve assembly results. With the development of sequencing technology, some of the high-throughput physical mapping technologies have emerged, such as Bionano, chromosome conformation capture (Hi-C), and 10X Genomics, which compensate the shortcomings of traditional genetic and physical mapping techniques (Li et al. 2018).

Different from the method of BAC physical maps, Bionano technology labels long DNA molecules at a specific recognition site that is widely distributed in the genome, then linearizes and images labeled DNA molecules by Saphyr instrument to construct the physical maps. The Bionano optical map can improve the contiguity of genome assemblies by ordering and orienting contigs, but also can correct potential chimeric contigs in genome assemblies and estimate the gap size between adjacent contigs. A certain plant genome assembly has been improved by Bionano in terms of their accuracy and contiguity without the laborious and expensive construction of BAC physical maps, such as wheat sequenced by Illumina and PacBio reads (Zimin et al. 2017), sorghum using Nanopore reads (Deschamps et al. 2018), and maize using PacBio reads (Jiao et al. 2017b). The scaffold N50 of *Spirodela polyrhiza* 9509 was improved to 7.7 Mb with the integration of Bionano optical map, almost two-fold of the MP version (4.3 Mb), and reached the assembly level of *Spirodela polyrhiza 7498* with the scaffolds N50 was 7.6 Mb (Wang et al. 2014).

10X Genomics technology barcodes each DNA molecule (>50 kb) with the Linked-Reads (Phillippy 2017) and combines the Illumina sequencing to extend scaffolds. It is another long-range scaffolding technology with less cost but high throughput compared to the traditional BAC fingerprint technologies. For instance, the scaffold N50 sizes were significantly increased from 359.12 Kb to 1.217 Mb for *P. equestris* and from 391.46 Kb to 1.055 Mb for *D. catenatum* with the addition of 10× Genomics Linked-Reads (Zhang et al. 2017). There are currently no duckweed genomes using 10× Genomics technology for the scaffold extension.

The technique of Hi-C is able to capture the genome conformation based on the chromosomal interaction rules to reconstruct chromosome-scale genomes (Burton et al. 2013). Hi-C technique avoids complicated experiments in genetic and BAC-based physical maps, resulting in its extensive applications in the following fields:

1. Improvement of draft genome. The initial genome assembly can be further improved to the chromosome level by Hi-C information. By using chromosome conformation capture data, 6347 super scaffolds of the barley genome were ordered, and 4.54 Gb (~90%) of the genomic sequence were mapped to the precise chromosomal location in the Hi-C map (Mascher et al. 2017).
2. Improvement of the highly heterozygous plant genome. Because each chromosome occupies a unique territory, even for homologous chromosomes, it has an important role in distinguishing heterozygous chromosomes. For example, durian is a highly heterozygous genome. With the contact maps of CHiCAGO (in vitro chromatin reconstitution of high-molecular-weight DNA) and Hi-C (in vivo fixation of chromosomes), the final reference assembly reached 30 chromosome-scale pseudomolecules longer than 10 Mb and covered 95% of the 712 Mb assembly (Teh et al. 2017).
3. Improvement of the polyploid genome. Using single-molecule real-time (SMRT) sequencing technology from PacBio and optical maps from Bionano, the initial assembly of *Chenopodium quinoa* (polyploid species) contains 4014 scaffolds, with a scaffold N50 of 2.45 Mb. After using chromosome-intact data from Dovetail Genomics, the number of scaffolds was reducing to 3486, with a scaffold N50 of 3.84 Mb. There were 439 scaffolds that covered 90% of the assembled genome (Jarvis et al. 2017).

The chromosomally integrated genome of *Spirodela polyrhiza* was constructed by using mcFISH with 96 BACs probes from 20 chromosome pairs (Cao et al. 2016). Hi-C technique could be an alternative way to reconstruct the chromosomes of duckweed genomes.

Still, the technologies of Bionano, Hi-C, and 10X Genomics are lack of the fine resolution to improve contig length compared with PacBio or Nanopore sequencing. Currently, PacBio or Nanopore sequencing combined with Bionano optical map or 10X Genomics and Hi-C technology will be the best choice to optimize the assembly of complex duckweed genomes.

## 6.2 Tools for Assembling Duckweeds

### 6.2.1 De Novo Assembling Tools for Illumina Paired-End Reads

The selection of sequencing platforms for duckweeds is determined by their characteristics of the genomes. In addition, the bioinformatics programs for assembling genomes are also needed to customize. Here, we will introduce the most common programs in terms of duckweed genome assembly by using next-generation sequencing.

SOAPdenovo (http://soap.genomics.org.cn/soapdenovo.html) (Luo et al. 2012) is a de novo assembly software developed by BGI, based on a *de Bruijn* graph-based algorithm. It is featured in the fast speed of genome assembly and the long scaffold N50 value, but the error rate is higher than other programs. Therefore, SOAPdenovo is widely used for assembly of large genomes, such as barley (Mascher et al. 2017), maize (Hirsch et al. 2016), and quinoa (Jarvis et al. 2017).

ALLPATHS-LG [(http://software.broadinstitute.org/allpaths-lg/blog/) (Gnerre et al. 2011)], the short-read genome assembler from the Computational Research and Development group at the Broad Institute, also based on a *de Bruijn* graph-based algorithm, has a high assembly accuracy, but consumes a significant memory and CPU resource (Henson et al. 2012). The following issues need to be considered.

ALLPATHS-LG needs sequence data from multiple libraries with various insert sizes, including at least one paired reads from an "overlapping" fragment library. For example, the paired reads are overlapped with a read length of ~100 bp from inserted fragments of ~180 bp. Genome assembly requires the sequence coverage of 100X given the raw reads data (before error correction and filtering). ALLPATHS-LG does not support distributed computing using MPI, but makes use of shared memory parallelization. The usage of peak memory is roughly 1.7 bytes per read base, resulting in 17 G of memory for handling 10 Gb of input data. Different from other de novo assembly software, ALLPATHS-LG could determine the optimal K value after a series of self-trainings during the run (http://software.broadinstitute.org/allpaths-lg/blog/?page_id=336). ALLPATHS-LG supports the hybrid assembly with PacBio long reads, but it is still limited to assemble the small microbial genomes (Koren et al. 2012; Shibata et al. 2013; Koren and Phillippy 2015), which is not applied to any animal and plant genomes yet.

MaSuRCA (Zimin et al. 2013) was derived from the Celera Assembler (Myers et al. 2000), combining the algorithm of the *de Bruijn* graph and overlap–layout–consensus (OLC) approaches. MaSuRCA supports not only Illumina short reads, but also a hybrid of short and long reads. The MaSuRCA genome assembler has been widely used in the field of large animal and plant genomes (Chibucos et al. 2013) (Zimin et al. 2014; Zimin et al. 2017).

To achieve the most continuous assembly of *Lemna minor* 5500 genome, three programs were evaluated including SOAPdenovo2, CLC bio, and MaSuRCA. The draft genome generated by MaSuRCA is the best compared to that of SOAPdenovo2 and CLC bio (Van Hoeck et al. 2015). A high-quality draft genome of *Spirodela polyrhiza* 9509 containing 774 scaffolds with an N50 length of 4.3 Mb and a contig N50 length of 19 kb was reached using a combination of

ALLPATHS-LG and SSPACE (Boetzer et al. 2011).

### 6.2.2 De Novo Assembling Tools for PacBio or Nanopore Long Reads

There are still no released duckweed genomes sequenced by the third-generation sequencing of PacBio or Nanopore platforms that generate long reads. Here, we recommend four popular de novo assembling tools for long reads, and expect they can benefit the assembly of duckweed genomes in the near future.

Certain alignment algorithms have been developed to effectively discover overlaps among noisy long reads, such as DALIGNER (Myers 2014), BLASR (Chaisson and Tesler 2012), MHAP (Berlin et al. 2015), GraphMap (Sovic et al. 2016), and Minimap (Li 2016). DALIGNER was the first tool designed specifically for finding the overlaps between noisy long reads with the filtering or removals of k-mers redundancy. The program could increase the computation speed, decrease memory usage, and mitigate the effect of repetitive sequences. Given the risk of filtering important k-mers, the parameters (-k, -w, -h, -t) need to be customized. Another method to improve the alignment efficiency is to split the large data set into small blocks based on the total number of base pairs and read lengths by using the DBsplit utility (Myers 2014).

FALCON is an overlap–layout–consensus (OLC) genome assembler based on DALIGNER, which only supports PacBio data. The pipeline includes six steps to construct contigs, while the correction and polish of raw reads are the steps that consume the most computational resources. The script "fc_run.py" with a configuration file can complete the whole assembly process. The configuration file contains input files, optimal parameters, and computation resources. Several parameters need to be well considered due to their greater impacts on genome assembly, for instance, length_cutoff which controls the threshold during the error correction, and length_cutoff_pr which sets the cutoff value used for the later assembly overlapping step. Other parameters (-k, -w, -h, -t) are used to optimize k-mers. The parameters are determined by the sequencing depth and the characteristics of the sequenced species. The recommendations can be found in this Web site (https://pb-falcon.readthedocs.io/en/latest/parameters.html#parameters). It is suggested to choose a smaller length_cutoff in the initial computation run, then adjust length_cutoff_pr for a better assembly. If the coverage of the corrected high-quality reads longer than the cutoff length is more than 20x, we are recommended to set the min_cov to 5, max_cov to three times of coverage and the max_diff to twice of coverage (Chin 2016). Several studies have shown that FALCON has advantages in assembling highly complex plant genomes, such as maize and opium (Jiao et al. 2017a, b; Guo et al. 2018). It needs to be considered that FALCON requires a high computational cost to complete the long-read correction and the overlapping detection due to its alignment algorithm.

Hierarchical genome assembly process (HGAP) is developed from FALCON with the integration of the polish step by using arrow or quiver. A small genome with hundreds of Mb could be assembled via a web-based graphical user interface of HGAP, while a large genome needs to be run through the environment of the UNIX command line. HGAP has been widely used in the assemblies of multiple plant genomes, as well as of small genomes (VanBuren et al. 2015; Jiao et al. 2017b; Lan et al. 2017).

Canu, a successor of the Celera Assembler (Denisov et al. 2008), can assemble both PacBio and Nanopore sequencing reads. By the fact of the optimized algorithms in the initial overlapping and correction process, Canu is often able to generate a complete plant assembly less time than FALCON (Koren et al. 2017). The genome size is very critical parameter in Canu, which decides how sensitive the mhap overlapper should be. The rawErrorRate and correctedErrorRate are another two main parameters which are involved in overlap detection. A more accurate assembly will be achieved with a preferably smaller parameter than the default

**Table 6.1** Proposed best sequencing strategy for duckweed genomes

| | Genome size (Mb) | Strategy | Sequencing platform | Sequencing depth | Scaffold technology | Bioinformatic tools | Description |
|---|---|---|---|---|---|---|---|
| Spirodela | 150–167 | Reference-guided assembly | Illumina | 50X | Bionano | ALLPATHS-LG | Boetzer et al. (2011) |
| Landoltia | 372–397 | De novo assembly | Illumina | 200X | Bionano | ALLPATHS-LG | No ref available |
| Lemna | 323–760 | Reference-guided assembly | Illumina | 200X | Bionano | MaSuRCA | Van Hoeck et al. (2015) |
| Wolffiella | 623–973 | De novo assembly | Pacbio | 50X | Bionano | FALCON, Canu and MECAT | No ref available |
| Wolffia | 357–1881 | De novo assembly | Pacbio | 50X | Bionano | FALCON, Canu and MECAT | No ref available |

value. MECAT (Xiao et al. 2017) (https://github.com/xiaochuanle/MECAT) is also an ultra-fast mapping, error correction and de novo assembly tool for PacBio and Nanopore sequencing reads. MECAT employs novel alignment and error correction algorithms that are much faster than the state of art of aligners and error correction tools. MECAT exhibits a faster speed and better assembly results in the model plant Arabidopsis; however, the genomes are preferred to be assembled by mecat2cns and mecat2canu rather than only by MECAT (Xiao et al. 2017; Guo et al. 2018; Xia et al. 2018). When the computing resource is limited for a laboratory,Canu and MECAT both would be better choices for the complex duckweed genomes.

## 6.3 Summary

A single software cannot get the best assembly for a given genome; whereas, multiple softwares, parameters, and databases need to be tested to get the improved results. In summary, we can directly use FALCON, HGAP, or Canu to de novo assemble the PacBio or Nanopore long reads for small duckweed genomes. On the other side, we need use MECAT or Canu to get high-quality reads, then assemble them by FALCON to get a better result for large duckweed genomes.

Until now, only *Spirodela polyrhiza* and *Lemna gibba* genomes have been released (Table 6.1). By re-sequencing these two species, the Illumina paired-end sequencing combining with Bionano will be an affordable and effective strategy. But for large genome size duckweeds like *Wolffiella* and *Wolffia*, it would be better to choose long-read sequencing platform to achieve a good assembly results.

## References


Al-Dous EK, George B, Al-Mahmoud ME, Al-Jaber MY, Wang H et al (2011) De novo genome sequencing and comparative genomics of date palm (*Phoenix dactylifera*). Nat Biotechnol 29:521–527

Berlin K, Koren S, Chin CS, Drake JP, Landolin JM et al (2015) Assembling large genomes with single-molecule sequencing and locality-sensitive hashing. Nat Biotechnol 33:623–630

Boetzer M, Henkel CV, Jansen HJ, Butler D, Pirovano W (2011) Scaffolding pre-assembled contigs using SSPACE. Bioinformatics 27:578–579

Burton JN, Adey A, Patwardhan RP, Qiu R, Kitzman JO et al (2013) Chromosome-scale scaffolding of de novo genome assemblies based on chromatin interactions. Nat Biotechnol 31:1119–1125

Cao HX, Vu GT, Wang W, Appenroth KJ, Messing J et al (2016) The map-based genome sequence of *Spirodela polyrhiza* aligned with its chromosomes, a reference for karyotype evolution. New Phytol 209:354–363

Chaisson MJ, Tesler G (2012) Mapping single molecule sequencing reads using basic local alignment with successive refinement (BLASR): application and theory. BMC Bioinformatics 13:238

Chibucos MC, Crabtree J, NagarajS, Chaturvedi S, Chaturvedi V (2013) Draft genome sequences of human pathogenic fungus Geomyces pannorum sensu lato and bat white nose syndrome pathogen Geomyces

(*Pseudogymnoascus*) destructans. Genome Announcements 1: e01045–01013
Chin J (2016) Falcon Genome Assembly Tool Kit Manual
Denisov G, Walenz B, Halpern AL, Miller J, Axelrod N et al (2008) Consensus generation and variant detection by Celera Assembler. Bioinformatics 24:1035–1040
Deschamps S, Zhang Y, Llaca V, Ye L, May G et al (2018) A chromosome-scale assembly of the sorghum genome using nanopore sequencing and optical mapping. bioRxiv: 327817
Ernst E (2016) Status of the Lemna gibba 7742a and Lemna minor 8627 genomes, pp. 9–10. 3
Gnerre S, Maccallum I, Przybylski D, Ribeiro FJ, Burton JN et al (2011) High-quality draft assemblies of mammalian genomes from massively parallel sequence data. Proc Natl Acad Sci U S A 108:1513–1518
Guo L, Winzer T, Yang X, Li Y, Ning Z et al (2018) The opium poppy genome and morphinan production. Science
Guo S, Zhang J, Sun H, Salse J, Lucas WJ et al (2013) The draft genome of watermelon (*Citrullus lanatus*) and resequencing of 20 diverse accessions. Nat Genet 45:51–58
Henson J, Tischler G, Ning Z (2012) Next-generation sequencing and large genome assemblies. Pharmacogenomics 13:901–915
Hirsch CN, Hirsch CD, Brohammer AB, Bowman MJ, Soifer I et al (2016) Draft assembly of elite Inbred Line PH207 provides insights into genomic and transcriptome diversity in maize. Plant Cell 28:2700–2714
Huang S, Li R, Zhang Z, Li L, Gu X et al (2009) The genome of the cucumber, Cucumis sativus L. Nat Genet 41:1275–1281
Jain M, Koren S, Miga KH, Quick J, Rand AC et al (2018) Nanopore sequencing and assembly of a human genome with ultra-long reads. Nat Biotechnol 36:338–345
Jarvis DE, Ho YS, Lightfoot DJ, Schmöckel SM, Li B et al (2017) The genome of Chenopodium quinoa. Nature: 1–20
Jiao W-B, Accinelli GG, Hartwig B, Kiefer C, Baker D et al (2017a) Improving and correcting the contiguity of long-read genome assemblies of three plant species using optical mapping and chromosome conformation capture data. Genome research
Jiao Y, Peluso P, Shi J, Liang T, Stitzer MC et al (2017b) Improved maize reference genome with single-molecule technologies. Nature 546:524–527
Kang C, Darwish O, Geretz A, Shahan R, Alkharouf N et al (2013) Genome-scale transcriptomic insights into early-stage fruit development in woodland strawberry Fragaria vesca. Plant Cell 25:1960–1978
Koren S, Phillippy AM (2015) One chromosome, one contig: complete microbial genomes from long-read sequencing and assembly. Curr Opin Microbiol 23:110–120
Koren S, Schatz MC, Walenz BP, Martin J, Howard JT et al (2012) Hybrid error correction and de novo assembly of single-molecule sequencing reads. Nat Biotechnol 30:693–700
Koren S, Walenz BP, Berlin K, Miller JR, Bergman NH et al (2017) Canu: scalable and accurate long-read assembly via adaptive k-mer weighting and repeat separation. Genome Res 27:722–736
Kumar A, Bennetzen JL (1999) Plant retrotransposons. Annu Rev Genet 33:479–532
Lan T, Renner T, Ibarra-Laclette E, Farr KM, Chang T-H et al (2017) Long-read sequencing uncovers the adaptive topography of a carnivorous plant genome. Proc Natl Acad Sci 114:E4435–E4441
Li C, Lin F, An D, Wang W, Huang R (2018) Genome sequencing and assembly by long reads in plants. Genes 9:6
Li H (2016) Minimap and miniasm: fast mapping and de novo assembly for noisy long sequences. Bioinformatics 32:2103–2110
Luo R, Liu B, Xie Y, Li Z, Huang W et al (2012) SOAP denovo2: an empirically improved memory-efficient short-read de novo assembler. Gigascience 1:18
Mascher M, Gundlach H, Himmelbach A, Beier S, Twardziok SO et al (2017) A chromosome conformation capture ordered sequence of the barley genome. Nature 544:427–433
Michael TP, Bryant D, Gutierrez R, Borisjuk N, Chu P et al (2017) Comprehensive definition of genome features in *Spirodela polyrhiza* by high-depth physical mapping and short-read DNA sequencing strategies. Plant J 89:617–635
Michael TP, VanBuren R (2015) Progress, challenges and the future of crop genomes. Curr Opin Plant Biol 24:71–81
Ming R, Hou S, Feng Y, Yu Q, Dionne-Laporte A et al (2008) The draft genome of the transgenic tropical fruit tree papaya (*Carica papaya Linnaeus*). Nature 452:991–996
Myers EW, Sutton GG, Delcher AL, Dew IM, Fasulo DP et al (2000) A whole-genome assembly of Drosophila. Science 287:2196–2204
Myers G (2014) Efficient local alignment discovery amongst noisy long reads. In: International workshop on algorithms in bioinformatics. Springer, pp 52–67
Pacific B (2018a) Produce exceptional results with high-quality long reads. Pacific Biosciences
Pacific B (2018b) Sequel System: high-throughput, cost-effective access to SMRT Sequencing
Phillippy AM (2017) New advances in sequence assembly. Genome Res 27:xi–xiii
Schadt EE, Turner S, Kasarskis A (2010) A window into third-generation sequencing. Hum Mol Genet 19: R227–240
Shibata TF, Maeda T, Nikoh N, Yamaguchi K, Oshima K et al (2013) Complete Genome Sequence of Burkholderia sp. Strain RPE64, Bacterial Symbiont of the Bean Bug Riptortus pedestris. Genome Announc 1

Sovic I, Sikic M, Wilm A, Fenlon SN, Chen S et al (2016) Fast and sensitive mapping of nanopore sequencing reads with GraphMap. Nat Commun 7:11307

Teh BT, Lim K, Yong CH, Ng CCY, Rao SR et al (2017) The draft genome of tropical fruit durian (*Durio zibethinus*). Nat Genet 49:1633–1641

Van Hoeck A, Horemans N, Monsieurs P, Cao HX, Vandenhove H et al (2015) The first draft genome of the aquatic model plant Lemna minor opens the route for future stress physiology research and biotechnological applications. Biotechnol Biofuels 8:188

VanBuren R, Bryant D, Edger PP, Tang H, Burgess D et al (2015) Single-molecule sequencing of the desiccation- tolerant grass Oropetium thomaeum. Nature 527:508–511

Wang W, Haberer G, Gundlach H, Glasser C, Nussbaumer T et al (2014) The *Spirodela polyrhiza* genome reveals insights into its neotenous reduction fast growth and aquatic lifestyle. Nat Commun 5:3311

Wang W, Kerstetter RA, Michael TP (2011) Evolution of genome size in duckweeds (*Lemnaceae*). J Botany

Weirather JL, de Cesare M, Wang Y, Piazza P, Sebastiano V et al (2017) Comprehensive comparison of pacific biosciences and oxford nanopore technologies and their applications to transcriptome analysis. F1000Res 6:100

Willing EM, Rawat V, Mandakova T, Maumus F, James GV et al (2015) Genome expansion of Arabis alpina linked with retrotransposition and reduced symmetric DNA methylation. Nature Plants 1:14023

Xia M, Han X, He H, Yu R, Zhen G et al (2018) Improved de novo genome assembly and analysis of the Chinese cucurbit *Siraitia grosvenorii*, also known as monk fruit or luo-han-guo. GigaScience 7:giy067

Xiao CL, Chen Y, Xie SQ, Chen KN, Wang Y et al (2017) MECAT: fast mapping, error correction, and de novo assembly for single-molecule sequencing reads. Nat Methods 14:1072–1074

Zhang GQ, Liu KW, Li Z, Lohaus R, Hsiao YY et al (2017) The Apostasia genome and the evolution of orchids. Nature 549:379–383

Zimin A, Stevens KA, Crepeau MW, Holtz-Morris A, Koriabine M et al (2014) Sequencing and assembly of the 22-gb loblolly pine genome. Genetics 196:875–890

Zimin AV, Marçais G, Puiu D, Roberts M, Salzberg SL et al (2013) The MaSuRCA genome assembler. Bioinformatics 29:2669–2677

Zimin AV, Puiu D, Luo M-C, Zhu T, Koren S et al (2017) Hybrid assembly of the large and highly repetitive genome of *Aegilops tauschii*, a progenitor of bread wheat, with the MaSuRCA mega-reads algorithm. Genome research

# The Journey of Spirodela Whole-Genome Sequencing

7

Dong An and Wenqin Wang

**Abstract**

The Spirodela shows great potential in the fields of high-protein animal feed, biofuel, bioreactor, and wastewater remediator, whereas the deficiency of genome greatly impedes the advancement of molecular biology and industrial applications. Thus, sequencing and annotating Spirodela genome is the prerequisite for opening new frontiers in the study of aquatic plants. There are extraordinary resequencing efforts for more Spirodela genotypes, which would help fully interpret the genome and facilitate the functional genomic studies. The sequence information will allow the genetic dissection of the characters involved in the aquatic adaptation and the breeding strategies for the improvement of biofuel, as well as stopping environmental damage.

## 7.1 Introduction

The greater duckweeds, *Spirodela polyrhiza*, are the aquatic flowering plants with the simplest morphology and the smallest size (Wang et al. 2011). Spirodela is widely distributed over the world, stretching at nearly any altitude from frigid to torrid zones (Xue et al. 2012; Wang 1990). Some of the current uses of Spirodela are promising biofuel candidates (Cui and Cheng 2015), bioengineering protein factories (Stomp 2005), wastewater remediation (Rahman and Hasegawa 2011; Naumann et al. 2007), toxicity testing organism (Wang 1990), and animal feed (Culley and Epps 1973) (Fig. 7.1).

The *Spirodela polyrhiza* (Greater duckweed) is comprised of leave-like fronds and roots. The daughter fronds can be induced into the dormant stage, called turions, under the stimulations of low temperature, poor nutrition, and hormone of abscisic acid. The Spirodela contains rich features that is greatly utilized in molecular biology and biotechnology industry. The characteristics of fast growth, heavy metal absorption, clonal propagation, and starch accumulation contribute to their potential applications in the fields of biomass, bioremediator, bioreactor, and biofuel, respectively.

## 7.2 Fast-Growing, Forever-Young Plants as a Promising Biofuel

Spirodela fronds resemble cotyledons, and embryonic leaves inside plant seeds become the first leaves after germination. Unlike other plants that develop different kinds of leaves as they mature, Spirodela is arrested in the junior stage

D. An · W. Wang (✉)
School of Agriculture and Biology, Shanghai Jiao Tong University, Shanghai, China
e-mail: wang2015@sjtu.edu.cn





X. H. Cao et al. (eds.), *The Duckweed Genomes*, Compendium of Plant Genomes,
https://doi.org/10.1007/978-3-030-11045-1_7

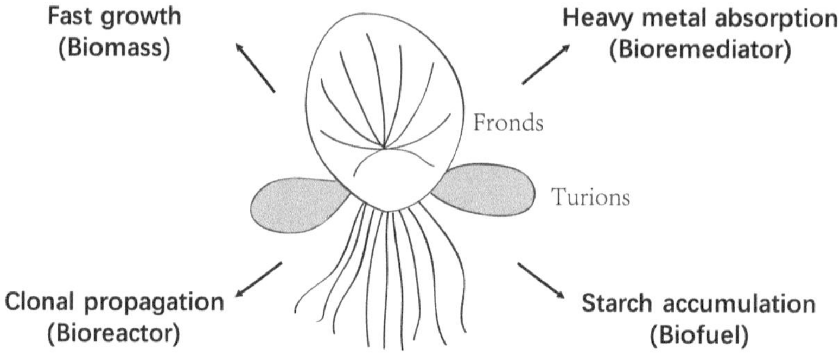


**Fig. 7.1** Spirodela biological features and potential applications

without the organ differentiation and continuously produces cotyledon leaves. This prolonging of juvenile traits is called "neoteny" (Wang et al. 2014).

The Spirodela doubles its biomass every two to four days, faster than any other flowering plants. Spirodela is floating on the surface of water that is different from land plants and is not necessary to hold themselves upright. Spirodela is absent of woody materials from feedstock, allowing it to be easily digested into ethanol (Ma et al. 2018). The carbohydrate in duckweed biomass is readily converted to fermentable sugars by using commercially available enzymes developed for corn-based ethanol production. The high starch content (up to ~70% of dry biomass) under the stress stimulation indicates a potential for ethanol production. Thus, the fast-growing speed, the high biomass yield, and the easy ethanol conversion have made Spirodela become a great alternative feedstock for the biofuel production (Cui and Cheng 2015).

## 7.3 A Valuable Plant for Biomanufacturing

Despite its rapid growth, the unique features including a simple architecture and unusual metabolic characteristics, lack of genetic tools in the duckweeds, have impeded the full implementation of this organism as model for biological research (Yamamoto et al. 2001). Many attempts have been made to develop a technology of genetic engineering of exogenous genes into nuclear genome through agrobacterium-mediated transformation and regeneration from tissue culture (Li et al. 2004; Vunsh et al. 2007). This technology not only allows expressing recombinant protein, polymer, small molecules in duckweed system (Stomp 2005; Yamamoto et al. 2001), but also facilitates functional gene studies in duckweeds (Yamamoto et al. 2001). For instances, a high-efficient bioreactor was developed to produce human monoclonal antibody (Cox et al. 2006) and interferon in duckweeds (De Leede et al. 2008).

## 7.4 An Aquatic Plant in Wastewater Treatment

The utility of duckweed species for bioremediation is sustainable because they recycle the nutrient from the wastewater and recover the aquatic ecosystem efficiently. Duckweeds absorb excess nitrogen and phosphate pollutants from agricultural and municipal wastewater and reproduce their biomass in a competent way (Cheng and Stomp 2009). Duckweed growth on ponds effectively inhibits algal growth, restrains mosquito larvae, concentrates heavy metals, and sequesters harmful organic and phenolic compounds. *Lemna minor* has been most extensively used in phytotoxicity testing (Ozengin and Elmaci 2007; Caicedo et al. 2000), and there are several standard methods which have been adopted by major international standardization agencies, e.g., US Environmental Protection Agency (Brain and Solomon 2007).

Duckweed species show variable sensitivity to the heavy metals. It was reported that the dormant organ of Spirodela before the germination was more tolerant to heavy metals than normal fronds (Olah et al. 2016). *Spirodela intermedia* and *Lemna minor* were proved to effectively remove several heavy metals by their phytoremediation capacity (Miretzky et al. 2004). A significant decrease of Cr and Cd level in polluted water was investigated by Spirodela polyrhiza with the 15-day treatment, showing their strong removal ability of heavy metals (Rai et al. 1995).

The use of dyes and pigments in industries has posed severe problems in wastewater that reduce light penetration and affect plant photosynthesis. It causes fatal impact on aquatic plants and animals, breaking the balance of ecosystem. The dried *Spirodela polyrhiza* was examined to be an efficient adsorbent to take away the basic dye of methylene blue from aqueous solution. As the amount of the dried Spirodela increased, more percentage of dye was eliminated accordingly (Waranusantigul et al. 2003).

## 7.5 Why Sequence Spirodela Genome

The previous studies were intensively concentrated on the classical botany, plant physiology, or biochemistry on duckweeds due to the lack of Spirodela genome. The Spirodela genome sequence could elucidate the remarkable potential of a rapidly growing speed, biomass accumulation, and biofuel production. The further functional genomics of Spirodela would gain insight into the mechanisms of the high production of starch and protein. Undoubtedly, the accessibility of Spirodela genome would set a milestone and represent significant advance in the fields of molecular biology and plant evolution.

## 7.6 History and Consortium

With the driving to develop sustainable and clean feedstocks for biofuel production, researchers from Rutgers University initiated the Spirodela genome sequencing project led by Dr. Joachim Messing together with Dr. Todd P. Michael and funded by the US Department of Energy Joint Genome Institute (JGI) (Fig. 7.2). Several other facilities also got involved in the project (Fig. 7.2). Dr. Joachim Messing's laboratory was in charge of high-quality DNA and RNA preparation, data integration, and team coordination. JGI was dedicated to genomic sequencing including 454, BAC-end sequencing, and RNA-sequencing, as well as genome assembly. Dr. Klaus Mayer's group from MIPS/IBIS, Helmholtz Center Munich, Germany, was responsible for computational analysis of genome annotation and comparative genomics. Dr. Ming-chen Luo's team from the University of California constructed the physical map by fingerprinting of 15,360 BAC clones.

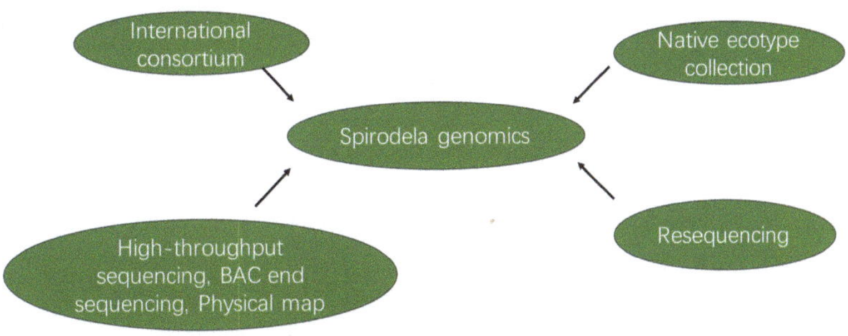


**Fig. 7.2** Spirodela genomics

The Spirodela genome is completed by the cooperation of the international consortium. The availability of whole-genome sequence relies on the technologies of high-throughput sequencing, BAC-end sequencing, and the physical map. The duckweed collection provides the germplasm with the broadly genetic diversity. Resequencing more Spirodela genomes would decipher the genomic variation associated with the phenotypic change.

## 7.7 Sequencing Overview

Spirodela polyrhiza 7498 (Sp7498) has a small plant genome of 158 Mb (Wang et al. 2011), similar to Arabidopsis (Bennett et al. 2003) but nearly half as many as the rice genome (Matsumoto et al. 2005) that was subjected to whole-genome shotgun sequence by using next-generation sequencing technology of Roche 454 (Fig. 7.2) (Wang et al. 2014). Nine million single-end and 1.4 million paired-end reads were generated, which resulted in 21 times of genome coverage. In addition, the Sanger sequencing produced one sequencing depth of BAC-end sequencing from the 15,360 BAC clones (Fig. 7.2). The reads were de novo assembled by Newbler version 2.6 with default parameters after trimming poor bases from ends and masking vector sequences. The contig N50 was 18,927 bp, and scaffold N50 was 3,759,109 bp, respectively. Of the 158 Mb genome, as measured by flow cytometry, 90% was assembled into contigs, 97% of which was assembled in 252 scaffolds, and 94.1% in the top 50 largest scaffolds (Wang et al. 2014).

## 7.8 Construction of Physical Map

A BAC library with 100 Kb insertion was constructed with 40 times of genome coverage. A total of 15,360 clones were subjected to DNA fingerprinting, generating a physical map (Fig. 7.2). We also used BAC-end sequences (BES) to anchor the assembly with the physical map, leading to the scaffolds joined into 32 pseudomolecules (Fig. 7.2). Gaps in sequences like centromeres amounted to 10.7% of the genome and remained in unnamed bases (Ns) (Wang et al. 2014). To examine how the 32 pseudomolecules relate to the 20 chromosomes of the haploid genome of Spirodela polyrhiza, a fluorescence in situ hybridization (FISH) was conducted by selecting those BACs that were low in repeat sequences (Cao et al. 2016). A cytogenetic map with an average distance of 0.89 Mb was constructed by consecutive FISH analyses. Seven ancestral blocks emerged from duplicated chromosome segments of 19 Spirodela chromosomes were elucidated. The chromosomally integrated genome of Sp7498 established a framework for comparative genomics and karyotype evolution of duckweed species (Cao et al. 2016).

## 7.9 Spirodela Genes

Whole-genome sequence provides access to the total number of genes that contribute to the growth, development, and stress response. Increasing studies of individual genes and their corresponding gene families have elucidated their functions in diverse molecular, physiological, and biological processes and have provided novel clues on their regulation and gene expression. A number of 19,623 protein-coding genes were annotated by an integrated pipeline, showing 28% less copies than Arabidopsis (27,416) (Bennett et al. 2003) and 50% less copies than rice (37,544) (Matsumoto et al. 2005). The small duckweed genome turns out many missing genes, including those for plant maturation and production of cellulose and lignin, whereas it retains more genes for starch production than comparable genomes. The most surprising finding was the insight into the molecular basis involved in maintaining a forever-young lifestyle. Spirodela had fewer genes to promote and more genes to repress the switch from juvenile to mature growth.

Spirodela appears to have a significantly lower number of tandem gene clusters (948) than rice (2,602) (Matsumoto et al. 2005), tomato

(2,340) (Sato et al. 2012), and Arabidopsis (1,938) (Lamesch et al. 2012), but it is surprisingly close to banana (1,048) (D'Hont et al. 2012), which has 1.9 times of the Spirodela gene number (Wang et al. 2014).

## 7.10 Resequencing of Spirodela Genome

The investigation for specific turion (dormant stage) yields appeared that many factors including phosphate deficiency and temperature could affect the dormant organ development (Appenroth and Adamec 2015). Spirodela polyrhiza 9509 (Sp9509) was shown to have a low turion yield. To obtain genome-wide information on intraspecific variations between different Spirodela populations, Sp9509 was resequenced by using high-depth short-read sequencing and high-throughput genome mapping technologies (Fig. 7.2) (Michael et al. 2017). The draft genome of Sp9509 was assembled and was further defined into 20 chromosomes with the BioNano physical map.

The genome comparison between Sp7498 and Sp9509 revealed conflicts and identified potential misassembled sites in each genome, indicating that PCR validations or long reads spanning over the junctions were required. There were 96 high-confidence structure variations (SVs) with the range of 1000–100,000 bp between the two BioNano genome maps of Sp9509 and Sp7498. The copy number of the rDNA repeats units in Sp9509, as well as four different accessions of *S. polyrhiza* was significantly shrunk less than 100, which was even fewer than that of yeast (Michael et al. 2017). There was 25.25% repeat content with 271 full-length long terminal repeats (LTRs) in the Sp9509 genome, compared with $\sim$17% of Sp7498 (Wang et al. 2014). The transposon similarity in Sp9509 was very low against other species of Brachypodium, rice, and sorghum, indicating a large evolutionary distance between them. The overall DNA methylation level in Spirodela was the lowest (9%) among the tested plants of A. thaliana (32%), rice (39%), Setaria italica (44%), and B. distachyon (54%) (Michael et al. 2017). The high copy number of tandem repeats (TRs) generally occurs in the chromosome centromeres (Melters et al. 2013). The Sp7498 genome was predicted to have a 138 bp centromere repeat-like sequence, whereas Sp9509 was found a 119 bp TR on 19 out of the 20 chromosomes with high DNA methylation levels (Michael et al. 2017). The distribution of the 119 bp centromere repeat across some of the Spirodela chromosomes suggested that they were holocentric. This result was consistent with a dispersed heterochromatin signal observed in cytological studies (Cao et al. 2015). The bioinformatics analysis predicted that there were 59 conserved microRNAs (miRNAs) of 22 families and 25 novel miRNAs. The small RNA-sequencing validated 29 Spirodela-specific miRNAs in the genome of Sp9509. The sequence-based annotation identified five and three loci for miRNA156 and miRNA159 in Sp9509, respectively (Michael et al. 2017). In contrast, the Sp7498 genome included 24 loci encoded for miRNA156 and one locus encoded for miRNA159 (Wang et al. 2014).

## 7.11 Going Back to the Native Ecotypes

Genetic diversity represents a great resource for the improvement of breeding. Dr. Landolt collected more than 1,000 native ecotypes all over the world and shifted the biggest collection to Rutgers duckweed stock cooperative (http://www.ruduckweed.org) (Fig. 7.2). There are five genera of duckweeds including 37 species. The best marker to identify duckweed species is the atpF-atpH intergenic region (Wang et al. 2010). Intriguing the large native collection would shed new light on their charming and thus make the most use of duckweeds as biofeed, biofuel, and bioremediator.

## 7.12 Data Availability

The assembled genome sequence of Sp7498 is deposited in DDBJ/EMBL/GenBank nucleotide core database under the accession code ATDW 00000000 with the BioSample of SAMN0298 1544 and the BioProject of PRJNA205940 (Wang et al. 2014). The sequence reads done by JGI are assigned an NCBI bioproject (https://www.ncbi.nlm.nih.gov/bioproject/46611). BAC-end sequences have been deposited in the GenBank GSS database under accession codes JY978532 to KG007076. Fosmid sequences have been deposited in the GenBank nucleotide core database under accession codes AC254537 to AC254559. The EST project data is available in NCBI Sequence Read Archive (https://www.ncbi.nlm.nih.gov/sra/SRX148325). The GBrowse and Blast search of *Spirodela polyrhiza* v2 is accessible at JGI Phytozome 12: https://phytozome.jgi.doe.gov/pz/portal.html#!info?alias=Org_Spolyrhiza. The bulk data can be also downloaded with a new account registration.

The genome sequence for Sp9509 is deposited with an accession number of GCA_001981405. The project can be retrieved at: (https://www.ncbi.nlm.nih.gov/bioproject/PRJNA308109/). The raw reads from Illumina platform with the libraries of 200, 500 bp, 2, and 5 Kb can be also found under this bioproject (Michael et al. 2017).

## 7.13 Conclusions

The initiation of Spirodela as an aquatic plant for dissecting the physiological, evolutionary, and architectural traits of asexual plants was accelerated by the release of whole-genome sequence information. This has accentuated the nearly neglected research of basal monocots and promoted the plant genomics into richer genetic and genomic resources. Attempts to generate large-scale and unrestricted genomic and transcriptomic data via web-based databases would certainly benefit the duckweed community and accelerate functional genomics studies and molecular breeding in the fields of biofeed, biofuel, bioremediator, and bioreactor. Furthermore, the potential abiotic stress tolerance in duckweeds has encouraged scientists to explore the molecular mechanism of strong adaption and vitality over the world that would improve stress tolerance not only in themselves but other crops.

## References

Appenroth KJ, Adamec L (2015) Specific turion yields of different clones of Spirodela polyrhiza depend on external phosphate thresholds. Plant Biol (Stuttg) 17:125–129

Bennett MD, Leitch IJ, Price HJ, Johnston JS (2003) Comparisons with Caenorhabditis (similar to 100 Mb) and Drosophila (similar to 175 Mb) using flow cytometry show genome size in Arabidopsis to be similar to 157 Mb and thus similar to 25% larger than the Arabidopsis genome initiative estimate of similar to 125 Mb. Ann Bot-London 91(5):547–557

Brain RA, Solomon KR (2007) A protocol for conducting 7-day daily renewal tests with Lemna gibba. Nat Protoc 2(4):979–987

Caicedo JR, Van der Steen NP, Arce O, Gijzen HJ (2000) Effect of total ammonia nitrogen concentration and pH on growth rates of duckweed (Spirodela polyrrhiza). Water Res 34(15):3829–3835

Cao HX, Vu GTH, Wang W, Messing J, Schubert I (2015) Chromatin organisation in duckweed interphase nuclei in relation to the nuclear DNA content. Plant Biol (Stuttg) 17:120–124

Cao HX, Vu GTH, Wang WQ, Appenroth KJ, Messing J, Schubert I (2016) The map-based genome sequence of Spirodela polyrhiza aligned with its chromosomes, a reference for karyotype evolution. New Phytol 209(1):354–363

Cheng JJ, Stomp AM (2009) Growing duckweed to recover nutrients from wastewaters and for production of fuel ethanol and animal feed. Clean-Soil Air Water 37(1):17–26

Cox KM, Sterling JD, Regan JT, Gasdaska JR, Frantz KK, Peele CG, Black A, Passmore D, Moldovan-Loomis C, Srinivasan M, Cuison S, Cardarelli PM, Dickey LF (2006) Glycan optimization of a human monoclonal antibody in the aquatic plant Lemna minor. Nat Biotechnol 24(12):1591–1597

Cui W, Cheng JJ (2015) Growing duckweed for biofuel production: a review. Plant Biol (Stuttg) 17(Suppl 1):16–23

Culley DD, Epps EA (1973) Use of Duckweed for waste treatment and animal feed. J Water Pollut Con F 45(2):337–347

D'Hont A, Denoeud F, Aury JM, Baurens FC, Carreel F, Garsmeur O, Noel B, Bocs S, Droc G, Rouard M, Da Silva C, Jabbari K, Cardi C, Poulain J, Souquet M,

Labadie K, Jourda C, Lengelle J, Rodier-Goud M, Alberti A, Bernard M, Correa M, Ayyampalayam S, Mckain MR, Leebens-Mack J, Burgess D, Freeling M, Mbeguie-A-Mbeguie D, Chabannes M, Wicker T, Panaud O, Barbosa J, Hribova E, Heslop-Harrison P, Habas R, Rivallan R, Francois P, Poiron C, Kilian A, Burthia D, Jenny C, Bakry F, Brown S, Guignon V, Kema G, Dita M, Waalwijk C, Joseph S, Dievart A, Jaillon O, Leclercq J, Argout X, Lyons E, Almeida A, Jeridi M, Dolezel J, Roux N, Risterucci AM, Weissenbach J, Ruiz M, Glaszmann JC, Quetier F, Yahiaoui N, Wincker P (2012) The banana (Musa acuminata) genome and the evolution of monocotyledonous plants. Nature 488(7410):213–217

De Leede LG, Humphries JE, Bechet AC, Van Hoogdalem EJ, Verrijk R, Spencer DG (2008) Novel controlled-release Lemna-derived IFN-alpha2b (Locteron): pharmacokinetics, pharmacodynamics, and tolerability in a phase I clinical trial. J Interferon Cytokine Res 28(2):113–122

Lamesch P, Berardini TZ, Li DH, Swarbreck D, Wilks C, Sasidharan R, Muller R, Dreher K, Alexander DL, Garcia-Hernandez M, Karthikeyan AS, Lee CH, Nelson WD, Ploetz L, Singh S, Wensel A, Huala E (2012) The arabidopsis information resource (TAIR): improved gene annotation and new tools. Nucleic Acids Res 40(D1):D1202–D1210

Li J, Jain M, Vunsh R, Vishnevetsky J, Hanania U, Flaishman M, Perl A, Edelman M (2004) Callus induction and regeneration in Spirodela and Lemna. Plant Cell Rep 22(7):457–464

Ma YB, Zhu M, Yu CJ, Wang Y, Liu Y, Li ML, Sun YD, Zhao JS, Zhou GK (2018) Large-scale screening and characterisation of Lemna aequinoctialis and Spirodela polyrhiza strains for starch production. Plant Biol (Stuttg) 20(2):357–364

Matsumoto T, Wu JZ, Kanamori H, Katayose Y, Fujisawa M, Namiki N, Mizuno H, Yamamoto K, Antonio BA, Baba T, Sakata K, Nagamura Y, Aoki H, Arikawa K, Arita K, Bito T, Chiden Y, Fujitsuka N, Fukunaka R, Hamada M, Harada C, Hayashi A, Hijishita S, Honda M, Hosokawa S, Ichikawa Y, Idonuma A, Iijima M, Ikeda M, Ikeno M, Ito K, Ito S, Ito T, Ito Y, Ito Y, Iwabuchi A, Kamiya K, Karasawa W, Kurita K, Katagiri S, Kikuta A, Kobayashi H, Kobayashi N, Machita K, Maehara T, Masukawa M, Mizubayashi T, Mukai Y, Nagasaki H, Nagata Y, Naito S, Nakashima M, Nakama Y, Nakamichi Y, Nakamura M, Meguro A, Negishi M, Ohta I, Ohta T, Okamoto M, Ono N, Saji S, Sakaguchi M, Sakai K, Shibata M, Shimokawa T, Song JY, Takazaki Y, Terasawa K, Tsugane M, Tsuji K, Ueda S, Waki K, Yamagata H, Yamamoto M, Yamamoto S, Yamane H, Yoshiki S, Yoshihara R, Yukawa K, Zhong HS, Yano M, Sasaki T, Yuan QP, Shu OT, Liu J, Jones KM, Gansberger K, Moffat K, Hill J, Bera J, Fadrosh D, Jin SH, Johri S, Kim M, Overton L, Reardon M, Tsitrin T, Vuong H, Weaver B, Ciecko A, Tallon L, Jackson J, Pai G, Van Aken S, Utterback T, Reidmuller S, Feldblyum T, Hsiao J, Zismann V, Iobst S, de Vazeille AR, Buell CR, Ying K, Li Y, Lu TT, Huang YC, Zhao Q, Feng Q, Zhang L, Zhu JJ, Weng QJ, Mu J, Lu YQ, Fan DL, Liu YL, Guan JP, Zhang YJ, Yu SL, Liu XH, Zhang Y, Hong GF, Han B, Choisne N, Demange N, Orjeda G, Samain S, Cattolico L, Pelletier E, Couloux A, Segurens B, Wincker P, D'Hont A, Scarpelli C, Weissenbach J, Salanoubat M, Quetier F, Yu Y, Kim HR, Rambo T, Currie J, Collura K, Luo MZ, Yang TJ, Ammiraju JSS, Engler F, Soderlund C, Wing RA, Palmer LE, de la Bastide M, Spiegel L, Nascimento L, Zutavern T, O'Shaughnessy A, Dike S, Dedhia N, Preston R, Balija V, McCombie WR, Chow TY, Chen HH, Chung MC, Chen CS, Shaw JF, Wu HP, Hsiao KJ, Chao YT, Chu MK, Cheng CH, Hour AL, Lee PF, Lin SJ, Lin YC, Liou JY, Liu SM, Hsing YI, Raghuvanshi S, Mohanty A, Bharti AK, Gaur A, Gupta V, Kumar D, Ravi V, Vij S, Kapur A, Khurana P, Khurana P, Khurana JP, Tyagi AK, Gaikwad K, Singh A, Dalal V, Srivastava S, Dixit A, Pal AK, Ghazi IA, Yadav M, Pandit A, Bhargava A, Sureshbabu K, Batra K, Sharma TR, Mohapatra T, Singh NK, Messing J, Nelson AB, Fuks G, Kavchok S, Keizer G, Llaca ELV, Song RT, Tanyolac B, Young S, Il KH, Hahn JH, Sangsakoo G, Vanavichit A, de Mattos LAT, Zimmer PD, Malone G, Dellagostin O, de Oliveira AC, Bevan M, Bancroft I, Minx P, Cordum H, Wilson R, Cheng ZK, Jin WW, Jiang JM, Leong SA, Iwama H, Gojobori T, Itoh T, Niimura Y, Fujii Y, Habara T, Sakai H, Sato Y, Wilson G, Kumar K, McCouch S, Juretic N, Hoen D, Wright S, Bruskiewich R, Bureau T, Miyao A, Hirochika H, Nishikawa T, Kadowaki K, Sugiura M (2005) Project IRGS. The map-based sequence of the rice genome. Nature 436(7052):793–800

Melters DP, Bradnam KR, Young HA, Telis N, May MR, Ruby JG, Sebra R, Peluso P, Eid J, Rank D, Garcia JF, De Risi JL, Smith T, Tobias C, Ross-Ibarra J, Korf I, Chan SW (2013) Comparative analysis of tandem repeats from hundreds of species reveals unique insights into centromere evolution. Genome Biol 14 (1):R10

Michael TP, Bryant D, Gutierrez R, Borisjuk N, Chu P, Zhang HZ, Xia J, Zhou JF, Peng H, El Baidouri M, ten Hallers B, Hastie AR, Liang T, Acosta K, Gilbert S, McEntee C, Jackson SA, Mockler TC, Zhang WX, Lam E (2017) Comprehensive definition of genome features in Spirodela polyrhiza by high-depth physical mapping and short-read DNA sequencing strategies. Plant J 89(3):617–635

Miretzky P, Saralegui A, Cirelli AF (2004) Aquatic macrophytes potential for the simultaneous removal of heavy metals (Buenos Aires, Argentina). Chemosphere 57(8):997–1005

Naumann B, Eberius M, Appenroth KJ (2007) Growth rate based dose-response relationships and EC-values of ten heavy metals using the duckweed growth inhibition test (ISO 20079) with Lemna minor L. clone St. J Plant Physiol 164(12):1656–1664

Olah V, Hepp A, Meszaros I (2016) Assessment of giant duckweed (Spirodela polyrhiza L. Schleiden) turions as model objects in ecotoxicological applications. Bull Environ Contam Toxicol 96(5):596–601

Ozengin N, Elmaci A (2007) Performance of duckweed (Lemna minor L.) on different types of wastewater treatment. J Environ Biol 28(2):307–314

Rahman MA, Hasegawa H (2011) Aquatic arsenic: phytoremediation using floating macrophytes. Chemosphere 83(5):633–646

Rai UN, Sinha S, Tripathi RD, Chandra P (1995) Waste-water treatability potential of some aquatic macrophytes—removal of heavy-metals. Ecol Eng 5 (1):5–12

Sato S, Tabata S, Hirakawa H, Asamizu E, Shirasawa K, Isobe S, Kaneko T, Nakamura Y, Shibata D, Aoki K, Egholm M, Knight J, Bogden R, Li CB, Shuang Y, Xu X, Pan SK, Cheng SF, Liu X, Ren YY, Wang J, Albiero A, Dal Pero F, Todesco S, Van Eck J, Buels RM, Bombarely A, Gosselin JR, Huang MY, Leto JA, Menda N, Strickler S, Mao LY, Gao S, Tecle IY, York T, Zheng Y, Vrebalov JT, Lee J, Zhong SL, Mueller LA, Stiekema WJ, Ribeca P, Alioto T, Yang WC, Huang SW, Du YC, Zhang ZH, Gao JC, Guo YM, Wang XX, Li Y, He J, Li CY, Cheng ZK, Zuo JR, Ren JF, Zhao JH, Yan LH, Jiang HL, Wang B, Li HS, Li ZJ, Fu FY, Chen BT, Han B, Feng Q, Fan DL, Wang Y, Ling HQ, Xue YBA, Ware D, McCombie WR, Lippman ZB, Chia JM, Jiang K, Pasternak S, Gelley L, Kramer M, Anderson LK, Chang SB, Royer SM, Shearer LA, Stack SM, Rose JKC, Xu YM, Eannetta N, Matas AJ, McQuinn R, Tanksley SD, Camara F, Guigo R, Rombauts S, Fawcett J, Van de Peer Y, Zamir D, Liang CB, Spannagl M, Gundlach H, Bruggmann R, Mayer K, Jia ZQ, Zhang JH, Ye ZBA, Bishop GJ, Butcher S, Lopez-Cobollo R, Buchan D, Filippis I, Abbott J, Dixit R, Singh M, Singh A, Pal JK, Pandit A, Singh PK, Mahato AK, Dogra V, Gaikwad K, Sharma TR, Mohapatra T, Singh NK, Causse M, Rothan C, Schiex T, Noirot C, Bellec A, Klopp C, Delalande C, Berges H, Mariette J, Frasse P, Vautrin S, Zouine M, Latche A, Rousseau C, Regad F, Pech JC, Philippot M, Bouzayen M, Pericard P, Osorio S, del Carmen AF, Monforte A, Granell A, Fernandez-Munoz R, Conte M, Lichtenstein G, Carrari F, De Bellis G, Fuligni F, Peano C, Grandillo S, Termolino P, Pietrella M, Fantini E, Falcone G, Fiore A, Giuliano G, Lopez L, Facella P, Perrotta G, Daddiego L, Bryan G, Orozco M, Pastor X, Torrents D, van Schriek KNVMGM, Feron RMC, van Oeveren J, de Heer P, daPonte L, Jacobs-Oomen S, Cariaso M, Prins M, van Eijk MJT, Janssen A, van Haaren MJJ, Jo SH, Kim J, Kwon SY, Kim S, Koo DH, Lee S, Hur CG, Clouser C, Rico A, Hallab A, Gebhardt C, Klee K, Jocker A, Warfsmann J, Gobel U, Kawamura S, Yano K, Sherman JD, Fukuoka H, Negoro S, Bhutty S, Chowdhury P, Chattopadhyay D, Datema E, Smit S, Schijlen EWM, van de Belt J, van Haarst JC, Peters SA, van Staveren MJ, Henkens MHC, Mooyman PJW, Hesselink T, van Ham RCHJ, Jiang GY, Droege M, Choi D, Kang BC, Kim BD, Park M, Kim S, Yeom SI, Lee YH, Choi YD, Li GC, Gao JW, Liu YS, Huang SX, Fernandez-Pedrosa V, Collado C, Zuniga S, Wang GP, Cade R, Dietrich RA, Rogers J, Knapp S, Fei ZJ, White RA, Thannhauser TW, Giovannoni JJ, Botella MA, Gilbert L, Gonzalez R, Goicoechea JL, Yu Y, Kudrna D, Collura K, Wissotski M, Wing R, Schoof H, Meyers BC, Gurazada AB, Green PJ, Mathur S, Vyas S, Solanke AU, Kumar R, Gupta V, Sharma AK, Khurana P, Khurana JP, Tyagi AK, Dalmay T, Mohorianu I, Walts B, Chamala S, Barbazuk WB, Li JP, Guo H, Lee TH, Wang YP, Zhang D, Paterson AH, Wang XY, Tang HB, Barone A, Chiusano ML, Ercolano MR, D'Agostino N, Di Filippo M, Traini A, Sanseverino W, Frusciante L, Seymour GB, Elharam M, Fu Y, Hua A, Kenton S, Lewis J, Lin SP, Najar F, Lai HS, Qin BF, Qu CM, Shi RH, White D, White J, Xing YB, Yang KQ, Yi J, Yao ZY, Zhou LP, Roe BA, Vezzi A, D'Angelo M, Zimbello R, Schiavon R, Caniato E, Rigobello C, Campagna D, Vitulo N, Valle G, Nelson DR, De Paoli E, Szinay D, de Jong HH, Bai YL, Visser RGF, Lankhorst RMK, Beasley H, McLaren K, Nicholson C, Riddle C, Gianese G, Consortium TG (2012) The tomato genome sequence provides insights into fleshy fruit evolution. Nature 485(7400):635–641

Stomp AM (2005) The duckweeds: a valuable plant for biomanufacturing. Biotechnol Annu Rev 11:69–99

Vunsh R, Li J, Hanania U, Edelman M, Flaishman M, Perl A, Wisniewski JP, Freyssinet G (2007) High expression of transgene protein in Spirodela. Plant Cell Rep 26(9):1511–1519

Wang W (1990) Literature review on duckweed toxicity testing. Environ Res 52(1):7–22

Wang W, Kerstetter R, Michael T (2011) Evolution of genome size in duckweeds (Lemnaceae). J Bot 2:1

Wang W, Haberer G, Gundlach H, Glasser C, Nussbaumer T, Luo MC, Lomsadze A, Borodovsky M, Kerstetter RA, Shanklin J, Byrant DW, Mockler TC, Appenroth KJ, Grimwood J, Jenkins J, Chow J, Choi C, Adam C, Cao XH, Fuchs J, Schubert I, Rokhsar D, Schmutz J, Michael TP, Mayer KFX, Messing J (2014) The Spirodela polyrhiza genome reveals insights into its neotenous reduction fast growth and aquatic lifestyle. Nat Commun, 5

Wang WQ, Wu YR, Yan YH, Ermakova M, Kerstetter R, Messing J (2010) DNA barcoding of the Lemnaceae, a family of aquatic monocots. BMC Plant Biol 10

Waranusantigul P, Pokethitiyook P, Kruatrachue M, Upatham ES (2003) Kinetics of basic dye (methylene blue) biosorption by giant duckweed (Spirodela polyrrhiza). Environ Pollut 125(3):385–392

Xue HL, Xiao Y, Jin YL, Li XB, Fang Y, Zhao H, Zhao Y, Guan JF (2012) Genetic diversity and geographic differentiation analysis of duckweed using inter-simple sequence repeat markers. Mol Biol Rep 39(1):547–554

Yamamoto YT, Rajbhandari N, Lin XH, Bergmann BA, Nishimura Y, Stomp AM (2001) Genetic transformation of duckweed Lemna gibba and Lemna minor. Vitro Cell Dev-Pl 37(3):349–353

# Repetitive Sequences: Impacts and Uses in the Spirodela Genome

# 8

Paul Fourounjian

**Abstract**

Repetitive DNA, consisting of small and large satellite repeats and transposable elements, comprises over 50% of most plant genomes. The Lemnaceae family demonstrates a ~12-fold difference in genome size and relatively similar number of genes, indicating a wide variability in repeat content. The best studied genome of the family *Spirodela polyrhiza* had a normal total satellite DNA content, yet a surprisingly high 50% of those were dinucleotide microsatellite repeats. The telomeres and 119 bp centromere repeats were typical, although ribosomal repeats appear scarce. Genomic studies showed a small number of 24nt heterochromatic siRNAs accompanied by the lowest rate of DNA methylation seen in any plant sequenced at 9% and low rates of heterochromatin formation. Despite this low level of regulation, the transposable elements are unexpectedly rare and old. In fact, they even show high rates of DNA methylation and high rates of inactivation through illegitimate recombination. This suggests that the scarce 24nt siRNAs are surprisingly effective and an intriguing topic of further research.

P. Fourounjian (✉)
Waksman Institute of Microbiology, Rutgers University, Piscataway 08854, USA
e-mail: pjf99@scarletmail.rutgers.edu

In the early years of DNA and chromosome research, structural components of chromosomes were noticed as patterns in DNA and protein stains, often in the centromeric or telomeric regions. Once DNA sequencing began it was uncovered that virtually all eukaryotic genomes contain significant portions of repetitive DNA, previously thought of as "junk DNA" (Biscotti et al. 2015). In plants, repetitive elements comprise the majority of most genomes sequenced, ranging from a mere 14% in the grain teff to 85% in maize (Wendel et al. 2016). These repetitive elements can be categorized into tandem repeats which aid in chromosome structure, and longer interspersed repeats derived from transposable elements (TEs). As of 2018 there are two published sequences for *Spirodela polyrhiza* clones 7498 and 9509, and the *Lemna minor* 5500, along with draft genomes of two *Lemna* species *minor* and *gibba* and the *Wolffia* species *australiana* (Unpublished), (Wang et al. 2014; Van Hoeck et al. 2015; Ernst and Martienssen 2016; Michael et al. 2017). Similar to other angiosperms as a whole, these genomes vary considerably in size, but not significantly in gene number (Table 8.1). The Lemnaceae family displays a 12-fold difference in genome size from the smallest sequenced monocot *Spirodela polyrhiza* to the 1881 megabase *Wolffia arrhiza* (Wang et al. 2011). A recent review on plant genome architecture summarized that these size variations between genomes are due to common whole genome duplication, followed by



X. H. Cao et al. (eds.), *The Duckweed Genomes*, Compendium of Plant Genomes,
https://doi.org/10.1007/978-3-030-11045-1_8

**Table 8.1** Lemnaceae genome size and gene content

| Species, clone | Genome size (Megabases) | Gene copy # |
|---|---|---|
| *S. polyrhiza*, 7498 | 158 | 19,623 |
| *S. polyrhiza*, 9509 | 158 | 18,507 |
| *L. minor*, 5500 | 481 | 22,382 |
| *L. minor*, 8627 | 800 | NA |
| *L. gibba*, 7742 | 450 | 21,830 |
| *W. australiana* | ~380 | NA |

reduction of coding genes, and proliferation of transposable elements (Wendel et al. 2016). Taken in summary, these repeats play a large role in genomic size and composition and chromosomal structure, in the duckweeds and eukaryotes as a whole.

When DNA was separated by density gradient centrifugation tandem repeats with differential AT/GC content created satellite bands above and below the majority of DNA eventually leading to the name satellite DNA. These tandem repeats range in size from the 180 bp corresponding to a nucleosome to tiny 2 nucleotide microsatellite repeats. They were found to have structural implications in centromeres and telomeres where they maintain heterochromatic structure, and disruptions of their expression have been shown to lead to genomic instability and cancer (Biscotti et al. 2015). The strain 7498 genome study showed that the small *Spirodela polyrhiza* genome had a normal number of satellite DNA repeats, at 1.3% of the genome. Yet while most plants have 10–100 bp minisatellites making up roughly half of the total satellite DNA, strain 7498 satellite DNA was 50% microsatellite repeats, largely comprised of GA repeats, which may have been mutated from methylated CG heterochromatin sequences (Wang et al. 2014; Michael et al. 2017). For the *Lemna minor* 5500 genome, we know that satellite and microsatellite repeats made up 0.6 and 3% of the genome, indicating a similar enrichment of microsatellite repeats (Van Hoeck et al. 2015). In a follow-up study assembling the 32 pseudo-molecules into 20 chromosomes relied on the telomeric repeats of TTTAGGG and the suspected centromeric repeats to help support the confidence of the pseudomolecule assembly (Cao et al. 2016). Another analysis of the 7498 and 9509 strains of *Spirodela* was run using longer reads for better resolution of repeat regions and found a high homology with few indels and less than 0.06% heterozygosity in SNPs. They found that a previously reported 138 bp centromeric repeat was found at 1 centromere and that 19 of 20 chromosomes contained large numbers of a 119 bp centromeric repeat (Melters et al. 2013; Michael et al. 2017). Additionally, they found an extremely low ribosomal DNA copy number of 81 compared to 570 in the similarly sized *Arabidopsis thaliana* genome. In summary, while the centromeres and telomeres of *Spirodela polyrhiza* are consistent with other plant genomes, the microsatellite repeats are very abundant and the ribosomal repeats are very rare.

Probably, the most interesting repeat elements are the transposable elements (TEs), which include DNA copying transposons, RNA copying retrotransposons with autonomous versions capable of replicating themselves and non-autonomous versions of each. Thanks to this replication potential, these selfish genes are always attempting to proliferate, while the plant host genome is perpetually suppressing them and removing them through illegitimate recombination. This push and pull occurring in countless plant species shows that of our crop plants TEs can comprise as little as 14% of the genome in teff and as much as 85% in maize (Wendel et al. 2016). In the annotation of the 7498 genome, LTR retrotransposons were annotated based on homology and found to be 15.5% of the genome, which agreed with its size, while the transposons were too distant from their homologs in their genomes an unable to be annotated (Wang et al. 2014). This lack of homology is due to the age of the transposons, which mutate over time. In *Spirodela,* the relatively few LTRs (264) had an average age of 4.3 million years, while the average in Brachypodium and rice was found to be 1.8 and 0.7 million years, respectively. In the later analysis of the 9509 genome, TEs were annotated by homology to other known TEs, and by mapping 22–24nt siRNAs known to regulate them through methylation. This showed that the genome is 25% TEs, with a Gypsy/Copia ratio of 1.5. In accordance with the age of the LTRs, the *Spirodela* genome was found to be purging them through illegitimate recombination resulting in the highest ratio of deactivated solo to intact LTRs seen in any plant genome.

After the *Spirodela* 7498 genome was published, the draft genome of *Lemna minor* 5500 was published due to its importance in ecotoxicological studies (Van Hoeck et al. 2015). While *Lemna minor* strains vary in genome size from 323 to 760 Mb strain 5500 is 481 Mb in size and only has 14% more annotated genes than *Spirodela polyrhiza* 7498 (Table 8.1). Compared to *Spirodela* 94.5% of the difference in genome size is due to repeats. These repeats make up 61% of the genome and 36% of the genome is TEs, mainly retrotransposons, which is slightly higher than *Spirodela.* The count of LTRs increased ~10-fold to 210,531. There was a final category of unclassified repeats that made up 21% of the genome. In strain, 7498 DNA-based transposons were difficult to annotate based on their old age and low homology, and in strain, 9509 the annotation relied on siRNAs. Therefore, the unclassified repeats may include many ancient unannotated transposons.

The relative lack of TEs in *Spirodela* brought attention to the RNA directed DNA methylation (RdDM) pathway. This is a mechanism of silencing transposons through siRNAs where Pol IV creates a ssRNA transcript and RDR2 makes it a dsRNA (Matzke et al. 2015). Then, DCL3 cleaves it into 24nt het-siRNAs (heterochromatic) that are loaded onto AGO4, which binds to DRM2 and RDM1 proteins that methylate the 5' end of cytosine in GC, CHG, and CHH sequences. To finish the process a collection of proteins in a histone-modifying complex converts the methylated TE sequence to silenced heterochromatin. This pathway is highly conserved across all land plants, with the notable outlier of the Norway Spruce, which has relatively few 24nt het-siRNAs, mainly localized to reproductive organs (Matzke et al. 2015).

In *Spirodela polyrhiza,* it was noticed that 24nt sRNAs were rare, comprising 7.3% of the small RNAs in strain LT5a and 1% in strain 7498 (Fourounjian et al. 2019). While the 9509 genome had the lowest DNA methylation rate of any plant sequenced at 9%, the TEs had an average methylation rate of 20% (Michael et al. 2017). Furthermore, older TEs were annotated based on the mapping of 22–24nt siRNAs, suggesting that they were expressed and active. The *Spirodela* genome also revealed a low number of old TEs suggesting that it has been very successful at halting their proliferation (Wang et al. 2014; Michael et al. 2017). Taken together it looks like the RdDM pathway is working with little to no 24nt het-siRNAs. This could be similar to the results seen in Norway spruce where 24nt het-siRNAs are localized to flowers, which are very rare in *Spirodela,* or perhaps other mechanisms may be at play. The mystery of how the Lemnaceae, particularly *Spirodela,* regulate their TEs is an exciting field of research that is still currently unfolding.

**Acknowledgements** Thank you to Dr. Alex Harkess for reviewing this chapter to confirm its accurate, but not complete description of the RdDM pathway.

## References

Biscotti MA, Olmo E, Heslop-Harrison JS (2015) Repetitive DNA in eukaryotic genomes. Chromosom Res 23:415–420. https://doi.org/10.1007/s10577-015-9499-z

Cao HX, Vu GTH, Wang W et al (2016) The map-based genome sequence of Spirodela polyrhiza aligned with its chromosomes, a reference for karyotype evolution. New Phytol 209:354–363. https://doi.org/10.1111/nph.13592

Ernst E, Martienssen R (2016) Status of the Lemna gibba 7742a and Lemna minor 8627 genomes. Duckweed Forum, 12

Fourounjian P, Tang J, Tanyolac B, Feng Y, Gelfand B, Kakrana A, Tu M, Wakim C, Meyers BC, Ma J, Messing J (2019) Post-transcriptional adaptation of the aquatic plant under stress and hormonal stimuli. Plant J

Matzke MA, Kanno T, Matzke AJM (2015) RNA-directed DNA methylation: the evolution of a complex epigenetic pathway in flowering plants. Annu Rev Plant Biol 66:243–267. https://doi.org/10.1146/annurev-arplant-043014-114633

Melters DP, Bradnam KR, Young HA et al (2013) Comparative analysis of tandem repeats from hundreds of species reveals unique insights into centromere evolution. Genome Biol 14:R10. https://doi.org/10.1186/gb-2013-14-1-r10

Michael TP, Bryant D, Gutierrez R et al (2017) Comprehensive definition of genome features in Spirodela polyrhiza by high-depth physical mapping and short-read DNA sequencing strategies. Plant J 89:617–635. https://doi.org/10.1111/tpj.13400

Van Hoeck A, Horemans N, Monsieurs P et al (2015) The first draft genome of the aquatic model plant Lemna minor opens the route for future stress physiology research and biotechnological applications. Biotechnol Biofuels 8:188. https://doi.org/10.1186/s13068-015-0381-1

Wang W, Haberer G, Gundlach H et al (2014) The Spirodela polyrhiza genome reveals insights into its neotenous reduction fast growth and aquatic lifestyle. Nat Commun 5:1–13. https://doi.org/10.1038/ncomms4311

Wang W, Kerstetter RA, Michael TP (2011) Evolution of genome size in Duckweeds (Lemnaceae). J Bot 2011:1–9. https://doi.org/10.1155/2011/570319

Wendel JF, Jackson SA, Meyers BC, Wing RA (2016) Evolution of plant genome architecture. Genome Biol 17:1–14. https://doi.org/10.1186/s13059-016-0908-1

# 9 Stranger than Fiction: Loss of MADS-Box Genes During Evolutionary Miniaturization of the Duckweed Body Plan

## Loss of MADS-Box Genes in Duckweeds

Lydia Gramzow and Günter Theißen

**Abstract**

Duckweeds (Lemnoideae or Lemnaceae) are unusual flowering plants. They all lack a stem, many species even do not have roots, and most of them flower only rarely, if at all. Evolution of duckweeds obviously involved extreme miniaturization and simplification of their structure and life cycle. This raises the question as to whether the evolutionary changes of the duckweed body plan were accompanied, and thus potentially causally linked, to a loss of function of genes that control the development of affected structures. MADS-box genes are involved in the control of many developmental processes in flowering plants, including root, flower, and fruit development. Their phylogeny is quite well known and reveals a fairly strong correlation between some gene clades and function. Therefore, we used the available genome sequences of several duckweed species to test as to whether the loss of specific MADS-box genes can be linked to the reduction of morphological structures. In all duckweed genomes analysed, 5 of the 17 clades of MIKC$^{C}$-group MADS-box genes that probably existed in the most recent common ancestor of extant flowering plants, appear to be absent. Our analyses thus demonstrate that duckweeds are the plant taxon with the lowest number of MIKC-type MADS-box gene clades of all flowering plants that were investigated so far. While *AGL15*-like genes have probably been lost in the stem group of extant monocots already, and *FLC*-like genes have been lost in the stem group of the order Alismatales (to which duckweeds belong), *AGL9*- (*SEP3*-), *AGL12*- and *OsMADS32*-like genes were lost very likely in the stem group of extant duckweeds. The potential functional relevance of our findings is discussed. For example, the loss of *AGL12*-like genes might be linked to the vestigial or absent root formation in duckweeds.

## 9.1 Small Is Beautiful: Evolutionary Developmental Biology of the Duckweed Body Plan

Duckweeds are strange plants. People easily assume that they are algae, ferns or mosses, but in fact, they are flowering plants (angiosperms), even though many of them flower only occasionally or never. Actually, duckweeds are the smallest and morphologically simplest flowering

L. Gramzow · G. Theißen (✉)
Matthias Schleiden Institute - Genetics, Friedrich Schiller University, Philosophenweg 12, 07743 Jena, Germany
e-mail: guenter.theissen@uni-jena.de


X. H. Cao et al. (eds.), *The Duckweed Genomes*, Compendium of Plant Genomes,
https://doi.org/10.1007/978-3-030-11045-1_9

plants that currently exist. However, they are not simply miniature versions of regular angiosperms, but rather represent a highly modified structural organization, involving the simplification, and loss of many anatomical features. The peculiar morphology of duckweeds obviously reflects an extreme adaptation to aquatic life as free-floating plants (Bogner 2009).

Several lines of evidence, including molecular data, suggest that duckweeds are an early offshoot of the family Araceae (arum family, aroids), subfamily Lemnoideae. Traditionally, however, duckweeds have been treated as a separate family (Lemnaceae), a view that is still quite fashionable (Bogner 2009; Michael et al. 2017; Hoang et al. 2018). In any case, duckweeds are relatively basal representatives of the Alismatales, the most basal order of monocotyledonous plants except Acorales (Wang et al. 2011, 2014; Olsen et al. 2016).

There are currently 37 duckweed species in five genera recognized, *Spirodela*, *Landoltia*, *Lemna*, *Wolfiella* and *Wolffia* (Hoang et al. 2018). The first three genera constitute the tribe Lemneae, the latter two the Wolfieae (Bogner 2009). Among duckweeds, *Spirodela* represents the most ancestral and *Wollfia* the most derived lineage (Wang et al. 2011). There is a successive reduction of morphological structures and size in parallel with evolutionary advancement within Lemnaceae, ranging from the 1.5 cm long *Spirodela polyrhiza* to the less than 1 mm long *Wolffia globosa* (Wang et al. 2011). Interestingly, the DNA content continuously increases in duckweeds parallel to the reduction in body size, from a genome size of 150 million base pairs (Mbp) in case of the ancestral *Spirodela polyrhiza* to 1881 Mbp in case of the highly derived and miniaturized *Wolffia arrhiza* (Wang et al. 2011).

The duckweed plant body is organized as a 'thalloid' or 'frond' lacking a stem. Lemneae have only simple roots; the Wolffieae are even rootless; the usually very short stipe of the new fronds of the Lemneae has been interpreted as a stolon. The tribes Lemneae and Wolffieae show, in their evolution, a clear line to the reduction of their organs. Historically, three morphological interpretations have been proposed for the frond of duckweeds, (a) that it corresponds to a leaf, (b) that it corresponds to a shoot of leaf-like shape, (c) that the basal part of the frond represents a shoot, and the distal part a foliar organ (phyllome) (reviewed by Bogner 2009). The flower-like structures have previously been interpreted as a single flower, or alternatively, as an inflorescence in which all flowers are reduced to either a single stamen or a single gynoecium (reviewed by Bogner 2009). Recently, the detailed analysis of both extant and fossil plants revealed in some cases clear trends of reduction and allowed the identification of homologies. According to Bogner (2009), the pouch of the *Spirodela*, *Landoltia*, and *Lemna* frond is homologous to the petiole sheath of typical Araceae. The frond's distal part is homologous to the veined leaf blade. The shoot is reduced to a vegetation point that generates both new fronds and inflorescences. The membranous envelope around the inflorescence in the Lemneae was interpreted as the spathe (lacking in Wolffieae); the spadix is reduced to a single bisexual flower in the Wolffieae and to one bisexual flower and one male flower (with one stamen only) in the Lemneae (Bogner 2009).

Nevertheless, duckweed evolution involved not only miniaturization, simplification, and reduction. Some duckweeds have even evolved novel structures, such as a protrusion on the ventral surface termed a 'pseudoroot' in case of *Wolffia microscopica* (Sree et al. 2015).

Duckweeds flower only occasionally, with remarkable exceptions such as the frequently flowering but very rare *Wolffia microscopica* (Sree et al. 2015). Duckweeds usually reproduce by vegetative daughter fronds initiated from the mother frond. They do so in a very efficient way, with doubling time of the fastest growing species under optimal growth conditions of less than 30 h, nearly twice as fast as other 'fast-growing' flowering plants and more than double that of conventional crops (Wang et al. 2014).

Duckweeds develop flowers after rapid vegetative development on juvenile tissues. The considerable modification of an Araceae body plan during the origin of duckweeds thus obviously represents an extreme case of heterochrony, a change in the relative timing of developmental events. Because the timing of reproductive relative to vegetative development is changed, so that a juvenile stage becomes reproductively mature, this represents a clear case of paedomorphosis. This kind of heterochrony comes in two different flavours, neoteny and progression (Arthur 2011). Duckweeds have been considered as cases of neoteny, the retention of juvenile traits into the adult form as a result of retardation of somatic development (Bogner 2009; Wang et al. 2014). However, since duckweeds develop flowers on juvenile tissue, they might be better considered as cases of progenesis, the acceleration of developmental processes such that the juvenile form becomes a sexually mature adult.

The evolution of duckweeds involved dramatic deviations from regular Araceae in terms of the developmental trajectory, growth habit, and body plan. Did this dramatic developmental change leave footprints in the duckweed genome? More specifically, were the evolutionary changes of the duckweed body plan accompanied, and thus potentially causally linked, to a loss of function of genes that control the development of structures that were subjected to miniaturization, simplification or even loss? To address these questions, we focus here on a gene family that brought about numerous developmental control genes during plant evolution, the MADS-box genes.

## 9.2 MADS About Development: MTF Phylogeny and the Ontogeny of Angiosperms

Many aspects of angiosperm development are controlled by MADS-box genes, encoding MADS-domain transcription factors (MTFs) (for a review, see Gramzow and Theißen 2010; Smaczniak et al. 2012; Theißen et al. 2000, 2016). MTFs are characterized by a highly conserved DNA-binding domain, the MADS domain, named after the four founding members of this family: MINICHROMOSOME MAINTENANCE FACTOR1 (MCM1) from baker's yeast (*Saccharomyces cerevisiae*), AGAMOUS (AG) from thale cress (*Arabidopsis thaliana*), DEFICIENS (DEF) from snapdragon (*Antirrhinum majus*) and SERUM RESPONSE FACTOR (SRF) from human (*Homo sapiens*). MTFs appear to be absent from prokaryotes, but two types existed probably already in the most recent common ancestor (MRCA) of extant eukaryotes, termed Type I and Type II MTFs (Gramzow et al. 2010). Plant Type II MTFs acquired a characteristic domain structure very likely in the stem group of extant streptophytes (charophyte green algae and land plants). These streptophyte Type II proteins exhibit a domain structure in which the MADS domain is followed by an Intervening, a Keratin-like and a C-terminal domain, and have hence been coined MIKC-type proteins (Theißen et al. 2000).

Type I MTFs in flowering plants are further subdivided in three groups, Mα, Mβ, and Mγ (Gramzow and Theißen 2010). The genes encoding angiosperm Type I MTFs have higher birth and death rates than Type II genes. In line with this, many of them have only quite subtle functions in female gametophyte, embryo and seed development, and some might even be pseudogenes (reviewed by Gramzow and Theißen 2010). The Type II (MIKC-type) genes of land plants, including angiosperms, are subdivided into MIKC$^{C}$-group and MIKC*-group genes based on phylogeny reconstructions and structural features (reviewed by Gramzow and Theißen 2010). In euphyllophytes (ferns and their allies, and seed plants), two clades of MIKC*-type genes exist, termed S and P clade (Gramzow et al. 2012). In angiosperms, MIKC*-group genes appear to have a conserved role in pollen development (Liu et al. 2013).

In contrast to the relatively limited interest that Type I and MIKC*-group MTFs have found so far, MIKC$^{C}$-group genes took plant biology by storm, not least due to the spectacular homeotic and heterochronic phenotypes of some mutants. The most well-known MIKC$^{C}$-group genes include those genes that confer floral organ identity and hence lead to homeotic phenotypes upon mutation (for a review, see Gramzow and Theißen 2010; Smaczniak et al. 2012; Theißen et al. 2016). Five classes (A–E) of floral organ identity genes have been identified by mutant analysis, with class A + E genes specifying sepals, A + B + E petals, B + C + E stamens, C + E carpels, and C + D + E ovules (Theißen et al. 2016). Almost all of these genes encode MIKC$^{C}$-group MTFs.

The phylogeny of MIKC$^{C}$-group genes is characterized by preferential retention of duplicate genes after whole genome duplications, followed by sequence divergence, sub- and neo-functionalization of the paralogs (Gramzow and Theißen 2013, 2015, 2016). Increases in gene numbers occurred independently in different groups of land plants. Interestingly, the floral homeotic genes are all members of gene clades that are seed plant- or flowering plant-specific. Phylogeny reconstructions revealed that all the MIKC$^{C}$-group genes of angiosperms are members of 11 seed plant-specific superclades which had been established already in the MRCA of extant seed plants (spermatophytes) about 300 million years ago (MYA). None of these gene clades may have existed already in the MRCA of extant euphyllophytes, i.e. monilophytes (ferns and their allies such as horsetails) and seed plants (gymnosperms + angiosperms) about 400 MYA (Gramzow et al. 2014). Among these, superclades are those containing the genes providing the floral homeotic A-function (*FLC/SQUA*-like, or *FLC/AP1*-like genes), containing class B genes (*DEF/GLO/OsMADS32*-like, or *AP3/PI/OsMADS32*-like genes), containing class C and D genes (*AG*-like genes), and containing class E genes (*SEP/AGL6*-like genes). Due to gene duplications in the stem group of angiosperms, these genes evolved into 17 clades that had already been established in the MRCA of extant angiosperms (Fig. 9.1). These angiosperm-specific clades include distinct *DEF*- (*AP3*-) and *GLO*- (*PI*-) like genes (class B), the *AG*-like and *STK*-like genes (class C and D), the *AGL2*-like (*SEP1*-like), *AGL9*-like (*SEP3*-like) and *AGL6*-like genes (class E), and the *SQUA*- (*AP1*-) like genes (class A) (Gramzow et al. 2014). Genes of the remaining clades are involved in diverse developmental processes ranging from root to fruit development. These clades comprise *AGL12*-like, *AGL15*-like, *AGL17*-like, *FLC*-like, *GGM13*-like ($B_{sister}$), *OsMADS32*-like, *StMADS11*-like (*SVP*-like), *TM3*-like (*SOC1*-like) and *TM8*-like genes (alternative clade names given in brackets) (Smaczniak et al. 2012).

## 9.3 When Less Is More: Loss of MADS-Box Genes During Evolution

Somewhat similar to living beings also genes are born (e.g. by gene duplication or de novo from non-genic DNA), and eventually, they die (even though that may take millions of years). Mutational gene death (nonfunctionalization) and loss is probably the most frequent fate of duplicated genes, but compared to gene birth, it has found relatively little scientific interest (Panchy et al. 2016). This ignorance has probably at least two reasons, a scientific and a technical one. On the one hand, gene loss might be easily viewed as the trivial outcome of random mutations in sequences without a function on which hence purifying selection is not acting anymore. On the other hand, gene loss is much more difficult to demonstrate than gene birth; while one can conclude that a gene birth must have happened at some time from the simple presence of a gene in an organismic lineage, demonstration of gene loss in a rigorous way requires solid evidence of the absence of the gene in an organismic lineage and hence typically depends on reliable whole-genome information. Comparably little is known, therefore, about the mechanisms and dynamics of the loss of genes during evolution.

Nevertheless, the interest in gene loss has rapidly increased recently. One reason is the availability of rising numbers of whole-genome sequences of high quality from diverse taxa, which facilitates the identification of gene loss. Another reason is mounting evidence that gene loss can be of considerable adaptive value (for a review, see Albalat and Cañestro 2016; Hoffmeier et al. 2018).

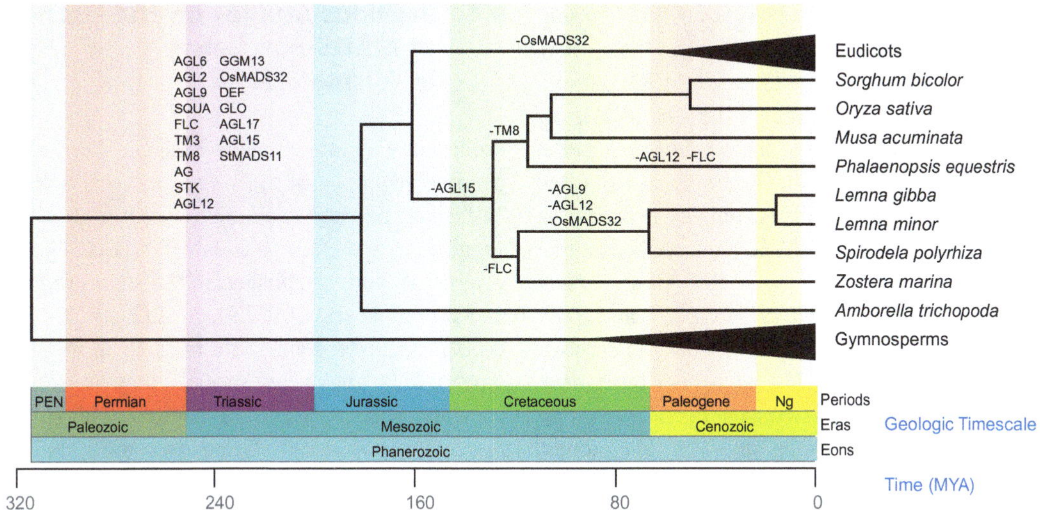


**Fig. 9.1** Loss of MADS-box gene clades in the evolution of flowering plants. The 17 clades of MIKC$^{C}$-group genes that originated prior to the divergence of extant angiosperms (AGL6–StMADS11) are listed at the corresponding branch. Clades that have been lost in particular lineages are indicated on the corresponding branches. The phylogeny was drawn using TimeTree (Kumar et al. 2017). All species named except the basal angiosperm *Amborella trichopoda* are monocotyledonous plants. *Lemna* and *Spirodela* represent duckweeds, together with *Zostera* they are Alismatales. *Phalaenopsis* is an orchid, *Musa* a banana species, *Sorghum* and *Oryza* are grasses (Poaceae). Concerning the evolution of *FLC*-like and *TM3*-like genes in Alismatales, only one of two possible scenarios is shown (for details, see text)

Gene loss is the most frequent fate of young paralogs after gene duplication, and that certainly also applies to MADS-box genes in plants, especially to Type I genes with their high death rates (Gramzow and Theißen 2013). However, MIKC-type MADS-box genes do not only show the preferential retention after whole-genome duplications typical for genes encoding transcription factors, but also the establishment and strong conservation of some gene clades in stem groups of plants that have been of utmost importance for plant biodiversity on land, most importantly seed and flowering plants. Nevertheless, phylogenomic analyses in recent years revealed that even ancient and relatively strongly conserved clades of genes can be completely lost in some organismic lineages. The earlier that happens during evolution, the more extant species are typically affected.

Closely related paralogs within one and the same clade of MIKC$^{C}$-group genes have often partially redundant functions and then may have a high probability to get lost. On the other hand, there is a considerable correlation between clade membership and function (Theißen et al. 1996), so that loss of a complete gene clade in a plant has a quite high probability to be of functional relevance (Gramzow and Theißen 2015). Accordingly, most clades of MADS-box genes appear to have never been lost completely during angiosperm evolution in any lineage, at least as far as this can be determined already based on a quite limited sampling. For example, in a previous study involving 27 angiosperm genome sequences, of the 17 clades of MIKC$^{C}$-group genes that probably existed in the MRCA of extant angiosperms, ten have not been wiped out in any of the investigated species. Besides the clades of floral organ identity genes (*AGL2*-, *AGL6*-, *SQUA*-, *DEF*-, *GLO*-, and *AG*-like genes), these include also *TM3*-like, *StMADS11*-like, *AGL17*-like, and *GGM13*-like ($B_{sister}$) genes. This finding suggests that also the less well-studied (non-homeotic) gene clades provide functions that are more important than has been recognized previously.

Nevertheless, some clades of MIKC$^C$-group genes were lost, some even more than once. Among the at least 12 clades of MIKC$^C$-group genes that had been established in the MRCA of extant seed plants, one, named *GpMADS4*-like genes, may have been lost in the lineage that led to extant flowering plants (Gramzow et al. 2014). Of the 17 clades of MIKC$^C$-group genes that probably existed in the MRCA of extant angiosperms, three (*AGL12*-, *AGL15*-, and *STK*-like genes) have been lost at least once, two (*AGL9*- and *OsMADS32*-like genes) have been lost at least twice, and two have been lost at least five (*TM8*-like genes) or six (*FLC*-like genes) times independently throughout angiosperm evolution (Gramzow and Theißen 2015). Because of the limited genome sampling, these are very conservative estimates, and the real numbers are probably much higher. Interestingly, a subclade of $B_{sister}$ genes, termed *GORDITA*-like genes, has been lost several times in parallel in crucifers alone (Hoffmeier et al. 2018).

The reasons why some clades of MIKC$^C$-group genes become dispensable during evolution and hence get lost in some lineages are all but clear. Which roles do genetic drift based on low purifying selection on the one hand and adaptive advantage on the other hand play?

It would be interesting to relate gene loss to the organismal context in which it happens. A well-known example is gene loss as a consequence of parasitism, since many parasites receive functions from their hosts rather than to encode them in their own genomes. Indeed, it has been reported that the dodder species *Cuscuta australis*, a root- and leafless-parasitic plant from the family Convolvulaceae, has no *FLC*-, *StMADS11*-(*SVP*-), and *AGL17*-like MIKC$^C$-group genes (Sun et al. 2018). Whether the loss of these genes can be attributed to the loss of major plant organs remains to be seen. However, not only some kinds of parasitism, but also other evolutionary processes in angiosperms involved the loss of organs. An obvious case in point is duckweeds.

## 9.4 Previous Studies on the Loss of MADS-Box Genes in Alismatales

Preliminary analyses made possible because a genome sequence of *Spirodela polyrhiza* (clone 7498) had become available (Wang et al. 2014) suggested that four clades of MIKC$^C$-group genes that were established in the stem group of extant angiosperms (*AGL9*-, *AGL12*-, *FLC*-, and *OsMADS32*-like genes), are missing in the genome of this duckweed species; they thus have possibly been lost in the lineage that led to *Spirodela polyrhiza* (Gramzow and Theißen 2015). An independent analysis confirmed the absence of these gene clades, and reported also the absence of MIKC*-group genes (which had not been investigated in the previous study) (Olsen et al. 2016). Thus, *Spirodela polyrhiza* may have lost the highest number of clades of MIKC-type genes of all investigated species (Gramzow and Theißen 2015). It was tempting to hypothesize, therefore, that the relatively high number of possibly lost clades is causally linked to the simplification of the duckweed body plan during evolution. Maybe duckweeds such as *Spirodela polyrhiza* lost some ancestral clades of MIKC-type genes because they were not required anymore to control developmental processes that had been significantly simplified or even abolished (Gramzow and Theißen 2015).

The only other species from Alismatales except duckweeds for which a genome sequence is currently available is *Zostera marina* (Zosteraceae), which is phylogenetically more derived than duckweeds (Olsen et al. 2016). The finding that *AGL12*-, *AGL9*-, *OsMADS32*-like, and MIKC*-group genes have been found in that species (Olsen et al. 2016) suggests that these gene clades have been lost in the lineage that led to duckweeds after the lineage that led to more derived Alismatales had branched-off. Probably these gene losses are duckweed-specific. In contrast, *FLC*-like genes were reported to be missing in both *Spirodela polyrhiza* and *Zostera* *marina*, suggesting that these genes were lost before the lineages that led to extant Zosteraceae and Lemnoideae separated. Given the relatively frequent loss of *FLC*-like genes throughout angiosperm evolution (Gramzow and Theißen 2015), however, it also appeared quite possible that the absence of *FLC*-like genes in both species traces back to two independent gene-loss events. However, even though evidence was provided, that the first genome sequence of a duckweed contained at least 90% of genic sequences (Wang et al. 2014), incompleteness and inaccuracies of any genome sequence make it difficult to exclude false negative results.

## 9.5 Lost in Miniaturization: Setback of MADS-Box Genes in Duckweed Genomes

To test our initial conclusions concerning the loss of MIKC-type genes in duckweeds, we here use the additional genomic resources that have become available since the time of our previous study. These novel data include the genome sequence of another strain of *Spirodela polyrhiza* (strain 9509), and the genomes of *Lemna gibba* and *Lemna minor* (Michael et al. 2017; van Hoeck et al. 2015; Ernst 2016). Furthermore, we reanalysed the *Zostera marina* MADS-box genes (Olsen et al. 2016) and extended our analyses to MIKC*-group genes. MADS-box genes were identified, their phylogeny was reconstructed and the presence and absence of MADS-box gene clades were determined essentially as previously described (Gramzow and Theißen 2015). Redundant genes from the same species were removed using Jalview and a redundancy threshold of 95 (Waterhouse et al. 2009).

### 9.5.1 Conserved Clades of MIKC-Type Genes in Alismatales

Our analyses identified 11 out of 17 clades of $MIKC^C$-group genes and both (S and P) clades of MIKC*-group genes in all duckweed species analysed and in *Zostera marina*, suggesting that these clades are conserved throughout Alismatales (Table 9.1; Fig. 9.1). The conserved clades of $MIKC^C$-group genes are essentially the ones also identified in our previous study of flowering plants (Gramzow and Theißen 2015) and include most of the clades which are formed by genes with important functions in flower development (*AG*-, *AGL2*-, *AGL6*-, *DEF*-, *GGM13*-($B_{sister}$), *GLO*-, *SQUA*-, *STK*-like genes). Furthermore, some clades comprised of genes that in functional terms are less well understood, *StMADS11*- and *AGL17*-like genes, are also conserved in Alismatales and hence in all other flowering plants studied. Additionally, *TM8*-like genes were found to be present throughout the Alismatales species studied (Table 9.1), but not throughout flowering plants in general (Gramzow and Theißen 2015). In our previous study, we found *TM3*-like genes a clade of $MIKC^C$-group genes that had not been lost in any of the species analysed. In our current study, the genes previously classified as *TM3*-like genes from *Spirodela polyrhiza* strain 7498 now cluster among the *FLC*-like genes, however, as do their closest relatives from the other duckweed species and strains. Hence, assignment of genes to the clade of *FLC*- or *TM3*-like genes is ambiguous, and additional analyses will be required to reveal whether the genes identified from duckweeds belong to the clade of *TM3*- or *FLC*-like genes. Only then one can determine which of the two clades was actually lost in duckweeds. So either *FLC*-like genes have been lost in the stem group of extant Alismatales (the scenario shown in Fig. 9.1), or *TM3*-like genes have been lost in the stem group of extant duckweeds.

### 9.5.2 Clades of $MIKC^C$-Group Genes Missing Completely in Alismatales

Representatives of one or two clades of $MIKC^C$-group genes may be absent from all Alismatales species analysed (Table 9.1). In contrast to previous analyses (Gramzow and Theißen 2015), in our current study, *AGL15*-like genes have not been identified in Alismatales (Table 9.1) and

**Table 9.1** Number of different MIKC-type MADS-box genes in Alismatales species

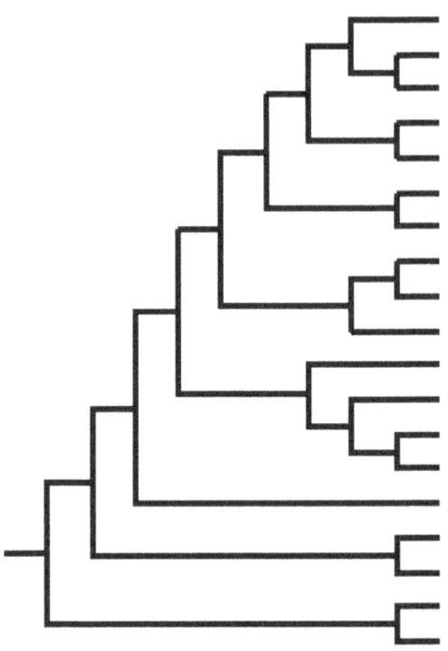

| Clade | *Zostera marina* | *Lemna gibba* | *Lemna minor* | *Spirodela polyrhiza strain 7498* | *Spirodela polyrhiza strain 9509* |
|---|---|---|---|---|---|
| AGL6 | 1 | 1 | 1 | 1 | 1 |
| AGL2 | 1 | 1 | 1 | 2 | 1 |
| AGL9 | 1 | 0 | 0 | 0 | 0 |
| SQUA | 4 | 2 | 2 | 2 | 1 |
| FLC | 0 | 0 | 0 | 0 | 0 |
| TM3 | 1 | 8[a] | 12[a] | 3[a] | 1[a] |
| TM8 | 2 | 1 | 1 | 1 | 1 |
| AG | 2 | 1 | 1 | 1 | 1 |
| STK | 2 | 1 | 1 | 1 | 1 |
| AGL12 | 2 | 0 | 0 | 0 | 0 |
| GGM13 | 1[b] | 1 | 1 | 1 | 1 |
| OsMADS32 | 1 | 0 | 0 | 0 | 0 |
| DEF | 1 | 1 | 3 | 1 | 1 |
| GLO | 1 | 1 | 2 | 2 | 1 |
| AGL17 | 1 | 1 | 1 | 1 | 1 |
| AGL15 | 0 | 0 | 0 | 0[c] | 0 |
| StMADS11 | 1 | 9 | 9 | 8 | 7 |
| MIKC* P | 4[d] | 1 | 2 | 1 | 1 |
| MIKC* S | 1 | 1 | 2 | 1 | 1 |

The table lists the 17 clades of MIKC$^C$-group genes that probably existed in the MRCA of extant angiosperms (AGL6–StMADS11) and the two clades of MIKC*-group genes that probably existed in the MRCA of euphyllophytes (P and S clade). The phylogeny on the left represents the relationships between the different clades according to Zhao et al. (2017). For each clade, the number of genes belonging to these clades in different species and strains of the Alismatales is given

[a]The corresponding genes are classified as *FLC*-like genes in some of our recent, preliminary phylogeny reconstructions
[b]The corresponding gene has been classified as Type I, clade Mα, in Olsen et al. (2016) but fell into the clade of *GGM13*-like genes in our phylogeny reconstruction; in line with this, a new gene prediction revealed an MIKC-type MADS-box gene
[c]The gene identified as *AGL15*-like gene from *Spirodela polyrhiza* strain 7498 (Olsen et al. 2016; Gramzow and Theißen 2015) belongs to the Type I genes according to our recent phylogeny reconstructions
[d]The corresponding genes were classified as Type I Mα in Olsen et al. (2016)`

*Musa acuminata*; thus, they may have actually been lost in the stem group that led to extant monocots (Fig. 9.1). Furthermore, as already mentioned above, depending on correct classification, *FLC*-like genes may have been lost in the MRCA of Alismatales (Fig. 9.1).

### 9.5.3 Clades of MIKC$^C$-Group Genes Lost in the Lineage Leading to Duckweeds

We did not identify *AGL9*-(*SEP3*-), *AGL12*-, and *OsMADS32*-like genes in any of the duckweed species studied, but in *Zostera marina* (Table 9.1). The fact that we did not find representatives of these clades in any of the four duckweed genomes studied makes effects of incomplete genome sequencing or bias due to sequencing of a single individual unlikely or impossible, respectively, and suggests genuine loss of the respective clades of MIKC$^C$-group genes. Hence, these clades appear to have been lost in the lineage that led to the MRCA of duckweeds (Fig. 9.1).

Duckweeds hence remain the plant family with the fewest number of conserved clades of MIKC$^C$-group genes. Five out of 17 clades are missing in duckweeds. Three clades alone, *AGL9*-, *AGL12*-, and *OsMADS32*-like genes, may have been lost in the lineage that led to the MRCA of extant duckweeds (Fig. 9.1). For the orchid *Phalaenopsis equestris,* we did not identify representative of four clades, *AGL15*-, *TM8*-, *AGL12*-, and *FLC*-like genes, in the current version of its genome sequence (data not shown;

Cai et al. 2015). Hence, all other analysed species are missing less clades of MIKC$^{C}$-group genes than duckweeds.

### 9.5.4 Link Between the Simplified Body Plan of Duckweeds and Loss of Clades of MIKC$^{C}$-Group Genes

Our analyses confirm some previous reports about the absence of clades of MIKC$^{C}$-group MADS-box genes in duckweed genomes. They thus provide a solid starting point for future investigations on the functional importance of MADS-box gene loss in flowering plants in general, and especially in duckweeds. Some of our findings provide already quite some food for thought. For example, the *AGL12*-like gene of *Arabidopsis thaliana* (also known as *XAANTAL1*, *XAL1*) has been shown to be involved in root development as a promoter of cell proliferation in the root apical meristem (García-Cruz et al. 2016). Considering that duckweeds only have simple or even no roots makes it tempting to speculate that the loss of this gene clade is linked to the loss of complex roots. Since *XAL1* plays also a role in the transition to flowering (Tapia-López et al. 2008), this involvement must be dispensable in duckweeds, possibly due to functional redundancy to other genes.

*OsMADS32* has been studied in rice where it has functions in flower development. In *OsMADS32* mutants, the lodicules (organs homologous to petals) are transformed into hull-like organs and the number of stamens is reduced (Wang et al. 2015), revealing a function similar to that of class B floral homeotic genes, to which *OsMADS32* is closely related (Gramzow et al. 2014). *OsMADS32*-like genes may have become dispensable during duckweed evolution as duckweeds have strongly reduced flowers without petals. Note, however, that also eudicots have lost their *OsMADS32*-like genes (Fig. 9.1; Gramzow et al. 2014), possibly because of redundancy to class B floral homeotic genes proper (specifying petal and stamen identity).

When *AGL9*-like (*SEP3*-like) genes are mutated in rice, the corresponding plants flower late and show homeotic changes of lodicules, stamens, and carpels into palea/lemma-like organs, as well as a loss of floral determinacy (Cui et al. 2010). Hence, the loss of *AGL9*-like genes may be correlated to the rare flowering of duckweeds. The fact that duckweeds still are able to produce fertile stamens and carpels may be explained by functional redundancy. *AGL9*-like genes form a superclade with *AGL2*- (*SEP1*-), and *AGL6*-like genes which are still present in duckweeds. For *AGL2*- and *AGL6*-like genes in petunia, it has been shown that they function redundantly in floral organ formation (Rijpkema et al. 2009), and the *SEP* genes *of Arabidopsis thaliana* are also largely redundant (Pelaz et al. 2000). SEP proteins may be able to substitute each others in the transcription factor complexes ('floral quartets') that specify organ identity in the flower (Theißen et al. 2016).

Taking together, our study suggests that the loss of specific MADS-box genes clades is correlated to the simplification of the duckweed body plan. Demonstrating a causal link between both observations in the one or other direction will require detailed studies in the future. In addition to investigations on gene loss it will also be interesting, however, to figure out as to why some clades, such as *AGL17*-like genes with a major function in root development, have been retained in duckweeds. Since some duckweed species (*Landoltia punctata*, *Lemna gibba*, and *L. minor*) can be transformed already, genome editing tools such as CRISPR-Cas9 could possibly be used to determine the function of the retained genes.

**Acknowledgements** We thank Hieu Cao for his kind invitation to write this article, and his patience.

## References

Albalat R, Cañestro C (2016) Evolution by gene loss. Nat Rev Genet 17:379–391

Arthur W (2011) Evolution—a developmental approach. Wiley-Blackwell

Bogner J (2009) The free-floating Aroids (Araceae)—living and fossil. Zitteliana 48:113–128

Cai J, Liu X, Vanneste K, Proost S, Tsai WC, Liu KW, Chen LJ, He Y, Xu Q, Bian C, Zheng Z, Sun F, Liu W, Hsiao YY, Pan ZJ, Hsu CC, Yang YP, Hsu YC, Chuang YC, Dievart A, Dufayard JF, Xu X, Wang JY, Wang J, Xiao XJ, Zhao XM, Du R, Zhang GQ, Wang M, Su YY, Xie GC, Liu GH, Li LQ, Huang LQ, Luo YB, Chen HH, Van de Peer Y, Liu ZJ (2015) The genome sequence of the orchid *Phalaenopsis equestris*. Nat Genet 47:65–72

Cui R, Han J, Zhao S, Su K, Wu F, Du X, Xu Q, Chong K, Theißen G, Meng Z (2010) Functional conservation and diversification of class E floral homeotic genes in rice (*Oryza sativa*). Plant J 61:767–781

Ernst E (2016) Status of the *Lemna gibba* 7742a and *Lemna minor* 8627 genomes. ISCDRA 3:9–10

García-Cruz KV, García-Ponce B, Garay-Arroyo A, Sanchez MP, Ugartechea-Chirino Y, Desvoyes B, Pacheco-Escobedo MA, Tapia-López R, Ransom-Rodríguez I, Gutierrez C, Alvarez-Buylla ER (2016) The MADS-box XAANTAL1 increases proliferation at the Arabidopsis root stem-cell niche and participates in transition to differentiation by regulating cell-cycle components. Ann Bot 4:787–796

Gramzow L, Theißen G (2010) A hitchhiker's guide to the MADS world of plants. Genome Biol 11:214

Gramzow L, Theißen G (2013) Phylogenomics of MADS-box genes in plants—two opposing life styles in one gene family. Biology 2:1150–1164

Gramzow L, Theißen G (2015) Phylogenomics reveals surprising sets of essential and dispensable clades of $MIKC^C$-group MADS-box genes in flowering plants. J Exp Zool (Mol Develop Evol) 324B:353–362

Gramzow L, Barker E, Schulz C, Ambrose B, Ashton N, Theißen G, Litt A (2012) *Selaginella* genome analysis—entering the "homoplasy heaven" of the MADS world. Front Plant Sci 3:214

Gramzow L, Ritz MS, Theißen G (2010) On the origin of MADS-domain transcription factors. Trends Genet 26:149–153

Gramzow L, Weilandt L, Theißen G (2014) MADS goes genomic in conifers: towards determining the ancestral set of MADS-box genes in seed plants. Ann Bot 114:1407–1429

Hoang PNT, Michael TP, Gilbert S, Chu P, Motley ST, Appenroth KJ, Schubert I, Lam E (2018) Generating a high-confidence reference genome map of the Greater Duckweed by integration of cytogenomic, optical mapping, and Oxford Nanopore technologies. Plant J 96:670–684

Hoffmeier A, Gramzow L, Bhide A, Kottenhagen N, Greifenstein A, Schubert O, Mummenhoff K, Becker A, Theißen G (2018) A dead gene walking: convergent degeneration of a clade of MADS-box genes in Crucifers. Mol Biol Evol 35:2618–2638

Kumar S, Stecher G, Suleski M, Hedges SB (2017) TimeTree: a resource for timelines, timetrees, and divergence times. Mol Biol Evol 34:1812–1819

Liu Y, Cui S, Wu F, Yan S, Lin X, Du X, Chong K, Schilling S, Theißen G, Meng Z (2013) Functional conservation of MIKC*-type MADS box genes in *Arabidopsis* and rice pollen maturation. Plant Cell 25:1288–1303

Michael TP, Bryant D, Gutierrez R, Borisjuk N, Chu P, Zhang H, Xia J, Zhou J, Peng H, El Baidouri M, ten Hallers B, Hastie AR, Liang T, Acosta K, Gilbert S, McEntee C, Jackson SA, Mockler TC, Zhang W, Lam E (2017) Comprehensive definition of genome features in *Spirodela polyrhiza* by high-depth physical mapping and short-read DNA sequencing strategies. Plant J 89:617–635

Olsen JL, Rouzé P, Verhelst B, Lin YC, Bayer T, Collen J, Dattolo E, De Paoli E, Dittami S, Maumus F, Michel G, Kersting A, Lauritano C, Lohaus R, Töpel M, Tonon T, Vanneste K, Amirebrahim M, Brakel J, Boström C, Chovatia M, Grimwood J, Jenkins JW, Jueterbock A, Mraz A, Stam WT, Tice H, Bornberg-Bauer E, Green PJ, Pearson GA, Procaccini G, Duarte CM, Schmutz J, Reusch TB, Van de Peer Y (2016) The genome of the seagrass *Zostera marina* reveals angiosperm adaptation to the sea. Nature 530:331–335

Panchy N, Lehti-Shiu M, Shiu S-H (2016) Evolution of gene duplication in plants. Plant Physiol 171:2294–2316

Pelaz S, Ditta GS, Baumann E, Wisman E, Yanofsky MF (2000) B and C floral organ identity functions require *SEPALLATA* MADS-box genes. Nature 405:200–203

Rijpkema AS, Zethof J, Gerats T, Vandenbussche M (2009) The petunia AGL6 gene has a SEPALLATA-like function in floral patterning. Plant J 60:1–9

Smaczniak C, Immink RGH, Angenent GC, Kaufmann K (2012) Developmental and evolutionary diversity of plant MADS-domain factors: insights from recent studies. Development 139:3081–3098

Sree KS, Maheshwari SC, Boka K, Khurana JP, Keresztes A, Appenroth K-J (2015) The duckweed *Wolffia microscopica*: a unique aquatic monocot. Flora 210:31–39

Sun G, Xu Y, Liu H, Sun T, Zhang J, Hettenhausen C, Shen G, Qi J, Qin Y, Li J, Wang L, Chang W, Guo Z, Baldwin IT, Wu J (2018) Large-scale gene losses underlie the genome evolution of parasitic plant *Cuscuta australis*. Nat Comm 9:1–8

Tapia-López R, Garcia-Ponce B, Dubrovsky JG, Garay-Arroyo A, Pérez-Ruiz RV, Kim SH, Acevedo F, Pelaz S, Alvarez-Buylla ER (2008) An AGAMOUS-related MADS-box gene, XAL1 (AGL12), regulates root meristem cell proliferation and flowering transition in Arabidopsis. Plant Physiol 146:1182–1192

Theißen G, Gramzow L (2016) Structure and evolution of plant MADS-domain transcription factors. In: Gonzalez DH (ed) Plant transcription factors: evolutionary, structural and functional aspects, pp 127–138. Elsevier Inc.

Theißen G, Becker A, Di Rosa A, Kanno A, Kim JT, Münster T, Winter K-U, Saedler H (2000) A short history of MADS-box genes in plants. Plant Mol Biol 42:115–149

Theißen G, Kim J, Saedler H (1996) Classification and phylogeny of the MADS-box multigene family suggest defined roles of MADS-box gene subfamilies in the morphological evolution of eukaryotes. J Mol Evol 43:484–516

Theißen G, Melzer R, Rümpler F (2016) MADS-domain transcription factors and the floral quartet model of flower development: linking plant development and evolution. Development 143:3259–3271

Van Hoeck A, Horemans N, Monsieurs P, Cao HX, Vandenhove H, Blust R (2015) The first draft genome of the aquatic model plant *Lemna minor* opens the route for future stress physiology research and biotechnological applications. Biotechnol Biofuels 8:188

Wang W, Haberer G, Gundlach H, Gläßer C, Nussbaumer C, Luo MC, Lomsadze A, Borodovsky M, Kerstetter RA, Shanklin J, Bryant DW, Mockler TC, Appenroth KJ, Grimwood J, Jenkins J, Chow J, Choi C, Adam C, Cao X-H, Fuchs J, Schubert I, Rokhsar D, Schmutz J, Michael TP, Mayer KFX, Messing J (2014) The *Spirodela polyrhiza* genome reveals insights into its neotenous reduction fast growth and aquatic lifestyle. Nat Commun 5:3311

Wang W, Kerstetter RA, Michael TP (2011) Evolution of genome size in duckweeds (Lemnaceae). J Botany, Article ID 570319, Hindawi Publishing

Wang H, Zhang L, Cai Q, Hu Y, Jin Z, Zhao X, Fan W, Huang Q, Luo Z, Chen M, Zhang D, Yuan Z (2015) OsMADS32 interacts with PI-like proteins and regulates rice flower development. J Integr Plant Biol 57:504–513

Waterhouse AM, Procter JB, Martin DMA, Clamp M, Barton GJ (2009) Jalview version 2—a multiple sequence alignment editor and analysis workbench. Bioinformatics 25:1189–1191

Zhao T, Holmer R, de Bruijn S, Angenent GC, van den Burg HA, Schranz ME (2017) Phylogenomic synteny network analysis of MADS-box transcription factor genes reveals lineage-specific transpositions, ancient tandem duplications, and deep positional conservation. Plant Cell 29:1278–1292

# Duckweed Chloroplast Genome Sequencing and Annotation

10

Yating Zhang and Wenqin Wang

**Abstract**

Duckweed chloroplasts make great contributions to their rapid growth and abundant starch accumulation. The availability of chloroplast genomes will greatly facilitate the understanding of duckweed traits and further genetic chloroplast bioengineering. Here, we summarized the progress of duckweed chloroplast genome sequencing, together with the developments of sequencing technology and bioinformatics tools. The high-quality chloroplast DNA preparation was specifically emphasized in this chapter due to the key of genome sequencing. We also highlighted the potential applications of chloroplast genomes in the phylogenetic studies and the improvement of plant desirable traits.

## 10.1 Introduction

*Lemnaceae* (duckweeds) is comprised of five genera of 37 species: *Spirodela*, *Landoltia*, *Lemna*, *Wolffiella*, and *Wolffia*. Their taxonomy belongs to the basal monocotyledon of flower plants. They have a list of appealing characteristics in plant science, attracting tremendous attentions in recent years. As one of the ideal experimental plants, duckweeds could survive in an aquatic environment with little nutrients and even limited space when the high density of individuals is arrived. They grow quickly often by the way of asexually reproduction and double their biomass within two days.

Duckweeds have shown invaluable utilization in wastewater treatment, source of biofeed and biofuel (Cheng and Stomp 2009; Stomp and El-Gewely 2005). The exploration of more native duckweed ecotypes with the diverse genetic background is critical in duckweed breeding. However, the duckweed is a minimalist with an extremely simple morphology containing few leaves (called fronds) with or without roots (Stomp and El-Gewely 2005). The frond size is less than one centimeter by average. Professor Elias Landolt from the ETH Zurich, Switzerland is the father of duckweeds, and he is the only person who could recognize all 37 species of duckweeds based on a morphological basis (Landolt 1986). An expert who does not get a deep training may hardly distinguish the duckweed sister species. Thus, the highly morphological degeneration becomes a bottleneck to identify duckweeds efficiently until the appearance of DNA barcode. The Consortium for the Barcode of Life (CBOL) plant-working group proposed seven leading candidate sequences all from the chloroplast genome were the best promising barcoding markers (Group CPW 2009). A comprehensive study was carried out to

Y. Zhang · W. Wang (✉)
School of Agriculture and Biology, Shanghai Jiao Tong University, 800 Dong Chuan Road, Shanghai 200240, China
e-mail: wang2015@sjtu.edu.cn



X. H. Cao et al. (eds.), *The Duckweed Genomes*, Compendium of Plant Genomes,
https://doi.org/10.1007/978-3-030-11045-1_10

evaluate the barcode efficiency to discriminate duckweed species for seven pairs of chloroplast-derived markers. It has shown that the best molecular marker is the *atpF-atpH* noncoding spacer that could correct identify 14 out of 19 species (Wang et al. 2010). Still, the maker could not reach the complete discriminations, especially for closely related species even with the combination of several markers. The whole chloroplast genome as a barcode marker has the power to provide higher resolution and more polymorphism to identify species. The use of whole chloroplast genomes as a barcode system was demonstrated in certain species (Douglas 1998), breaking through the previous limitations.

Genetic information in plants stores in three parts of nucleus, mitochondrion, and chloroplast (plastid). Nuclear genome is a hub coordinating the activities of both mitochondrion and chloroplast in spite of the cross-talk between them. The mitochondrion is a membrane-bound organelle that converts large molecules of carbohydrates, proteins, or lipids into the energy of adenosine triphosphate (ATP) by respiration. The chloroplast is a producer that allows plant to capture the sun energy into energy-rich molecules. Its genome is a circular, double-stranded DNA molecule with thousands of copies in a cell (Fig. 10.1). There are roughly 100 mitochondrial, 1000 chloroplast, and 2 nuclear genome copies in a duckweed cell (Wang et al. 2012). It is believed to have evolved from endosymbiosis of a cyanobacterium, together with a massive gene transfer from the chloroplast to the nucleus (Timmis et al. 2004). The chloroplast genome is responsible for encoding the key protein complexes involved in photosynthesis and other metabolic processes. The genome sizes are relatively constant with a range of 107–218 Kb (Daniell et al. 2016). The chloroplast genome is a circular molecule containing a large sing copy, a small single copy and two inverted repeats. The gene content and gene structure are also highly conserved, indicating their essential functions throughout the plant evolution (Daniell et al. 2016).

The advent of high-throughput sequencing technologies and bioinformatics tools has facilitated the rapid progress in the fields of chloroplast genetics and genomics. Until 2018, the database of NCBI Organelle Genomes section has collected 1975 annotated chloroplast/plastid genomes from land plants (https://www.ncbi.nlm.nih.gov/genome/organelle/). The first duckweed chloroplast genome of *Lemna minor* was released in 2008 by using traditional Sanger sequencing (Mardanov et al. 2008), and the other three duckweed chloroplast genomes were sequenced by next-generation sequencing (NGS) in 2011 (Wang and Messing 2011). All the sequences were deposited in NCBI database and could be fetched with the unique ID number.

Here, we reviewed the complete chloroplast genomes of four duckweed species, which gave insights into the overall evolutionary dynamics and phylogenetic relationship compared to other plants. We also present the strategy of genome assembly with the short reads from NGS and prospected the long reads produced by third-generation technology. The pipeline including DNA preparation, sequencing technology, and computational tools for genome assembly and annotation were informatively covered (Fig. 10.2). We expect the availability of more duckweed chloroplast genomes would help our understanding for the origins and features of duckweed species, but shed new lights on their evolution and biotechnological applications.

## 10.2 DNA Preparation

### 10.2.1 Pure cpDNA Isolation

#### 10.2.1.1 Gradient Centrifuge

The high-quality and decent amount of starting chloroplast DNA (cpDNA) is a prerequisite for sequencing chloroplast genome, whereas cpDNA isolation is tedious and time-consuming that has restricted the broad applications. It is well known that a plant cell contains three types of DNA derived from nucleus, mitochondrion and chloroplast. A regular CTAB way is to get the

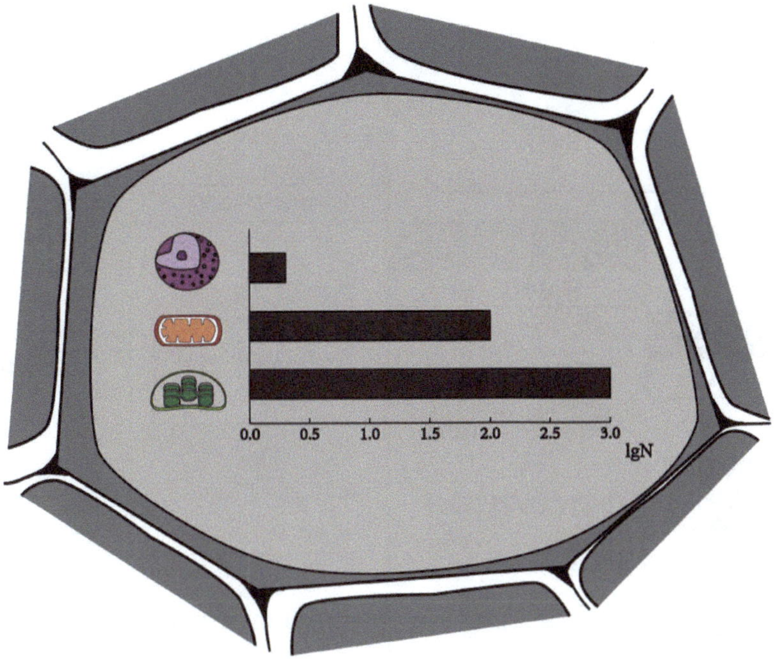


**Fig. 10.1** Genome copy numbers for nuclear, mitochondrial, and chloroplast genome in a duckweed cell. There are three DNA resources in a plant cell including nuclear, mitochondrial, and chloroplast DNA. It was reported that there were two of nuclear genome, 100 of mitochondrial and 1000 of chloroplast genome copies in a typical duckweed cell (Wang et al. 2012). Here, the *X*-axis means log value of copy number. The *Y*-axis represents nucleus, mitochondria, and chloroplast from top to bottom

mixture of total DNA (Murray and Thompson 1980) including not only cpDNA but also mitochondrial and nuclear DNA (Fig. 10.1; Table 10.1).

The most common methods to harvest pure cpDNA are to isolate intact chloroplasts by using sucrose or Percoll gradient centrifugation (Shi et al. 2012; Palmer 1986), which includes the steps of chloroplast separation from other organelles, the lysis of the chloroplasts and purification of DNA. The gradient centrifugation is feasible for most land plants to harvest enough quality and quantity of cpDNA. However, it is still challengeable to get completely pure cpDNA by the facts of DNA contamination derived from mitochondria and nucleus. A modified protocol based on sucrose gradients was developed to isolate the cpDNA for the species of *Oryza brachyantha*, *Leersia japonica,* and *Prinsepia utilis*. The mapped reads from Illumina sequencing against the chloroplast reference genomes accounted for only 40–50% of total reads (Shi et al. 2012).

#### 10.2.1.2 Enzyme Digestion

The method of DNase I treatment could extract cpDNA with the digestion of nuclear DNA (Kolodner and Tewari 1979) (Table 10.1). However, this treatment also degrades the exposed cpDNA that is outside of intact plastids that more start material is needed. A couple of plants was attempted to isolate cpDNA via the DNase I treatment and only *Lactuca sativa* and *Ginkgo biloba* recovered the decent results (Jansen et al. 2005).

#### 10.2.1.3 High Ionic Strength

The high ionic strength with high salt buffers diminishes the adherence of nucleus to the chloroplast membranes, yielding an enriched cpDNA with less nuclear contamination (Table 10.1). The cpDNA from pea was prepared in a high concentration of salts. The harvested cpDNA was successfully used for restriction enzyme mapping, Southern transfers, and cloning (Bookjans et al. 1984). The cpDNA from the

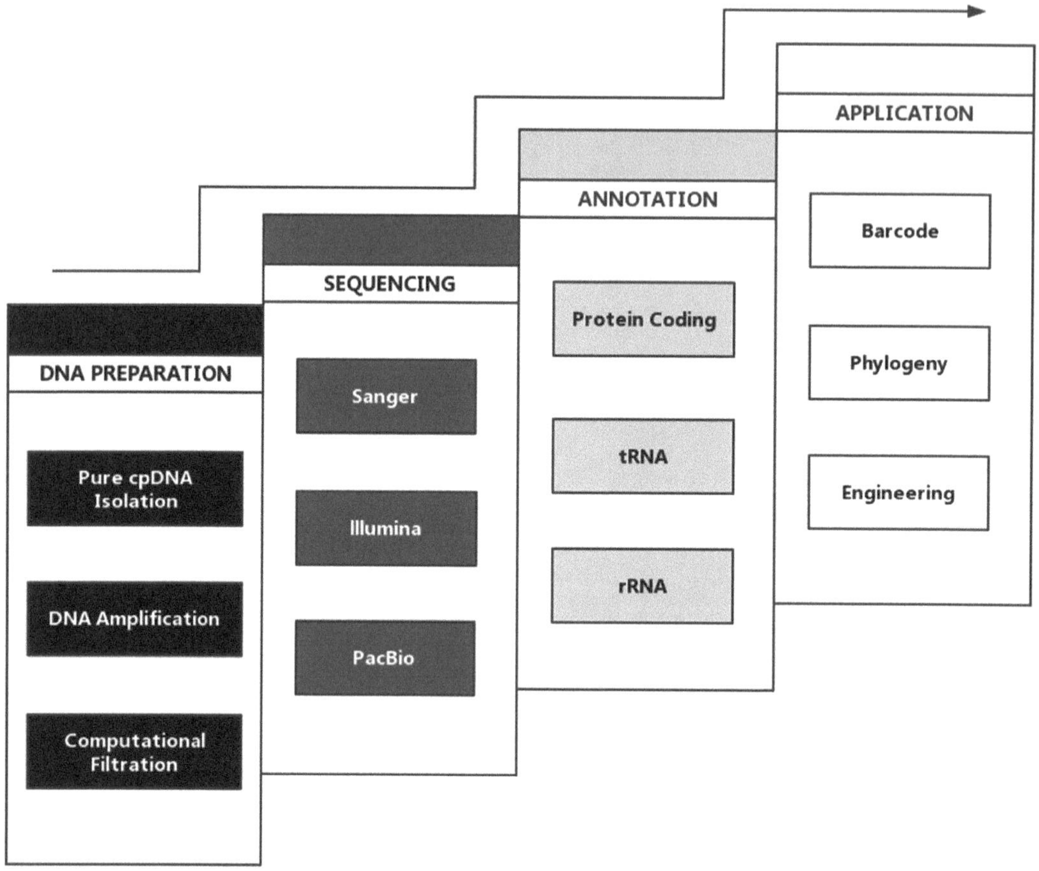


**Fig. 10.2** Workflow of chloroplast genome sequencing and annotation. The chloroplast genome sequencing starts with DNA preparation. The high-quality DNA is subjected to sequence by the platforms of Sanger, Illumina, and PacBio. The sequenced reads are assembled into chloroplast genome and then annotated by various bioinformatics tools. The chloroplast sequences are extraordinary applied in the fields of barcode, phylogeny, and biotechnology engineering

plant of *Ranunculus macranthus* was attempted, yielding sufficient purity and amount to run genomic sequencing (Jansen et al. 2005). However, the method is highly species dependent, resulting in very limited success in plants. The combination of high salt wash buffers with the sucrose/Percoll gradient technique improved the extraction of conifer cpDNA (Vieira Ldo et al. 2014). It was very cost-effective in terms of the balance between the quality and yield of cpDNA. The gradient allowed the increased chloroplast isolation, along with decreased the contamination of nuclear DNA and secondary metabolites in cpDNA by using high salt buffer.

## 10.2.2 DNA Amplification

### 10.2.2.1 Whole Genome Amplification

The mentioned protocols are not suitable to isolate pure cpDNA from all plants. Enough yield and sufficient purity are still restricted the downstream genome sequence and analysis. A bacteriophage Phi29 polymerase, which has the ability of amplifying more than 70 Kb without disassociating from the DNA template (Dean et al. 2002), provides a great opportunity to enrich the whole chloroplast genome using rolling circle amplification (RCA) (Table 10.1). It was found that the RCA approach worked

**Table 10.1** Methods of chloroplast DNA preparation

| Methods | Protocol name | Chemicals | Advantages | Drawbacks | References |
|---|---|---|---|---|---|
| Pure cpDNA isolation | Gradient centrifuge | Sucrose or Percoll | Suitable for most land plants | Time-consuming; massive amounts of start meterial | Palmer (1986) |
| | Enzyme digestion | DNAse I | High purity | Lower yield; suitable for few plants | Kolodner and Tewari (1979) |
| | High ionic strength | High salt buffers | No gradient centrifugation and less contamination | Suitable for few plants | Bookjans et al. (1984) |
| DNA amplification | Whole genome amplification | PCR reagent and primers | Suitable for most land plants | Low specificity of DNA amplification | Dean et al. (2002) |
| | Long PCR amplification | PCR reagent and primers | Little start material | Universal primer required; considerable gaps | Mardanov et al. (2008) |
| Computational filtration | Computational filtration | Computer programs | Easily operation | Gap existence | Wang et al. (2011) |

efficiently for many seed plants, such as Ginkgo and Podocarpus (Jansen et al. 2005). The only problem was that the low annealing temperature of the PCR reduced the specificity of the random hexamer primers. The development of genome-specific primers for chloroplast would advance the specificity of the amplification.

#### 10.2.2.2 Long PCR Amplification

To overcome the difficulty of isolating high-quality chloroplast DNA, long polymerase chain reaction (PCR) amplifying of large fragments (up to 40 Kb) of the genome using conserved chloroplast-specific primers gives an alternative way to generate chloroplast DNA, where it requires multiple reactions in order to gain overlapping fragments of the chloroplast genome (Table 10.1). It would produce long gaps as PCR reactions fail in some recalcitrant regions (Cronn et al. 2008). The primer design and PCR success are essential for the acquisition of complete chloroplast genome DNA. A study was conducted to propose that all known angiosperm chloroplast genomes can be amplified by using the nine universal primer pairs designed from the highly conserved regions (Yang et al. 2014). The primers showed a broad application in the tested 24 species from major clades of angiosperms, producing enough PCR products to construct the sequencing library (Yang et al. 2014). More universal primers were summarized to serve as a resource for the poorly described chloroplast genomes (Heinze 2007). The first sequenced duckweed chloroplast genome was *L. minor* by using this protocol. After the total DNA was extracted with CTAB-based method, the fragments of chloroplast DNA were amplified by long range PCR. The fragments of PCR products with the length of 1–8 Kb were overlapped (Table 10.2). The full coverage of chloroplast genome allowed to determine the complete nucleotide sequence without gaps (Mardanov et al. 2008).

### 10.2.3 Computational Filtration

With the advent of next-generation sequencing, it becomes feasible to simultaneously capture multiple chloroplast genomes from the single lane of next-generation sequencing (Cronn et al. 2008). Still, the separation of cpDNA from plant

**Table 10.2** Sequencing and features of duckweed chloroplast genomes

| Species | DNA preparation | Purification | Sequence | Assembly | Annotation | Genbank number | Genome size (bp) | IR size (bp) | Protein coding gene | tRNA | rRNA | References |
|---|---|---|---|---|---|---|---|---|---|---|---|---|
| *Lemna minor* | Total DNA | Long PCR amplification | Sanger | Gene studio program | BLAST; tRNAscan-SE program; DOGMA | DQ400350 | 165,955 | 31,223 | 78 | 30 | 4 | Mardanov et al. 2008 |
| *Spirodela polyrhiza 7498* | Total DNA | Computational filtration | NGS | SOLiD system de novo accessory tools and velvet assembly engine | BLAST; tRNAscan-SE program; DOGMA | JN160603 | 168,788 | 31,755 | 85 | 37 | 8 | Wang et al. (2011) |
| *Wolffiella lingulata 7289* | | | | | | JN160604 | 169,337 | 31,683 | 83 | 37 | 8 | |
| *Wolffia australiana 7733* | | | | | | JN160605 | 168,704 | 31,930 | 83 | 37 | 8 | |

cells can be extremely tedious. The use of multiple long PCR reactions or RCA is also intimidating with the gaps due to PCR failure. An innovative method was able to sequence total DNA including nuclear, mitochondrial, and chloroplast genome by skipping any experimental isolation of pure cpDNA (Nock et al. 2011). The separation occurred in the next step by using computational filtration to remove any contaminated DNA derived from nucleus and mitochondria. Three duckweed chloroplast genomes (*Spirodela polyrhiza*, *Wolffiella lingulata*, and *Wolffia australiana*) were sequenced by using such method and were assembled without any experimental purification (Table 10.2). The data have shown that the reads of total DNA sequencing from a quadrant slide without any purification could reach more than 1000-fold coverage (Wang and Messing 2011).

## 10.3 Chloroplast Genome Sequencing and Annotation

It is well known that a chloroplast genome size is relatively small (~107 to 218 Kb) (Daniell et al. 2016). After removing one copy of inverted repeat, the size becomes comparable to a BAC clone. The chloroplast genome is generally a by-product of a whole nuclear genome sequencing project that could be obtained from a fosmid or a BAC clone under Sanger sequencing platform (Sanger 1988). With the rapid development of high-throughput next-generation sequencing (NGS) (Schuster 2008), it substantially facilitates the releases of chloroplast genomes. However, the short reads (~100 bp) from NGS are GC biased and cannot fully span over the junctions of inverted repeats (IRs), resulting in incomplete genome and unsolved IRs. PacBio reads become promising to overcome the assembly challenge due to the long reads (~10 Kb) that can specifically determine the sequence location in the genome (Fig. 10.2) (Eid et al. 2009).

### 10.3.1 Chloroplast Genome of *L. minor* Sequenced by Sanger Technology

The chloroplast genome of *L. minor* (common duckweeds) was sequenced by Sanger sequencer in 2008 (Mardanov et al. 2008). The cpDNA was prepared by the amplified PCR fragments ranging from 1 to 8 Kb. Each fragment was automatically sequenced on ABI sequencers using the BigDye Terminator. All fragments were sequenced ~6 times on average with the way of primer walking. Assembling was performed with the Gene Studio program (http://www.genestudio.com). *L. minor* chloroplast genome had a size of 165,955 bp in a circular molecule, including a pair of 31,223-bp inverted repeat regions, an 89,906-bp large single copy, and a 13,603-bp small single copy (Table 10.2). *L. minor* had a tendency of expansion in terms of the inverted repeats in comparison with other monocots. The genes of *infA*, *ycf15*, and *ycf68* were absent from *L. minor*, but present in other plant chloroplast genomes. The tRNA types found in *L. minor* chloroplast genome that could recognize all plastid codons (Mardanov et al. 2008).

### 10.3.2 Chloroplast Genomes of *S. polyrhiza*, *W. lingulata* and *W. australiana* Sequenced by NGS Technology

The number of sequenced plant chloroplast genomes has exploded to more than 1000 species mostly due to the availability of NGS at cheaper sequencing cost (Jansen et al. 2005). A multiplex sequencing-by-synthesis approach using the Illumina Genome Analyzer sequenced the chloroplast genomes of *Picea sitchensis* and seven pine species simultaneously. The pooled PCR-amplified products were ligated and multiplexed to the adapters including 3-bp indexing tags. The efficient sequencing approach yields an average sequence depth of 55× to 186× and produced high-quality draft genomes with an estimation of 88–94% completion (Cronn et al. 2008). Three *Brassica rapa* accessions sequenced by Illumina short reads also generated complete chloroplast genomes from total DNA without any experimental purification (Wu et al. 2012).

Three more duckweed chloroplast genomes (*S. polyrhiza*, *W. lingulata,* and *W. australiana*) were sequenced by using SOLiD platform with a read length of 50 bp. The short reads derived from cpDNA were filtered electronically using the reference of *L. minor*. The genomes were de novo assembled using the SOLiD System de novo Accessory Tools 2.0 (http://solidsoftwaretools.com/gf/project/denovo/) in conjunction with the velvet assembly engine (Zerbino and Birney 2008). The remaining gaps were closed with Sanger sequencing of 13–29 additional PCR products. The chloroplast genome sizes of duckweeds had a range of 165,955–169,353 bp. They contained two copies of ~31-Kb of inverted repeats, separated by a ~90-Kb LSC and a ~10-Kb SSC (Table 10.2). The borders between IR and the single-copy region were variable in duckweeds. The accessibility of duckweed chloroplast genomes and the sequence divergence indicated their taxonomic and phylogenetic relationships, which would provide the references for further species identification and chloroplast genetic engineering.

The overall structure of the duckweed chloroplast genomes was conserved. They had similar gene copy number and gene order. Still, there were numbers of rearrangements and sequences polymorphisms, like INDELs (insertion/deletion) and single-nucleotide variations. The sequence alignment and comparison revealed multiple variation hotspots, like the 10-Kb regions between *rpoB* and *psbD* in the genomes. The rich sequence variation occurred in the noncoding intergenic regions, while IR regions showed lower sequence divergence than the single-copy regions. There was a 505-bp deletion in *S. polyrhiza* compared to the 100-bp deletion in *W. lingulata*. It was also found a 353-bp insertion occurred at 31 Kb of the intergenic *petN-psbM* region of *W. Australiana* (Wang and Messing 2011).

The genomes were annotated by DOGMA (Wyman et al. 2004). There were 83–85 protein-coding genes in three duckweeds, 37 for rRNA genes and 8 for tRNA (Table 10.2). Generally, the chloroplast genome was conserved in gene number and organization with respect to the reference genome of *L. minor*.

### 10.3.3 Other Chloroplast Genomes Sequenced by TGS Technology

Given the high-throughput and the increase of computational capacity for NGS, the number of released chloroplast genome will double in a very short period and accelerate the development of the chloroplast genomes. However, the reads produced by NGS are relatively short (~150 bp), leading to the fragmented assembly and requiring the inevitable steps of gap filling with PCR amplification and Sanger sequencing (Dohm et al. 2008). The technology of third-generation sequencing (TGS), also called single-molecule real-time sequencing (SMRT) was invented by Pacific Biosciences in 2009 (Eid et al. 2009; Koren et al. 2013). The way of sequencing-by-synthesis could produce long reads and cover the full template without amplification bias. The current average read length generated by PacBio Sequel is exceptionally long (>20 Kb), spanning over the repeat regions and increasing genome contiguity (Ardui et al. 2018). It has been widely used in plant sciences, even for large and complex nuclear genomes (Li et al. 2017). Several genomes in terms of continuity and accuracy showed improved quality, including *U. gibba* (82 Mb) (Lan et al. 2017), *O. thomaeum* (245 Mb) (VanBuren et al. 2015), *C. quinoa* (1500 Mb) (Jarvis et al. 2017), *Zea mays* (2300 Mb) (Jiao et al. 2017), and *H. annuus* (3000 Mb) (Badouin et al. 2017; Li et al. 2017).

The revolutionary PacBio long reads also provide the chloroplast genome sequencing a cost-effective and straightforward approach. The chloroplast genome of *Potentilla micrantha* was the first one to be sequenced by using PacBio long reads. The qualified reads after error-correction represent 320-fold chloroplast genome depth with a mean length of 1902 bp. A single contig was generated from long-read assembly covering the entire genome without any ambiguities. The chloroplast genome assembled from PacBio long reads were consistent with that of Illumina short reads, whereas the PacBio assembly was more continuous and resolved 187 ambiguities existed in Illumina assembly (Ferrarini et al. 2013). There was a stronger positive correlation between the coverage and GC content in NGS, whereas Pacbio long reads did not show any obvious GC-bias. Still, TGS has its own limitation given its high inherent error rate that used to require NGS reads to correct (Ferrarini et al. 2013). With the falling of sequencing cost and the random errors in TGS, the sequences could be revised by a consensus deduced from deep sequencing coverage. The accuracy could be achieved to 99.9% without any impact on the assembly. It is worth to mention that Pacbio greatly facilitated the integration of inverted repeats (IRs). The unique sequence from the junctions of IRs and the small variation in IRs permitted them to be assigned unambiguously, which were extremely challenging for NGS. The two IRs with the individual length of 25,530 bp in *P. micrantha* were almost identical but with three nucleotide difference that was enough to correctly locate the IRs. However, it was a non-trivial task for untangling IRs from short-read assembly that required tedious steps of manual operations (Ku et al. 2013; Ferrarini et al. 2013).

Another two chloroplast genomes from the species *Picea glauca* (a gymnosperm) and *Sinningia speciosa* (a eudicot angiosperm) were also assembled by taking advantage of long reads. The pipeline of Organelle_PBA was specifically designed to assemble chloroplast and mitochondrial genome by computationally selecting the organelle long reads from total DNA sequencing (Soorni et al. 2017). The simple-to-use program and the long reads promoted the organelle genome assembly performance and resolve the inverted repeats. Furthermore, the application of PacBio long reads would enhance our understanding of the complex structure and function of chloroplast genome. It is believed that more plant chloroplast genomes, including duckweeds, will be sequenced by the long-read sequencing technology in near future.

## 10.4 Chloroplast Genome Applications

### 10.4.1 Chloroplast Genome Sequences for Plant Barcode

Chloroplast DNA sequence data are a powerful tool for plant identification and deciphering genetic relationships among plant species. DNA barcode is an efficient molecular identification system to tell species apart by a universal marker. The chloroplast-derived markers are the most popular for identifying the genetic distances in plants (Fig. 10.2). However, no single locus could distinguish between all plant species. It is even more challenging in duckweeds to perform species identification and taxonomic studies due to the highly reduced morphology and small plant size (Wang et al. 2010).

The *atpF-atpH* chloroplast marker was proposed to be the best one with its high success of PCR amplification and high power to discriminate duckweeds (Wang et al. 2010). To further barcode sibling species that was lack of enough sequence variation and polymorphism, the combination of two plastid sequences *rpl16* and *rps16* recognized the species in *Wolffia* genus (Landolt 1994; Bog et al. 2013). However, it failed to delineate all 11 species of because of highly closely inter- and intra-specific genetic distances. The study of using two barcodes (*atpF-atpH* and *psbK-psbI* spacer regions) could distinguish 30 of the 37 duckweed species. The increase of resolution from the whole chloroplast genome as a single-locus DNA barcode becomes the most feasible way to discriminate plant species.

Thanks to the massively parallel sequencing of NGS, the obtaining of chloroplast genomes from five grass species was provided as a promising barcode marker. The number of assembly gaps increased with evolutionary distance from the reference of *Oryza sativa*. A number of 91 SNPs were identified between the closely related species of *O. meridionalis* and *O. sativa* japonica (Nock et al. 2011). The complete chloroplast genomes of eight Fritillaria species were found to contain multiple polymorphic SSR, large repeat sequences and highly variable regions, which was invaluable for the species identification and the relation establishment in Fritillaria (Bi et al. 2018).

### 10.4.2 Engineering the Chloroplast Genomes for Biotechnology Applications

Genetic engineering is the biotechnology of directly transferring foreign genes into target plants in order to improve its desirable traits, such as yield, nutrition enhancement, and resistance to pathogen. An efficient genetic transformation protocols were developed in *Lemna gibba* and *L. minor* with a binary vector containing b-glucuronidase and nptII expression cassettes (Yamamoto et al. 2000). Transgenic duckweed could be regenerated after three months of agrobacterium-mediated transformation. The addition of the poorly assimilated carbohydrates of galactose or sorbitol yielded high levels of callus (Li et al. 2004). The stable and transgenic *Spirodela oligorrhiza* showed that the transgene protein of GFP expression reached more than 25% of total soluble proteins (Vunsh et al. 2007). An artificial microRNA gene silencing system was generated in *L. minor* fronds that the expression of CH42 was significantly inhibited, resulting in the reduction of chlorophyll pigmentation (Canto-Pastor et al. 2015). However, all these studies were nuclear-level modification, and there was no reported study about chloroplast transgene in duckweeds yet.

The duckweeds as an alternative biofuel and bioremediator have attracted extremely interests in recent years. The complete chloroplast genomes of duckweeds provide the framework to explore their potential values and to accelerate the chloroplast genetic engineering (Fig. 10.2). The chloroplast could be also engineered to produce high-value agricultural and biomedical products with highly expression by the fact of abundant chloroplast genome copies in a plant cell. As the transgene is inserted within the IRs region, its copy number of transgenes will be doubled theoretically. However, achieving the homoplastic state of genetically modified chloroplast is a non-trivial task. It requires two or three rounds of selection to eliminate all untransformed copies (>1000 per cell) of the native chloroplast genome. A comprehensive summary demonstrated the power of chloroplast genetic engineering, including 114 transgenes, the integration sites, and engineered traits (Daniell et al. 2016). The principle of chloroplast genome engineering has to incorporate the foreign genes into intergenic spacer regions without disrupting the native chloroplast genes. The cassette usually contains gene(s) of interest, a selectable marker gene, and two chloroplast genes used as flanking sequences. However, most intergenic spacer region is not conserved and the ideal sites are still absent.

**Sequence Information**

The duckweed chloroplast genomes are deposited into GenBank with the ID of DQ400350 for *L. minor*. JN160603, JN160604 and JN160605.

## References

Ardui S, Ameur A, Vermeesch JR, Hestand MS (2018) Single molecule real-time (SMRT) sequencing comes of age: applications and utilities for medical diagnostics. Nucleic Acids Res 46(5):2159–2168

Badouin H, Gouzy J, Grassa CJ, Murat F, Staton SE, Cottret L, Lelandais-Briere C, Owens GL, Carrere S, Mayjonade B, Legrand L, Gill N, Kane NC, Bowers JE, Hubner S, Bellec A, Berard A, Berges H, Blanchet N, Boniface MC, Brunel D, Catrice O, Chaidir N, Claudel C, Donnadieu C, Faraut T, Fievet G, Helmstetter N, King M, Knapp SJ, Lai Z,

Le Paslier MC, Lippi Y, Lorenzon L, Mandel JR, Marage G, Marchand G, Marquand E, Bret-Mestries E, Morien E, Nambeesan S, Nguyen T, Pegot-Espagnet P, Pouilly N, Raftis F, Sallet E, Schiex T, Thomas J, Vandecasteele C, Vares D, Vear F, Vautrin S, Crespi M, Mangin B, Burke JM, Salse J, Munos S, Vincourt P, Rieseberg LH, Langlade NB (2017) The sunflower genome provides insights into oil metabolism, flowering and Asterid evolution. Nature 546(7656):148–152

Bi Y, Zhang MF, Xue J, Dong R, Du YP, Zhang XH (2018) Chloroplast genomic resources for phylogeny and DNA barcoding: a case study on Fritillaria. Sci Rep 8(1):1184

Bog M, Schneider P, Hellwig F, Sachse S, Kochieva EZ, Martyrosian E, Landolt E, Appenroth KJ (2013) Genetic characterization and barcoding of taxa in the genus Wolffia Horkel ex Schleid. (Lemnaceae) as revealed by two plastidic markers and amplified fragment length polymorphism (AFLP). Planta 237(1):1–13

Bookjans G, Stummann BM, Henningsen KW (1984) Preparation of chloroplast DNA from pea plastids isolated in a medium of high ionic strength. Anal Biochem 141(1):244–247

Canto-Pastor A, Molla-Morales A, Ernst E, Dahl W, Zhai J, Yan Y, Meyers BC, Shanklin J, Martienssen R (2015) Efficient transformation and artificial miRNA gene silencing in *Lemna minor*. Plant Biol (Stuttg) 17 (Suppl 1):59–65

Cheng JJ, Stomp AM (2009) Growing duckweed to recover nutrients from wastewaters and for production of fuel ethanol and animal feed. CLEAN Soil Air Water 37(1):17–26

Cronn R, Liston A, Parks M, Gernandt DS, Shen R, Mockler T (2008) Multiplex sequencing of plant chloroplast genomes using Solexa sequencing-by-synthesis technology. Nucleic Acids Res 36(19):e122

Daniell H, Lin CS, Yu M, Chang WJ (2016) Chloroplast genomes: diversity, evolution, and applications in genetic engineering. Genome Biol 17(1):134

Dean FB, Hosono S, Fang L, Wu X, Faruqi AF, Bray-Ward P, Sun Z, Zong Q, Du Y, Du J, Driscoll M, Song W, Kingsmore SF, Egholm M, Lasken RS (2002) Comprehensive human genome amplification using multiple displacement amplification. Proc Natl Acad Sci USA 99(8):5261–5266

Dohm JC, Lottaz C, Borodina T, Himmelbauer H (2008) Substantial biases in ultra-short read data sets from high-throughput DNA sequencing. Nucleic Acids Res 36(16):e105

Douglas SE (1998) Plastid evolution: origins, diversity, trends. Curr Opin Genet Dev 8(6):655–661

E Landolt (1994) Taxonomy and ecology of the section Wolffia of the genus Wolffia (Lemnaceae). Ber Geobot Inst ETH

Eid J, Fehr A, Gray J, Luong K, Lyle J, Otto G, Peluso P, Rank D, Baybayan P, Bettman B, Bibillo A, Bjornson K, Chaudhuri B, Christians F, Cicero R, Clark S, Dalal R, Dewinter A, Dixon J, Foquet M, Gaertner A, Hardenbol P, Heiner C, Hester K, Holden D, Kearns G, Kong X, Kuse R, Lacroix Y, Lin S, Lundquist P, Ma C, Marks P, Maxham M, Murphy D, Park I, Pham T, Phillips M, Roy J, Sebra R, Shen G, Sorenson J, Tomaney A, Travers K, Trulson M, Vieceli J, Wegener J, Wu D, Yang A, Zaccarin D, Zhao P, Zhong F, Korlach J, Turner S (2009) Real-time DNA sequencing from single polymerase molecules. Science (80) 323(5910):133–138

Ferrarini M, Moretto M, Ward JA, Surbanovski N, Stevanovic V, Giongo L, Viola R, Cavalieri D, Velasco R, Cestaro A, Sargent DJ (2013) An evaluation of the PacBio RS platform for sequencing and de novo assembly of a chloroplast genome. BMC Genom 14:670

Group CPW (2009) A DNA barcode for land plants. Proc Natl Acad Sci USA 106(31):12794–12797

Heinze B (2007) A database of PCR primers for the chloroplast genomes of higher plants. Plant Methods 3:4

Jansen RK, Raubeson LA, Boore JL, dePamphilis CW, Chumley TW, Haberle RC, Wyman SK, Alverson AJ, Peery R, Herman SJ, Fourcade HM, Kuehl JV, McNeal JR, Leebens-Mack J, Cui L (2005) Methods for obtaining and analyzing whole chloroplast genome sequences. Methods Enzymol 395:348–384

Jarvis DE, Ho YS, Lightfoot DJ, Schmöckel SM, Li B, Borm TJA, Ohyanagi H, Mineta K, Michell CT, Saber N, Kharbatia NM, Rupper RR, Sharp AR, Dally N, Boughton BA, Woo YH, Gao G, Schijlen EGWM, Guo X, Momin AA, Negrão S, Al-Babili S, Gehring C, Roessner U, Jung C, Murphy K, Arold ST, Gojobori T, van der Linden CG, van Loo EN, Jellen EN, Maughan PJ, Tester M (2017) The genome of Chenopodium quinoa. Nature:1–20

Jiao Y, Peluso P, Shi J, Liang T, Stitzer MC, Wang B, Campbell MS, Stein JC, Wei X, Chin C-S, Guill K, Regulski M, Kumari S, Olson A, Gent J, Schneider KL, Wolfgruber TK, May MR, Springer NM, Antoniou E, McCombie WR, Presting GG, McMullen M, Ross-Ibarra J, Dawe RK, Hastie A, Rank DR, Ware D (2017) Improved maize reference genome with single-molecule technologies. Nature 546:524–527

Kolodner R, Tewari KK (1979) Inverted repeats in chloroplast DNA from higher plants. Proc Natl Acad Sci USA 76(1):41–45

Koren S, Harhay GP, Smith TP, Bono JL, Harhay DM, McVey SD, Radune D, Bergman NH, Phillippy AM (2013) Reducing assembly complexity of microbial genomes with single-molecule sequencing. Genome Biol 14(9):R101

Ku C, Chung WC, Chen LL, Kuo CH (2013) The complete plastid genome sequence of Madagascar Periwinkle *Catharanthus roseus* (L.) G. Don: plastid genome evolution, molecular marker identification, and phylogenetic implications in Asterids. PLoS One 8(6):e68518

Lan T, Renner T, Ibarra-Laclette E, Farr KM, Chang TH, Cervantes-Perez SA, Zheng C, Sankoff D, Tang H,

Purbojati RW, Putra A, Drautz-Moses DI, Schuster SC, Herrera-Estrella L, Albert VA (2017) Long-read sequencing uncovers the adaptive topography of a carnivorous plant genome. Proc Natl Acad Sci USA 114(22):E4435–E4441

Landolt E (1986) The family of Lemnaceae - a monographic study, vol 1. Veroffentlichungen des Geobotanischen Institutes der Eidgenossischen Technischen Hochschule, Stiftung Rubel

Li C, Lin F, An D, Wang W, Huang R (2017) Genome sequencing and assembly by long reads in plants. Genes (Basel) 9(1)

Li J, Jain M, Vunsh R, Vishnevetsky J, Hanania U, Flaishman M, Perl A, Edelman M (2004) Callus induction and regeneration in Spirodela and Lemna. Plant Cell Rep 22(7):457–464

Mardanov AV, Ravin NV, Kuznetsov BB, Samigullin TH, Antonov AS, Kolganova TV, Skyabin KG (2008) Complete sequence of the duckweed (*Lemna minor*) chloroplast genome: structural organization and phylogenetic relationships to other angiosperms. J Mol Evol 66(6):555–564

Murray MG, Thompson WF (1980) Rapid isolation of high molecular weight plant DNA. Nucleic Acids Res 8(19):4321–4325

Nock CJ, Waters DL, Edwards MA, Bowen SG, Rice N, Cordeiro GM, Henry RJ (2011) Chloroplast genome sequences from total DNA for plant identification. Plant Biotechnol J 9(3):328–333

Palmer JD (1986) Isolation and structural analysis of chloroplast DNA. Methods Enzymol 118(2):167–186

Sanger F (1988) Sequences, sequences, and sequences. Annu Rev Biochem 57:1–28

Schuster SC (2008) Next-generation sequencing transforms today's biology. Nat Methods 5(1):16–18

Shi C, Hu N, Huang H, Gao J, Zhao YJ, Gao LZ (2012) An improved chloroplast DNA extraction procedure for whole plastid genome sequencing. PLoS ONE 7 (2):e31468

Soorni A, Haak D, Zaitlin D, Bombarely A (2017) Organelle_PBA, a pipeline for assembling chloroplast and mitochondrial genomes from PacBio DNA sequencing data. BMC Genom 18(1):49

Stomp A-M, El-Gewely MR (2005) The duckweeds: a valuable plant for biomanufacturing. In: Biotechnology annual review, vol 11. Elsevier, pp 69–99

Timmis JN, Ayliffe MA, Huang CY, Martin W (2004) Endosymbiotic gene transfer: organelle genomes forge eukaryotic chromosomes. Nat Rev Genet 5(2):123–135

VanBuren R, Bryant D, Edger PP, Tang H, Burgess D, Challabathula D, Spittle K, Hall R, Gu J, Lyons E, Freeling M, Bartels D, Ten Hallers B, Hastie A, Michael TP, Mockler TC (2015) Single-molecule sequencing of the desiccation-tolerant grass *Oropetium thomaeum*. Nature 527(7579):508–511

Vieira Ldo N, Faoro H, Fraga HP, Rogalski M, de Souza EM, de Oliveira Pedrosa F, Nodari RO, Guerra MP (2014) An improved protocol for intact chloroplasts and cpDNA isolation in conifers. PLoS ONE 9(1):e84792

Vunsh R, Li J, Hanania U, Edelman M, Flaishman M, Perl A, Wisniewski JP, Freyssinet G (2007) High expression of transgene protein in Spirodela. Plant Cell Rep 26(9):1511

Wang W, Messing J (2011) High-throughput sequencing of three Lemnoideae (duckweeds) chloroplast genomes from total DNA. PLoS ONE 6(9):e24670

Wang W, Wu Y, Messing J (2012) The mitochondrial genome of an aquatic plant *Spirodela polyrhiza*. PLoS One 7(10):e46747

Wang W, Wu Y, Yan Y, Ermakova M, Kerstetter R, Messing J (2010) DNA barcoding of the Lemnaceae, a family of aquatic monocots. BMC Plant Biol 10:205

Wu J, Liu B, Cheng F, Ramchiary N, Choi SR, Lim YP, Wang XW (2012) Sequencing of chloroplast genome using whole cellular DNA and solexa sequencing technology. Front Plant Sci 3:243

Wyman SK, Jansen RK, Boore JL (2004) Automatic annotation of organellar genomes with DOGMA. Bioinformatics 20(17):3252–3255

Yamamoto YT, Rajbhandari N, Lin X, Bergmann BA, Nishimura Y, Stomp AM (2000) Genetic transformation of duckweed *Lemna gibba* and *Lemna minor*. Vitro Cell Dev Plant 37:349–353

Yang JB, Li DZ, Li HT (2014) Highly effective sequencing whole chloroplast genomes of angiosperms by nine novel universal primer pairs. Mol Ecol Res 14(5):1024–1031

Zerbino DR, Birney E (2008) Velvet: algorithms for de novo short read assembly using de Bruijn graphs. Genome Res 18(5):821–829

# 11 Genotyping-by-Sequencing for Species Delimitation in *Lemna* Section *Uninerves* Hegelm. (Lemnaceae)

M. Bog, S. Xu, A. Himmelbach, R. Brandt, F. Wagner, K.-J. Appenroth and K. S. Sree

**Abstract**
*Lemna* section *Uninerves* presently consists of three species, *Lemna minuta* Kunth (common synonyms *L. minima* Chev., *L. minuscula* Herter), *L. valdiviana* Phil., and *L. yungensis* Landolt. The species *L. yungensis* was discovered by E. Landolt about 20 years ago. He mentioned that although this species specifically grows on wet rocks, it is closely related to *L. valdiviana* and is morphologically very similar to the other two species in this section. Therefore, we started with molecular taxonomic studies of this section by using genotyping-by-sequencing (GBS). It was found that *L. minuta* could be clearly separated from *L. valdiviana*/*L. yungensis*; however, no separation was found between *L. valdiviana* and *L. yungensis*, despite the fact that they occupy different ecological niches. These data revealed that *L. yungensis* is identical to *L. valdiviana* on the genetic basis. Based on these results, we suggest that the name *Lemna yungensis* should be synonymised. As the description of *L. valdiviana* Phil. is the older one, this would be the valid common name for all the clones defined before as *L. valdiviana* and *L. yungensis*. This decision reduces the number of Lemnaceae species to 36. Testing the method of GBS with the genus *Lemna* demonstrated its superiority in comparison with barcoding based on PCR amplifications.

## 11.1 Introduction

*Lemna* section *Uninerves* Hegelm. represents the most reduced group, based on morphology, within the genus *Lemna* L. (Lemnaceae) positioning it as the most derived one on evolutionary basis (Landolt 1986; Les et al. 1997). After the revision of the internal structure of *Lemna* by Les et al. (2002), all remaining sections are monophyletic, i.e. *L.* section *Lemna, L.* section *Alatae, L.* section *Biformes,* and *L.* section *Uninerves,* each with 100% bootstrap support. *Lemna*

M. Bog (✉)
Institute of Botany and Landscape Ecology, University of Greifswald, Greifswald, Germany
e-mail: manuela.bog@uni-greifswald.de

S. Xu
Institute for Evolution and Biodiversity, University of Münster, Münster, Germany

A. Himmelbach · R. Brandt
Leibniz Institute of Plant Genetics and Crop Plant Research (IPK), Gatersleben, Germany

F. Wagner
Institute of Plant Sciences, University of Regensburg, Regensburg, Germany

K.-J. Appenroth
University of Jena, Matthias Schleiden Institute, Plant Physiology, Jena, Germany

K. S. Sree
Department of Environmental Science, Central University of Kerala, Periye, India



X. H. Cao et al. (eds.), *The Duckweed Genomes*, Compendium of Plant Genomes,
https://doi.org/10.1007/978-3-030-11045-1_11

section *Uninerves* has morphological diagnostic features, i.e. a single nerve compared to 3–7 nerves in the other *Lemna* sections. Landolt (1986) originally reported two species under this section, *Lemna minuta* Kunth (in that time as *L. minuscula* Herter) and *L. valdiviana* Phil., but had later added the species *L. yungensis* Landolt (Landolt 1998), which was discovered growing on wet rocks in the province Nor-Yungas, Bolivia. The delimitation of these species on a morphological basis has been always challenging (Crawford et al. 1996), especially after *L. yungensis* had been added, which is even more similar to *L. valdiviana* (Landolt 1998). The separation of all three species as section by barcoding using plastidic/morphological (Les et al. 2002) and nuclear (Tippery et al. 2015) markers was indicated by bootstrap values of 100%. Bog et al. (2010) used manually scored amplified fragment length polymorphism (AFLP) to analyse the genus *Lemna*. *Lemna* section *Uninerves* could be separated by this method from other clades (bootstrap value 86%). Although *L. minuta* formed a defined clade (bootstrap value 100%), the other two species did not form any defined group. Moreover, one of the authors of the aforementioned studies (Elias Landolt) had doubts concerning this clear definition of *L. minuta*, owing to the reason that the biogeographical coordinates of the five analysed clones might not represent the actual diversity of this species in nature. Similar to the study by Wang et al. (2010), who investigated *L. minuta* and *L. valdiviana* using plastidic DNA markers and found no separation between them, Borisjuk et al. (2015), who used the markers *psb*K-*psb*I and *atp*H-*atp*F on all the three species, also considered that the differentiation of species in *L.* section *Uninerves* was not completely reliable. Using a large number of clones and the plastidic marker *psb*K-*psb*I, Bog et al. (2020) found that *L. minuta* could be separated from *L. valdiviana* and *L. yungensis*, but the latter two could not be distinguished.

Recently, next-generation sequencing made large-scale sequencing of genomes cheaper and faster. Therefore, we aimed to apply genotyping-by-sequencing (GBS; c.f. Elshire et al. 2011; Poland et al. 2012; Wendler et al. 2014) in order to address the question whether the species of the section *Uninerves* can be distinguished as (A) three independent species (*L. minuta*, *L. valdiviana*, and *L. yungensis*) or (B) *L. valdiviana* and *L. yungensis* should be synonymised and treated as one species next to *L. minuta*. Moreover, this also represents the first investigation of using GBS on the genus *Lemna* (Lemnaceae) to address the taxonomic challenge within *L.* section *Uninerves*.

## 11.2 Materials and Methods

### 11.2.1 Plant Material

All plants were taken from the stock collection of the University of Jena, Matthias Schleiden Institute-Plant Physiology supplemented by some clones from the Landolt Duckweed Collection, Zürich. The clones and their origin are given in Table 11.1. Plants were cultivated in N-medium (Appenroth et al. 1996) as described before (Bog et al. 2010), shock frozen after harvest in liquid nitrogen and freeze-dried.

### 11.2.2 Molecular Methods

DNA extraction based on silica columns was done according to Valledor et al. (2014), and subsequent construction of barcoded libraries for the GBS approach was essentially performed as described in Wendler et al. (2014). Sequencing-by-synthesis (single read, 1 × 107 cycles, index read 8 cycles) using the Illumina HiSeq 2500 device was done according to protocols provided by the manufacturer (Illumina Inc.).

### 11.2.3 Data Analysis

Raw reads were trimmed and cleaned using AdapterRemoval v2.1.7 (Schubert et al. 2016). All stretches of Ns and consecutive stretches of low-quality bases (minimum quality of 30) on

**Table 11.1** Species and clones (ID and accession numbers) of the genus *Lemna* used for genotyping-by-sequencing. The two species in bold were used as an outgroup for the genus

| Species | Clone ID | Accession number | Origin |
|---|---|---|---|
| *L. aequinoctialis* | 6746 | a1258832 | USA, Virginia |
| *L. disperma* | 7269 | a1258835 | USA, California |
| *L. gibba* | 8682 | a1258833 | Saudi-Arabia |
| *L. japonica* | 8695 | a1258836 | Japan, Kyoto |
| *L. minor* | 5500 | a1096785 | Ireland |
| *L. minuta* | 6717 | a1433919 | Guatemala, Chinaltenango |
| *L. minuta* | 7612 | a1433922 | Peru, Cutco |
| *L. minuta* | 8669 | a1258827 | Japan, Kyoto |
| *L. minuta* | 9484 | a1433921 | Greece |
| *L. obscura* | 9342 | a1258834 | Venezuela, Maracaibo |
| *L. perpusilla* | 8539 | a1258826 | USA, Virginia |
| *L. tenera* | 9020 | a1258830 | Australia, Northern Territory |
| *L. trisulca* | 9529 | a1258837 | Germany |
| *L. turionifera* | 9434 | a1258828 | Russia, Baikal Area |
| *L. valdiviana* | 9401 | a1433927 | Venezuela, Sucre |
| *L. valdiviana* | 9228 | a1433926 | Brazil, Bahia |
| *L. valdiviana* | 9442 | a1258831 | Brazil, Mato Grosso |
| *L. yungensis* | 9207 | a1433919 | Bolivia, La Paz |
| *L. yungensis* | 9208 | a1258829 | Bolivia, La Paz |
| *L. yungensis* | 9209 | a1433924 | Bolivia, La Paz |
| *L. yungensis* | 9210 | a1433925 | Bolivia, La Paz |
| ***Landoltia punctata*** | 9589 | a1096784 | India, New Delhi |
| ***Spirodela polyrhiza*** | 7498 | a1096783 | USA, North Carolina |

both 5′ and 3′ end were removed. Reads shorter than 36 bases were discarded. In a second step, the reads were mapped to the *Lemna minor* genome (Van Hoeck et al. 2015) using the BWA-MEM algorithm of the Burrows-Wheeler Alignment Tool (Li and Durbin 2009). All bam files were merged, sorted, and all reads that had low mapping quality ($q < 10$) were removed using SAMtools v1.3.1 (Li et al. 2009). Indels were called using GATK v3 (McKenna et al. 2010), and the reads around the indels were realigned. SNPs were annotated and called from the realigned bam file. VCFtools v0.1.13 (Danecek et al. 2011) was used to filter the SNP data.

For tree reconstruction, we removed all SNP loci from the final dataset that had more than 40% missing data. In total, 3005 loci were kept. Among them, 2233 homozygous sites were used to build the FASTA sequence alignment for reconstructing the phylogenetic tree. Tree reconstruction was done using MrBayes v3.2.6 (Huelsenbeck and Ronquist 2001). We used the General Time Reversible substitution model with gamma-distributed rate variation across sites since this was the best model found by jModel-test2 v.2.1.5 (Darriba et al. 2012). We made two independent runs with four chains each. The temperature of the heated chains was set to 0.2. The programme was run for 2,000,000 generations with a sample frequency of 2000. All other settings were kept at default values.

In order to test our two scenarios of species delimitation (three species vs. two species with *L. valdiviana* and *L. yungensis* combined), we reduced our final dataset to biallelic SNPs for the members of *L.* section *Uninerves*, which dramatically decreased the number of SNPs to 56. We then estimated Marginal Likelihood (log-ML) values with SNAPP (Bryant et al. 2012) implemented in BEAST v2.3.2 (Bouckaert et al. 2014) by conducting ten separate runs of path sampling for each scenario using 100 steps, a chain length of 100,000, and a preburn-in of 10,000. Priors for the Yule birth rate ($\lambda$), the population size parameter ($\theta$), and the backward and forward mutation rates ($u$, $v$) were accepted as given by the BEAST companion programme BEAUTI v2.3.2. We performed some basic statistic tests in R v3.3.3 (R Core Team 2017) to search for a significant trend to favour one of the species delimitation scenarios over the other from the log-ML values. First, we tested the data for normality using the Shapiro–Wilk normality test, and second, the data were tested for homoscedasticity using the Bartlett test. Since the data showed homoscedasticity but non-normality, we used the Kruskal–Wallis test to test the stochastic homogeneity of the log-ML values for the two scenarios.

## 11.3 Results

The GBS approach yielded between ca. 370,000 and 2.4 mio raw reads per investigated clone (Table 11.2). The mapping rates were the highest for members of *L.* section *Lemna* with an average value of 49%. For members of all other sections, the rate of reads that could be mapped to the *L. minor* reference genome was noticeably lower (Table 11.2). Nevertheless, our final dataset comprised of 2233 SNPs with an average number of 1693 SNPs per clone without missing data. The outgroup species *Spirodela polyrhiza* and *Landoltia punctata* had the least complete SNP data with 878 and 483 SNPs without missing data, respectively, which indicated a high divergence of these two species to the genus *Lemna*. The number of pairwise SNP differences is shown in Table 11.3. Each clone was represented by a unique SNP profile. As expected, pairwise SNP differences are bigger between clones of different sections than between clones belonging to the same section, that are again bigger than SNP differences between clones of the same species. Especially, the number of SNP differences between clones of *L. minuta* was relatively small.

The Bayesian majority-rule consensus tree for the 11 clones of *L. minuta*, *L. valdiviana*, and *L. yungensis*, i.e. all species of *L.* section *Uninerves*, showed a clear separation of *L. minuta* from the other two species, while *L. valdiviana* and *L. yungensis* could not be separated into distinct clades (Fig. 11.1). This is in coherence with our species delimitation scenario B. This was also supported by the log-ML values as estimated by SNAPP. Scenario A had a mean log-ML value of −315.45 (sd 23.20) and scenario B had a mean log-ML value of −281.14 (sd 41.14). The Kruskal-Wallis test revealed that the log-ML value for scenario B is significantly higher than that of scenario A ($X^2 = 4.8$, df = 1, $P$-value = 0.03), additionally supporting the synonymisation of *L. valdiviana* and *L. yungensis*.

In order to analyse the position of *L.* section *Uninerves* within the genus *Lemna*, the data were analysed in a tree, with *L. punctata* and *S. polyrhiza* as outgroups (Fig. 11.2). Each species was covered by only one clone outside of *L.* section *Uninerves*, which was separated with high posterior probability from all other clades. The conspicuous long branch lengths confirm the special position of this section within the genus. Within the section, the differences between the three species were so small that hardly any distinction in this rooted tree could be visualised (cf. scales in Figs. 11.1 and 11.2). With one exception (one subgroup of *L.* section *Lemna*), all other nodes had a high posterior probability support of 1. The main clades equal the common categorisation in sections, i.e. besides *L.* section *Uninerves*, the sections *L.* section *Alatae*, *L.* section *Biformes*, and *L.* section *Lemna*.

**Table 11.2** Read and mapping statistics of the GBS approach for the investigated duckweed clones

| Section | Clone | Read statistics | | Mapping | |
|---|---|---|---|---|---|
| | | Number of reads | Number of clean reads | Number of mapped reads | Mapping rate (%) |
| Outgroup | *L. punctata* 9589 | 815,740 | 753,242 | 60,962 | 8.1 |
| | *S. polyrhiza* 7498 | 1,100,768 | 1,057,872 | 75,405 | 7.1 |
| *Alatae* | *L. aequinoctialis* 6746 | 563,781 | 554,631 | 72,442 | 13.1 |
| | *L. perpusilla* 8539 | 702,751 | 694,346 | 87,530 | 12.1 |
| *Biformes* | *L. tenera* 9020 | 697,572 | 687,127 | 93,949 | 13.7 |
| *Lemna* | *L. disperma* 7269 | 748,863 | 743,626 | 327,188 | 44.0 |
| | *L. gibba* 8682 | 377,355 | 373,654 | 143,638 | 38.4 |
| | *L. japonica* 8695 | 601,316 | 593,021 | 352,401 | 59.4 |
| | *L. minor* 5500 | 1,097,006 | 1,040,068 | 780,792 | 75.1 |
| | *L. obscura* 9342 | 611,875 | 608,713 | 247,506 | 40.7 |
| | *L. trisulca* 9529 | 1,034,470 | 1,006,162 | 219,335 | 21.8 |
| | *L. turionifera* 9434 | 693,417 | 688,584 | 415,229 | 60.3 |
| *Uninerves* | *L. minuta* 6717 | 1,947,091 | 1,910,447 | 191,725 | 10.0 |
| | *L. minuta* 7612 | 2,449,941 | 2,319,265 | 17,039 | 0.77 |
| | *L. minuta* 8699 | 620,974 | 617,616 | 83,777 | 13.6 |
| | *L. minuta* 9484 | 153,9211 | 1,524,385 | 167,027 | 11.0 |
| | *L. valdiviana* 9228 | 1,231,914 | 1,218,381 | 124,613 | 10.2 |
| | *L. valdiviana* 9401 | 956,734 | 949,995 | 108,176 | 11.4 |
| | *L. valdiviana* 9442 | 537,565 | 528,903 | 67,288 | 12.7 |
| | *L. yungensis* 9207 | 518,298 | 512,585 | 53,931 | 10.5 |
| | *L. yungensis* 9208 | 448,269 | 433,653 | 50,283 | 11.6 |
| | *L. yungensis* 9209 | 846,404 | 817,976 | 50,450 | 6.2 |
| | *L. yungensis* 9210 | 1,241,086 | 1,210,575 | 105,112 | 8.7 |

## 11.4 Discussion

Genotyping-by-sequencing separated *L.* section *Uninerves* from all other species of the genus *Lemna* with high probability and demonstrated its character as monophyletic taxa. This is in agreement with the results of previous studies that were based on morphological, phytochemical, plastidic, and genomic markers (Bog et al. 2010; Les et al. 1997, 2002; Tippery et al. 2015).

Within *L.* section *Uninerves*, all clones of the species *L. minuta* were distinguished from *L. valdiviana* and *L. yungensis*. In contrast, the species *L. valdiviana* and *L. yungensis* did not form separate clades. This demonstrates the capacity of the method GBS, which additionally characterises intraspecific differences, i.e. even clones of the same species were characterised and separated. From this, we would conclude that *L. valdiviana* and *L. yungensis* do not represent independent species. We therefore consider

**Table 11.3** Number of pairwise SNP differences for the investigated duckweed clones. All ambiguous positions were removed for each SNP pair. There were a total of 2233 SNPs in the final dataset. The number of different SNPs within a section is marked in grey. Minimum (○) and maximum (●) number of SNPs within sections are highlighted, where applicable

| Section | Clone | *La. punctata* 9589 | *S. polyrhiza* 7498 | *L. aequinoctialis* 6746 | *L. perpusilla* 8539 | *L. tenera* 9020 | *L. disperma* 7269 | *L. gibba* 8682 | *L. japonica* 8695 | *L. minor* 5500 | *L. obscura* 9342 | *L. trisulca* 9529 | *L. turionifera* 9434 | *L. minuta* 6717 | *L. minuta* 7612 | *L. minuta* 8699 | *L. minuta* 9484 | *L. valdiviana* 9228 | *L. valdiviana* 9401 | *L. valdiviana* 9442 | *L. yungensis* 9207 | *L. yungensis* 9208 | *L. yungensis* 9209 |
|---|---|---|---|---|---|---|---|---|---|---|---|---|---|---|---|---|---|---|---|---|---|---|---|
| Outgroup | *S. polyrhiza* 7498 | 44 | | | | | | | | | | | | | | | | | | | | | |
| *Alatae* | *L. aequinoctialis* 6746 | 90 | 227 | | | | | | | | | | | | | | | | | | | | |
| | *L. perpusilla* 8539 | 86 | 215 | 6 | | | | | | | | | | | | | | | | | | | |
| *Biformes* | *L. tenera* 9020 | 53 | 74 | 78 | 75 | | | | | | | | | | | | | | | | | | |
| *Lemna* | *L. disperma* 7269 | 70 | 178 | 239 | 211 | 129 | | | | | | | | | | | | | | | | | |
| | *L. gibba* 8682 | 98 | 225 | 312 | 288 | 127 | 124 | | | | | | | | | | | | | | | | |
| | *L. japonica* 8695 | 80 | 161 | 253 | 220 | 124 | 193 | 256 | | | | | | | | | | | | | | | |
| | *L. minor* 5500 | 57 | 127 | 192 | 168 | 104 | 163 | 201 | 18○ | | | | | | | | | | | | | | |
| | *L. obscura* 9342 | 93 | 205 | 234 | 215 | 98 | 235 | 264 | 169 | 155 | | | | | | | | | | | | | |
| | *L. trisulca* 9529 | 69 | 146 | 199 | 179 | 98 | 161 | 209 | 89 | 76 | 147 | | | | | | | | | | | | |
| | *L. turionifdera* 9434 | 82 | 157 | 243 | 214 | 117 | 208 | 265● | 38 | 22 | 181 | 100 | | | | | | | | | | | |
| *Uninerves* | *L. minuta* 6717 | 144 | 243 | 330 | 299 | 140 | 544 | 630 | 666 | 547 | 554 | 519 | 641 | | | | | | | | | | |
| | *L. minuta* 7612 | 119 | 202 | 258 | 236 | 119 | 433 | 505 | 539 | 455 | 456 | 429 | 516 | 3 | | | | | | | | | |
| | *L. minuta* 8699 | 161 | 293 | 353 | 320 | 155 | 607 | 711 | 746 | 622 | 624 | 595 | 709 | 3 | 9 | | | | | | | | |
| | *L. minuta* 9484 | 170 | 325 | 388 | 352 | 162 | 648 | 757 | 782 | 644 | 670 | 625 | 743 | 2 | 7 | 3 | | | | | | | |
| | *L. valdiviana* 9228 | 165 | 317 | 382 | 346 | 159 | 645 | 755 | 782 | 645 | 673 | 628 | 744 | 23 | 19 | 27 | 25 | | | | | | |
| | *L. valdiviana* 9401 | 153 | 301 | 360 | 326 | 151 | 596 | 697 | 733 | 602 | 624 | 583 | 703 | 28 | 24 | 32● | 31 | 13 | | | | | |
| | *L. valdiviana* 9442 | 158 | 303 | 354 | 324 | 150 | 583 | 686 | 720 | 585 | 603 | 571 | 682 | 25 | 24 | 28 | 28 | 6 | 11 | | | | |
| | *L. yungensis* 9207 | 155 | 301 | 350 | 318 | 153 | 601 | 701 | 733 | 597 | 620 | 578 | 689 | 21 | 17 | 27 | 24 | 1○ | 12 | 7 | | | |
| | *L. yungensis* 9208 | 158 | 305 | 354 | 322 | 149 | 576 | 687 | 706 | 583 | 609 | 566 | 676 | 22 | 19 | 25 | 23 | 2 | 14 | 8 | 3 | | |
| | *L. yungensis* 9209 | 158 | 304 | 366 | 334 | 154 | 601 | 704 | 734 | 605 | 633 | 587 | 705 | 25 | 23 | 30 | 29 | 12 | 16 | 12 | 13 | 11 | |
| | *L. yungensis* 9210 | 166 | 317 | 377 | 344 | 159 | 633 | 749 | 766 | 630 | 656 | 614 | 732 | 25 | 22 | 30 | 29 | 14 | 16 | 14 | 14 | 11 | 1○ |

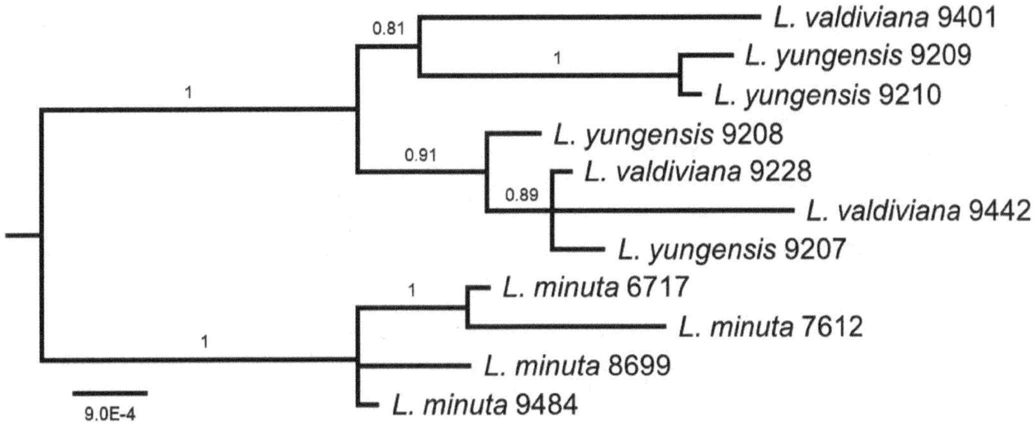


**Fig. 11.1** Bayesian majority-rule consensus tree for the 11 investigated clones of *L.* section *Uninerves* based on 2233 SNP loci as found by a genotyping-by-sequencing approach using two restriction enzymes. Posterior probabilities are given on the branches

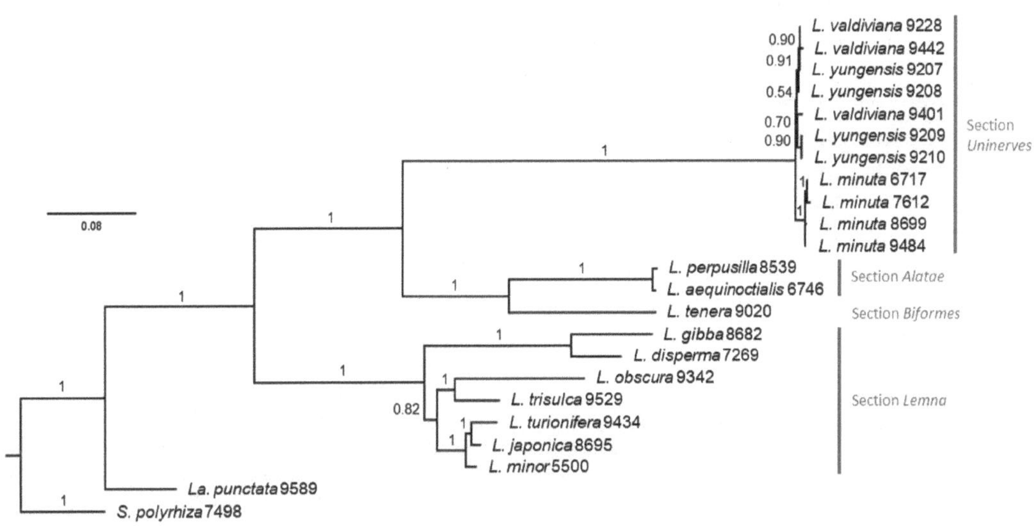


**Fig. 11.2** Bayesian majority-rule consensus tree for the 21 investigated clones of the genus *Lemna* based on 2233 SNP loci as found by a genotyping-by-sequencing approach using two restriction enzymes. Posterior probabilities are given on the branches. *Spirodela polyrhiza* and *Landoltia punctata* were used as outgroup species

*L. yungensis* as an ecotype of *L. valdiviana*. As *L. valdiviana* Phil. is the older name (Philippi 1864), we are suggesting that the name *L. yungensis* Landolt should be synonymised to *L. valdiviana*. This conclusion can be drawn on the basis of the powerful method of GBS and is also in agreement with the results of AFLP (Bog et al. 2010) and other molecular methods (Bog et al. 2020). This decision decreases the number of accepted species of Lemnaceae from 37 (Sree et al. 2016) to 36.

The special position of *L.* section *Uninerves* within the genus was also supported by very long branch lengths. This hints towards the point that the species of this section had separated from those of other sections for long time—in agreement with the previous conclusions of Landolt (1986) and Les et al. (2002). The alternative explanation is that the mutation rates in the species of this section are higher.

It is remarkable that the internal structure of the genus *Lemna* evaluated by GBS is almost identical to the structure evaluated on the basis of morphological and phytochemical markers (Les et al. 1997). All four sections were revealed as monophyletic taxa and evaluated by very high posterior probabilities. This confirms the conclusion of Les et al. (2002) which was drawn on the basis of different PCR fragments. The high resolution of the GBS markers in the genus *Lemna* also indicates that the section levels should be revised (Fig. 11.2). For example, to merge *L.* section *Alatae* and *L.* section *Biformes* or to separate *L.* section *Lemna* into several subsections. As the number of clones investigated per species in the present study remains small, the diversity within and between species might have some gaps. By the same reasons, it cannot be decided whether all ten species of the genus *Lemna* outside of *L.* section *Uninerves* should be considered as independent species. Therefore, to finalise the revision of the sections, we are expanding our current study by including more clones per species.

**Acknowledgements** The technical assistance of Susanne König, Ines Walde, Sandra Drießlein, and Christa Fricke (all IPK Gatersleben, Germany) is highly acknowledged. We also thank Ingo Schubert from IPK Gatersleben, who had provided access to the molecular methods and Walter Laemmler, Landolt Duckweed Collection, Zürich, for providing some clones.

## References

Appenroth KJ, Teller S, Horn M (1996) Photophysiology of turion formation and germination in *Spirodela polyrhiza*. Biol Plant 38:95–106. https://doi.org/10.1007/BF02879642

Bog M, Baumbach H, Schween U, Hellwig F, Landolt E, Appenroth KJ (2010) Genetic structure of the genus *Lemna* L. (Lemnaceae) as revealed by amplified fragment length polymorphism. Planta 232:609–619. https://doi.org/10.1007/s00425-010-1201-2

Bog M, Sree KS, Fuchs J, Hoang PTN, Schubert I, Kuever J, Rabenstein A, Paolacci S, Jansen MAK, Appenroth KJ (2020) A taxonomic revision of *Lemna* sect. *Uninerves* (Lemnaceae). Taxon (in press)

Borisjuk N, Chu P, Gutierrez R, Zhang H, Acosta K, Friesen N, Sree KS, Garcia C, Appenroth KJ, Lam E (2015) Assessment, validation and deployment strategy of a two-barcode protocol for facile genotyping of duckweed species. Plant Biol 17(Suppl. 1):42–49. https://doi.org/10.1111/plb.12229

Bouckaert R, Heled J, Kühnert D, Vaughan T, Wu CH, Xie D, Suchard MA, Rambaut A, Drummond AJ (2014) BEAST 2: a software platform for Bayesian evolutionary analysis. PLoS Comput Biol 10(4):e1003537. https://doi.org/10.1371/journal.pcbi.1003537

Bryant D, Bouckaert R, Felsenstein J, Rosenberg NA, RoyChoudhury A (2012) Inferring species trees directly from biallelic genetic markers: bypassing gene trees in a full coalescent analysis. Mol Biol Evol 29(8):1917–1932. https://doi.org/10.1093/molbev/mss086

Crawford DJ, Landolt E, Les DH (1996) An allozyme study of two sibling species of *Lemna* (Lemnaceae) with comments on their morphology, ecology and distribution. Bull Torrey Bot Club 123(1):1–6. https://doi.org/10.2307/2996300

Danecek P, Auton A, Abecasis G, Albers CA, Banks E, DePristo MA, Handsaker RE, Lunter G, Marth GT, Sherry ST, McVean G, Durbin R, 1000 Genomes Project Analysis Group (2011) The variant call format and VCFtools. Bioinformatics 27(15):2156–2158. https://doi.org/10.1093/bioinformatics/btr330

Darriba D, Taboada GL, Doallo R, Posada D (2012) jModelTest 2: more models, new heuristics and parallel computing. Nat Methods 9:772. https://doi.org/10.1038/nmeth.2109

Elshire RJ, Glaubitz JC, Sun Q, Poland JA, Kawamoto K, Buckler ES, Mitchell SE (2011) A robust, simple genotyping-by-sequencing (GBS) approach for high diversity species. PLoS ONE 6(5):e19379. https://doi.org/10.1371/journal.pone.0019379

Huelsenbeck JP, Ronquist F (2001) MrBayes: Bayesian inference of phylogenetic trees. Bioinformatics 17(8):754–755. https://doi.org/10.1093/bioinformatics/17.8.754

Landolt E (1986) The family of Lemnaceae—a monographic study. Vol. 1. Biosystematic Investigations in the Family of Duckweeds (Lemnaceae). Zürich: Stiftung Ruebel

Landolt E (1998) *Lemna yungensis*, a new duckweed species from rocks of the Andean Yungas in Bolivia. Bullet Geobotanical Inst ETH 64:15–21

Les DH, Crawford DJ, Landolt E, Gabel JD, Kimball RT (2002) Phylogeny and systematics of Lemnaceae, the duckweed family. Syst Bot 27(2):221–240

Les DH, Landolt E, Crawford DJ (1997) Systematics of the *Lemnaceae* (duckweeds): inferences from micromolecular and morphological data. Plant Syst Evol 204:161–177

Li H, Durbin R (2009) Fast and accurate short read alignment with Burrows–Wheeler transform. Bioinformatics 25(14):1754–1760. https://doi.org/10.1093/bioinformatics/btp324

Li H, Handsaker B, Wysoker A, Fennell T, Ruan J, Homer N, Marth G, Abecasis G, Durbin R, 1000 Genome Project Data Processing Subgroup (2009) The sequence alignment/map (SAM) format and SAMtools. Bioinformatics 25(16):2078–2079. https://doi.org/10.1093/bioinformatics/btp352

McKenna A, Hanna M, Banks E, Sivachenko A, Cibulskis K, Kernytsky A, Garimella K, Altschuler D, Gabriel S, Daly M, DePristo MA (2010) The genome analysis toolkit: a MapReduce framework for analyzing next-generation DNA sequencing data. Genome Res 20:1297–1303. https://doi.org/10.1101/gr.107524.110

Philippi RA (1864) Linnaea—Ein Journal für die Botanik in ihrem ganzen Umfange, vol 33. DFL von Schlechtendal, Halle, p 239

Poland JA, Brown PJ, Sorrells ME, Jannink JL (2012) Development of high-density genetic maps for barley and wheat using a novel two-enzyme genotyping-by-sequencing approach. PLoS ONE 7(2):e32253. https://doi.org/10.1371/journal.pone.0032253

R Core Team (2017) R: a language and environment for statistical computing. R Foundation for Statistical Computing, Vienna, Austria. https://www.R-project.org/

Schubert M, Lindgreen S, Orlando L (2016) AdapterRemoval v2: rapid adapter trimming, identification, and read merging. BMC Res Notes 9:88. https://doi.org/10.1186/s13104-016-1900-2

Sree KS, Bog M, Appenroth KJ (2016) Taxonomy of duckweeds (Lemnaceae), potential new crop plants. Emirates J Food Agric 28(5):291–302. https://doi.org/10.9755/ejfa.2016-01-038

Tippery NP, Les DH, Crawford DJ (2015) Evaluation of phylogenetic relationships in Lemnaceae using nuclear ribosomal data. Plant Biol 17(Suppl1):50–58. https://doi.org/10.1111/plb.12203

Valledor L, Escandón M, Meijón M, Nukarinen E, Canal MJ, Weckwerth W (2014) A universal protocol for the combined isolation of metabolites, DNA, long RNAs, small RNAs, and proteins from plants and microorganisms. Plant J 79:173–180. https://doi.org/10.1111/tpj.12546

Van Hoeck A, Horemans N, Monsieurs P, Cao HX, Vandenhove H, Blust R (2015) The first draft genome of the aquatic model plant *Lemna minor* opens the route for future stress physiology research and biotechnological applications. Biotechnol Biofuels 8:188. https://doi.org/10.1186/s13068-015-0381-1

Wang W, Wu Y, Yan Y, Ermakova M, Kerstetter R, Messing J (2010) DNA barcoding of the Lemnaceae, a family of aquatic monocots. BMC Plant Biol 10:205–214. https://doi.org/10.1186/1471-2229-10-205

Wendler N, Mascher M, Nöh C, Himmelbach A, Scholz U, Ruge-Wehling B, Stein N (2014) Unlocking the secondary gene-pool of barley with next-generation sequencing. Plant Biotechnol J 12(8):1122–1131. https://doi.org/10.1111/pbi.12219

# 12 The Transcriptome in *Landoltia punctata*

Yang Fang, Anping Du, Li Tan, Kaize He, Yanling Jin, Yanqiang Ding, Lin Guo and Hai Zhao

**Abstract**

Research on *Lemnaceae* (duckweeds) is hot spot recently due to its application potential in bioenergy production and phytoremediation. Among the five genera of duckweed, *Landoltia punctata* has great potential in starch production and heavy metal bio-extraction. Although starch accumulation and heavy metal absorption by *L. punctata* have been studied at biochemical and physiological levels for many years, the underlying mechanism has not been subjected to omics analysis until recently. Transcriptomics plays an important role in understanding gene expression regulation as response to changing environment. Several studies have been carried out to investigate the transcriptomic expression profiles of *L. punctata* to illustrate the mechanisms of sugar high starch content biosynthesis, cadmium hyperaccumulation, and flavonoids biosynthesis. Here, we reviewed the progress in transcriptome analysis of *L. punctata* to set framework and give the readers insights into the current status and future perspectives in researches and application potential of *L. punctata*.

## 12.1 Introduction

*Lemnaceae* (duckweed) is the smallest and fastest-growing aquatic flowering monocotyledonous plant in the world. It spreads a broad range of climates with 5 genera (*Spirodela*, *Landoltia*, *Lemna*, *Wolffiella*, and *Wolffia*), comprising 37 species around the world (Cao et al. 2014). It is able to reach a very high biomass yield (55 tons/ha/year dry weight) due to its asexual reproduction and rapid propagation (Zhao et al. 2012). Duckweed has been gained increasing attention due to its application potential in starch production, protein production, and phytoremediation.

Genus *Landoltia* is one of the widely distributed and practically applied duckweeds. It consists of one species, namely *Landoltia punctata*. Previously, *L. punctata* was a member of genus *Spirodela*, with old nomenclatures of *Spirodela oligorrhiza* and *Spirodela punctata* (Sree et al. 2016). It was recognized as a new genus since 1999 by Les. This generic name

Y. Fang · A. Du · L. Tan · K. He · Y. Jin · Y. Ding · L. Guo · H. Zhao (✉)
Key Laboratory of Environmental and Applied Microbiology, Chengdu Institute of Biology, Chinese Academy of Sciences, Chengdu 610041, China
e-mail: zhaohai@cib.ac.cn

Y. Fang · A. Du · L. Tan · K. He · Y. Jin · Y. Ding · L. Guo · H. Zhao
Environmental Microbiology Key Laboratory of Sichuan Province, Chengdu 610041, China

Y. Ding
University of Chinese Academy of Sciences, Beijing 100049, China


X. H. Cao et al. (eds.), *The Duckweed Genomes*, Compendium of Plant Genomes,
https://doi.org/10.1007/978-3-030-11045-1_12

*Landoltia* commemorates Elias Landoltia for his outstanding contributions to systematics and biology of *Lemnaceae*. *L. punctata* possesses 2–7 roots per frond, perforating the prophyllum. Its frond is ovate to lanceolate with 3–7 veins, and a red dorsal surface is often observed. Unlike *Spirodela*, *Landoltia* is turion-absent (Les and Crawford 1999).

The most application field of *L. punctata* is starch production. In 1970, Reid and Bieleski reported that *L. punctata* accumulated starch content to approximately 30% (dry weight) in 30 days cultivation on phosphorus-deficient complete mineral nutrient medium. The starch content was sharply increased to 75% (dry weight) with the presence of glucose (1%) in the medium (Reid and Bieleski 1970). In our previous studies, we evaluated various approaches, like nutrient starvation, uniconazole supplementation, and heavy metal (cobalt and nickel) stress, to induce starch accumulation by *L. punctata* (Tao et al. 2013; Liu et al. 2015b; Guo et al. 2017). The three approaches evaluated induced starch accumulation and resulted in approximately 50% (dry weight) of starch content in 10 days cultivation. Phytoremediation is another important application field for *L. punctata*. *L. punctata* is able to uptake nitrogen (N) and phosphorus (P) from water very quickly, even under relatively low N/P concentration. It is widely used to purify wastewater and eutrophic water bodies (Fang et al. 2007). Besides, *L. punctata* is resistant to heavy metals to some extent and is able to accumulate heavy metals like cadmium, cobalt, nickel, lead, uranium, and silver (Guo et al. 2017; Nie et al. 2016; Stegemeier et al. 2017; Tang et al. 2017; Fang et al. 2007). Also, it is reported that *L. punctata* is potential for phytoremediation on petroleum hydrocarbons (Ertekin et al. 2015). *L. punctata* is rich in flavonoids (Wang et al. 2014) and is used in traditional Chinese medicine. It is also potential for pharmaceutical drugs.

To further release and improve the potential applications, it is critical to interpret the physiological mechanism on how *L. punctata* response to changing environment. General central dogma states as "DNA makes RNA and RNA makes protein". RNA sequencing plays an important role in understanding gene expression regulation, and transcriptome analysis is able to reflect the global regulation. With the decreasing cost of next-generation sequencing, transcriptome analysis is accessible for individual laboratory and a few duckweed transcriptome studies were carried out in recent years (Table 12.1). Applying this deep-sequencing technology will set framework and stimulate novel potential of duckweeds. In this chapter, we introduced the primary researches on *L. punctata* and its transcriptome analysis. This will give the readers insights into the current status and future perspectives in researches and application potential of *L. punctata*.

## 12.2 Starch Production and Transcriptome Analysis

*L. punctata* is a potential bioenergy crop with high starch productivity and low lignin content. Several transcriptome studies in *L. punctata* had been carried out for understanding the mechanism of high starch content and low lignin content under abiotic stress.

The comparative transcriptome analysis was conducted to reveal the mechanism of high starch accumulation of *L. punctata* 0202 under nutrient starvation. *L. punctata* 0202 was transferred from nutrient-rich solution to distilled water and sampled in time course. Physiological measurements revealed that the activity of the key enzyme of starch biosynthesis, ADP-glucose pyrophosphorylase (AGPase), as well as the starch content increased continuously in *L. punctata* 0202 under nutrient starvation condition. Samples harvested at 0, 2, and 24 h were used for RNA-Seq, respectively. A comprehensive transcriptome, containing 74,797 contigs, was obtained by a de novo assembly of the RNA-Seq reads. Gene expression profiling showed that transcripts encoding key enzymes responsible for starch biosynthesis were up-regulated. Inversely, the expression of transcripts encoding enzymes involved in starch consumption and some photosynthesis-related transcripts were down-regulated. Specifically,

**Table 12.1** Summary of transcriptome studies in duckweeds

| Species | Condition | Time points | Main findings | References |
|---|---|---|---|---|
| *Landoltia punctata* 0202 | Nutrient starvation | 0, 2 and 24 h | Nutrient starvation down-regulated the global metabolic status, redirects metabolic flux of fixed $CO_2$ into starch synthesis branch in *L. punctata* | Tao et al. (2013) |
| *Spirodela polyrhiza* 7498 | ABA treatment | 3 days | Generated a model summarizing the signal transduction leading to *Spirodela* dormancy by comparing the transcriptome between fronds and developing turions | Wang et al. (2014) |
| *Landoltia punctata* 0202 | Uniconazole treatment | 0, 2, 5, 72, and 240 h | Uniconazole treatment altered endogenous hormone levels and enhanced chlorophyll content and net photosynthetic rate by regulating key enzymes involved in endogenous hormone and chlorophyll biosynthesis | Liu et al. (2015a) |
| *Landoltia punctata* 0202 | Uniconazole treatment | 0, 2, 5, 72, and 240 h | Uniconazole treatment altered endogenous hormone levels, thus resulted in starch accumulation | Liu et al. (2015b) |
| *Lemna minor* | Ammonium treatment | 7 days | The antioxidant enzyme system was activated under $NH4^+$ toxicity for ROS scavenging. The increased lignin biosynthesis might play an important role in $NH4^+$ toxicity resistance | Wang et al. (2016) |
| *Landoltia punctata* 0202 | NS, UT, FN | NS (0, 2, 24 h); UT (0, 2, 5, 72, 240 h); FN (2, 5, 72, 240 h) | Nutrient starvation is the best option to obtain high starch and flavonoid accumulation simultaneously in a short time | Tao et al. (2017) |
| *Lemna minor* | Ionising radiation | 7 days | Duckweeds shift from acclimation responses toward survival responses at increasing dose rates of ionising radiation | Van Hoeck et al. (2017) |
| *Lemna aequinoctialis* 6000 | Nitrogen starvation (applied sucrose) | 0, 3, and 7 days | Nitrogen starvation increased ADP-glucose and starch contents by regulating the gluconeogenesis and TCA pathways and lipids and pectin biosynthesis | Yu et al. (2017) |
| *Landoltia punctata* 6001 | $CdCl_2$ treatment | 0,1, 3, 6 days | Cd stress could affect the processes from DNA to protein metabolism, metabolic flux of carbohydrate, sulfur and ROS metabolism, and tonoplast transporter expression to cope with Cd cytotoxicity, tolerance, and detoxification | Xu et al. (2018) |

*Note NS* Nutrient starvation; *UT* Uniconazole treatment; *FN* Full nutrient (1/6 Hoagland solution)

the transcripts of some transporters were up-regulated within the first 2 h. Additionally, the expression of most transcripts encoding key enzymes involved in flavonoid biosynthesis was up-regulated drastically regardless of starvation, while the expression of the last rate-limiting enzyme of lignification, laccase, presented very low in all the three samples. Furthermore, the expression of genes involved in starch biosynthesis was confirmed by quantitative reverse transcription PCR (qRT-PCR) in *L. punctata* 0202. Another independent research on qRT-PCR assay for the expression of key starch biosynthesis enzymes, including AGPase, soluble starch synthase (SSS), starch degradation enzymes (SDE), and alpha- and beta-amylase, under nitrogen (N) and phosphorus (P) deficiency condition was consistent with this RNA-Seq data (Zhao et al. 2015). Those studies provided a comprehensive transcriptome analysis of *L. punctata* 0202 under nutrient starvation, which indicated that nutrient starvation down-regulated the global metabolic status and redirected metabolic flux of fixed $CO_2$ to starch biosynthesis pathway. It provided a valuable genomic resource for duckweed and paved the way for further molecular biological studies and the application of duckweed as a bioenergy crop (Tao et al. 2013).

Not only nutrient starvation but also uniconazole can improve the starch content. Uniconazole is a plant growth retardant, which can increase starch and biomass accumulation of *L. punctata* simultaneously under eutrophic conditions. The result of transcriptome sequencing of uniconazole application on fronds of *L. punctata* 0202 revealed the expression of genes involved in endogenous hormones and chlorophyll biosynthesis pathway changed responsively. The dry weight following the uniconazole treatment increased by 10% compared to the controls at 240 h, resulting from the uniconazole affecting endogenous hormones content, chlorophyll content, and the net photosynthetic rate. The dry weight starch content increased up to 48% compared to 15.7% in the control group after 240 h growth. Transcriptome sequencing revealed that the expression of regulatory elements of hormone signaling pathways that are involved in chlorophyll and starch metabolism changed correspondingly. Importantly, the expression of key enzymes responsible for starch biosynthesis was up-regulated, and transcript-encoding enzymes involved in starch degradation and other carbohydrate metabolic branches were down-regulated (Liu et al. 2015a, b).

According to current research on *L. punctata* 0202 starch accumulation, starch content could reach up to approximately 45% (dry weight) within 7 days under nutrient starvation condition (Tao et al. 2013; Huang et al. 2014) and approximately 48% (dry weight) within 10 days under uniconazole treatment (Liu et al. 2015a, b). A combined treatment of nutrient starvation and uniconazole application or others to *L. punctata* 0202 is considered a promising method to further improve the starch content. To support this, more tests should be carried out to verify the hypothesis in the future. According to the latest study, the starch content can reach 60% under nitrogen starvation in the presence of exogenously applied sucrose condition in *Lemna aequinoctialis* 6000 after treatment for 9 days. In this study, a combined transcriptome and metabolites analysis was carried out for metabolic flux in starch accumulation. The researchers evaluated expression of the genes involved in nitrogen metabolism, protein and amino acid metabolism, starch and sucrose metabolism, and lipid metabolism by sampling 7-day in time course under nitrogen starvation condition. The expression of genes encoding nitrate reductase, glutamine synthetase, and glutamate synthase was down-regulated; the expression of genes encoding enzymes involved in gluconeogenesis was up-regulated, whereas the majority of unigenes involved in glycolysis were down-regulated. The metabolome analysis revealed that more ADP-Glc was accumulated and lower levels of UDP-Glc were accumulated, which was consistent with the transcriptome results. The activity of AGPase involved in starch biosynthesis was significantly increased while the activity of UGPase was dramatically decreased. This study serves as an excellent candidate for functional transcriptome study and metabolic engineering to improve the production of next-generation biofuels in duckweeds (Yu et al. 2017).

## 12.3 Heavy Metal Accumulation and Transcriptome Analysis

Except for transcriptome analysis for starch accumulation, the gene expression response to cadmium stress in *L. punctata* 6001 was analyzed via RNA-Seq technique by Xu et al. *L. punctata* 6001, which is considered a promising candidate for Cd phytoremediation, was isolated by large-scale screening of over 200 duckweed clones. To understand the molecular mechanisms of Cd hyperaccumulation, a comprehensive transcriptome analysis was performed by RNA-Seq for samples treated with 20 μM $CdCl_2$ for 0, 1, 3, and 6 days. Xu et al. revealed that genes involved in DNA repair acted as an early response to Cd, and RNA and protein metabolism would be likely to respond as well. Furthermore, the carbohydrate metabolic flux tended to be modulated in response to Cd stress. Up-regulated genes involved in sulfur and reactive oxygen species (ROS) metabolism might contribute to Cd tolerance. Vacuolar sequestration most likely played an important role in Cd detoxification in *L. punctata* 6001. The novel findings provided important clues for molecular-assisted screening and breeding of Cd hyper-accumulating cultivars for phytoremediation (Xu et al. 2018).

## 12.4 Flavonoids Accumulation and Transcriptome Analysis

Flavonoids belong to phenolic compounds and are widely existed secondary metabolites in plants. It can be potentially exploited in the food and drug. The flavonoid content of duckweed can reach up to approximately 5.56%, and at least, 20 flavonoid compounds were found in duckweed (Tao et al. 2017).

Nutrient starvation also triggers high flavonoid accumulation in *L. punctata* 0202. A combined omics study was performed to investigate the biosynthesis of flavonoid and the metabolic flux changes in *L. punctata* 0202 grown in different culture media. To understand the global flavonoid and starch metabolite-related molecular response to nutrient starvation, *L. punctata* 0202 was cultivated in nutrient starvation, uniconazole treatment, and full nutrient, respectively. RNA-Seq analysis for three groups sample was carried out, and the proteome data obtained from iTRAQ-LC-MS/MS technology of previous studies was re-analyzed using the new transcriptome data as a reference database. The abundance of the most detected flavonoid-related proteins, including the phenylalanine ammonia-lyase (PAL), cinnamate 4-hydroxylase (C4H), 4-hydroxycinnamoyl-CoA ligase (4CL), chalcone synthase (CHS), chalcone isomerase (CHI), flavanone 3-hydroxylase (F3H), and anthocyanidin synthase (ANS) were detected to be improved in *L. punctata* 0202 when grown in distilled water. The proteome data was consistent with the transcriptome data in this study. Flavonoid content was measured, and purple coloration accumulation which is regarded as anthocyanin was observed on the dorsal fronds. The metabolome and morphology further verified the results of transcriptome and proteome analysis (Tao et al. 2017).

## 12.5 Other Transcriptomes Analysis

Several transcriptome studies in other duckweed species were also reported currently. *Lemna minor* can grow well in the high $NH_4^+$ environment but to some extent can also suffer toxic effects. To study the toxicity and tolerance of $NH_4^+$, the transcriptome study using RNA-Seq was reported in *L. minor*. The *L. minor* was cultured in the Hoagland solution of control and treated with 84 mg/L $NH_4^+$ and 840 mg/L $NH_4^+$. RNA-Seq generated 6.62 G nucleotides from the three distinct libraries. Bioinformatic analysis identified 70,728 unigenes and 14,207 differentially expressed genes (DEGs), most of which were down-regulated under $NH_4^+$ toxicity. Lignin biosynthesis-related genes in the phenylpropanoid biosynthesis pathway were up-regulated to enhance $NH_4^+$ toxicity resistance. The accumulation of ROS induced by $NH_4^+$ toxicity can cause oxidative damage leading to cell death in *L. minor*. The antioxidant enzyme system was also activated to scavenge

ROS and reduce the toxicity. The transcriptomic and physiological research of *L. minor* responding to high $NH_4^+$ may provide us a better understanding not only of toxic processed but also tolerance mechanisms (Wang et al. 2016).

Ionising radiation (IR) in the environment is considered harmful to plants and animals when conferring extremely high dose rates. To better understand the physiological response to plant exposure to ionising radiation, RNA-seq in *L. minor* treated ionising radiation for seven days in a dose rate-dependent manner was carried out. The gene expression data revealed that *L. minor* plants exposed at lower dose rates can tolerate the exposure by triggering acclimation responses. In contrast, the genes related to antioxidative defense systems in terms of DNA repair and cell cycle were high expressed at the highest dose rate. It indicated that plants can shift from acclimation responses toward survival responses at increasing dose rates of ionising radiation. Importantly, the photosynthetic process seems to be unaffected in *L. minor* plants among the tested dose rates (Van Hoeck et al. 2017).

## 12.6 Conclusion

In summary, in the absence of genomic data, RNA-seq of *L. punctata* was powerful in uncovering molecular mechanism under different treatments or stresses. Especially, the transcriptome of starch metabolism mapped the transcriptional profiles of high starch, high flavonoids, and low lignin metabolism pathways and revealed the molecular mechanisms of high starch accumulation under the particular conditions in duckweed. It provides new ideas for the study of gene regulation and genetic manipulation of starch metabolism in plants.

## References

Cao HX, Vu GT, Wang W et al (2014) Chromatin organisation in duckweed interphase nuclei in relation to the nuclear DNA content. Plant Biol (Stuttg) 17 (Suppl 1):120–124. https://doi.org/10.1111/plb.12194

Ertekin O, Kosesakal T, Unlu VS et al (2015) Phytoremediation potential of *Landoltia punctata* on petroleum hydrocarbons. Turk J Bot 39(1):23–29. https://doi.org/10.3906/bot-1403-42

Fang Y, Babourina O, Rengel Z et al (2007) Ammonium and nitrate uptake by the floating plant *Landoltia punctata*. Ann Bot 99(2):365–370

Guo L, Ding Y, Xu Y et al (2017) Responses of *Landoltia punctata* to cobalt and nickel: removal, growth, photosynthesis, antioxidant system and starch metabolism. Aquat Toxicol 190:87

Huang M, Fang Y, Xiao Y et al (2014) Proteomic analysis to investigate the high starch accumulation of duckweed (*Landoltia punctata*) under nutrient starvation. Ind Crops Prod 59:299–308. https://doi.org/10.1016/j.indcrop.2014.05.029

Les DH, Crawford DJ (1999) Landoltia (Lemnaceae), a new genus of duckweeds. Novon J Bot Nomenclature 9(4):530–533

Liu Y, Fang Y, Huang M et al (2015a) Uniconazole-induced starch accumulation in the bioenergy crop duckweed (*Landoltia punctata*) I: transcriptome analysis of the effects of uniconazole on chlorophyll and endogenous hormone biosynthesis. Biotechnol Biofuels 8:57. https://doi.org/10.1186/s13068-015-0246-7

Liu Y, Fang Y, Huang M et al (2015b) Uniconazole-induced starch accumulation in the bioenergy crop duckweed (*Landoltia punctata*) II: transcriptome alterations of pathways involved in carbohydrate metabolism and endogenous hormone crosstalk. Biotechnol Biofuels 8:64. https://doi.org/10.1186/s13068-015-0245-8

Nie X, Dong F, Bian L et al (2016) Uranium binding on *Landoltia punctata* as a result of formation of insoluble nano-U (VI) and U (IV) phosphate minerals. ACS Sustain Chem Eng 5(2)

Reid MS, Bieleski RL (1970) Response of spirodela-oligorrhiza to phosphorus deficiency. Plant Physiol 46 (4):609. https://doi.org/10.1104/pp.46.4.609

Sree KS, Bog M, Appenroth KJ (2016) Taxonomy of duckweeds (Lemnaceae), potential new crop plants. Emirates J Food Agric 28(5):1

Stegemeier JP, Colman BP, Schwab F et al (2017) Uptake and distribution of silver in the aquatic plant *Landoltia punctata* (Duckweed) exposed to silver and silver sulfide nanoparticles. Environ Sci Technol 51(9)

Tang J, Li Y, Wang X et al (2017) Effective adsorption of aqueous $Pb^{2+}$ by dried biomass of *Landoltia punctata* and *Spirodela polyrhiza*. J Clean Prod 145:25–34

Tao X, Fang Y, Xiao Y et al (2013) Comparative transcriptome analysis to investigate the high starch accumulation of duckweed (*Landoltia punctata*) under nutrient starvation. Biotechnol Biofuels 6(1):72

Tao X, Fang Y, Huang MJ et al (2017) High flavonoid accompanied with high starch accumulation triggered by nutrient starvation in bioenergy crop duckweed (*Landoltia punctata*). BMC Genom 18(1):166. https://doi.org/10.1186/s12864-017-3559-z

Van Hoeck A, Horemans N, Nauts R et al (2017) Lemna minor plants chronically exposed to ionising radiation:

RNA-seq analysis indicates a dose rate dependent shift from acclimation to survival strategies. Plant Sci 257:84–95. https://doi.org/10.1016/j.plantsci.2017.01.010
Wang NN, Xu GB, Fang Y et al (2014) New flavanol and cycloartane glucosides from *Landoltia punctata*. Molecules 19(5):6623–6634. https://doi.org/10.3390/molecules19056623
Wang W, Li R, Zhu Q et al (2016) Transcriptomic and physiological analysis of common duckweed Lemna minor responses to $NH4^+$ toxicity. BMC Plant Biol 16:92. https://doi.org/10.1186/s12870-016-0774-8
Xu H, Yu C, Xia X et al (2018) Comparative transcriptome analysis of duckweed (*Landoltia punctata*) in response to cadmium provides insights into molecular mechanisms underlying hyperaccumulation. Chemosphere 190:154–165. https://doi.org/10.1016/j.chemosphere.2017.09.146
Yu C, Zhao X, Qi G et al (2017) Integrated analysis of transcriptome and metabolites reveals an essential role of metabolic flux in starch accumulation under nitrogen starvation in duckweed. Biotechnol Biofuels 10:167. https://doi.org/10.1186/s13068-017-0851-8
Zhao H, Appenroth K, Landesman L et al (2012) Duckweed rising at Chengdu: summary of the 1st international conference on duckweed application and research. Plant Mol Biol 78(6):627–632. https://doi.org/10.1007/s11103-012-9889-y
Zhao Z, Shi HJ, Wang ML et al (2015) Effect of nitrogen and phosphorus deficiency on transcriptional regulation of genes encoding key enzymes of starch metabolism in duckweed (*Landoltia punctata*). Plant Physiol Biochem 86:72–81. https://doi.org/10.1016/j.plaphy.2014.11.007

# Transcriptome Responses of *Spirodela polyrhiza*

# 13

Paul Fourounjian

**Abstract**

In order to analyze the transcriptome of any species, RNA-seq has become the gold standard and evolved into a variety of library preparations and sequencing platforms to study more than mRNA abundance. This chapter reviews the transcriptional studies of *Spirodela polyrhiza*, the best-characterized member of the Lemnaceae family in a genomic sense. To date, there have been three studies of its transcriptome. The first two analyzed ribosomal RNA depleted total RNA of fronds and fronds developing into turions after exposure to abscisic acid. The first study analyzed 154 down-regulated genes involved in growth and 208 upregulated genes involved in starch, anthocyanin production, and seed development. The second study found 66 sites where chloroplast mRNAs were edited to create a functional protein, supporting the hypothesis that mRNA editing was evolved once, and the conservation of editing sites was phylogenetically correlated. The third study, also performed in the 7498 ecotypes, was sequencing of the uncapped polyadenylated transcripts. While the main aim was to observe miRNA induced cleavage, differences in the post-transcriptional regulation or abundance of degraded transcripts across the eight sequencing conditions can be observed. Taken together, these studies cover mRNA expression, post-transcriptional editing, and finally degradation.

P. Fourounjian (✉)
Waksman Institute of Microbiology, Rutgers University, Piscataway 08854, USA
e-mail: pjf99@scarletmail.rutgers.edu

Scientists have been interested in gene expression ever since discovering the central dogma of biology and have developed a number of methods over the decades to measure RNA quantity. After reverse transcription was discovered in 1970, Northern blot and Sanger sequencing followed in 1977, qPCR came out in the late 1980s, and then in 2005, the Roche 454 sequencing platform applied shotgun genome sequencing technology to massively parallel RNA-sequencing and quantification (Cieślik and Chinnaiyan 2017). The data from these Roche and Illumina sequencers were typically thousands to millions of 50–200 nt reads that need to be mapped to the genome and reassembled to determine splicing patterns and gene expression as fragments per kilobase per million (FPKM). Joining the next-generation sequencers are the high throughput, long-read sequencers like PacBio and Oxford Nanopore systems available in 2011 which often produce 20–200 kB reads that can easily span entire mRNA and long non-coding RNA transcripts (1–2 kb), thereby eliminating the reassembly steps to more precisely map the transcriptome with its splicing patterns and alternative polyadenylation sites. Unfortunately, the present challenge with these


X. H. Cao et al. (eds.), *The Duckweed Genomes*, Compendium of Plant Genomes,
https://doi.org/10.1007/978-3-030-11045-1_13

reads is their indels and mismatches which can be corrected via deeper PacBio sequencing of the same read, alignment with the reference genome, and alignment with short-read sequencing, all of which can be combined (An et al. 2018a). With these technologies, it is now possible to accurately characterize and measure the transcriptome of virtually any species with a reference genome or through *de novo* assembly.

When looking at the RNA-seq studies of the duckweed family, we see that most of the gene expression analyses have been done in the recently sequenced *Lemna minor*, or in *Landoltia punctata* and *Lemna aequinoctialis* with *de novo* transcriptomes assembled from small reads (An et al. 2018b). These species have genome sizes ranging from 379 to 650 Mb and de novo transcriptomes of 74,797 and 72,105 unique contigs, while *Lemna minor* has 22,382 annotated genes, and *Spirodela* has 19,623 and 18,507 in strains 7498 and 9509, respectively, suggesting a wide variety of possible mRNAs from the roughly 20,000 genes found in Lemnaceae genomes (Tao et al. 2013; Wang et al. 2014a, 2016; Van Hoeck et al. 2015; Michael et al. 2017; Yu et al. 2017). In addition to the transcriptomic analysis of ABA-induced turion formation, there have been a couple of other RNA-seq experiments within *Spirodela polyrhiza* that measure aspects other than gene expression to understand the full complexity of these plant transcriptomes. A reinvestigation of turion development RNA-seq identified the chloroplast genes that undergo mRNA editing and how this relates to the rest of the monocots. Another study was the sequencing of the degraded RNAs in *Spirodela* within eight conditions as a measure of miRNA induced cleavage, which could also show a rough measure of expression. These three studies provide researchers a chance to witness mRNA expression, editing, and degradation.

One unique aspect of the duckweed lifecycle is the formation of turions. Their role as an asexual organ of perennation makes them analogous to both seeds, which sexually give rise to an entire organism, and tubers or buds, which asexually survive winter and other unfavorable conditions. In 2014, an RNA-seq study was performed on *Spirodela* during the development of turions, making it the first genome wide-study of gene expression in a plant tuber, which was followed by a study of potatoes in 2015 (Wang et al. 2014b). It was previously discovered that 3 days of exposure to 10 μM abscisic acid, ABA, induced irreversible turion development and an increase of two enzymes involved in starch and cell wall production. This study, therefore, compared four biological replicates of *Spirodela* fronds with and without 3 days of exposure to ABA by sequencing 15–41 million 75 bp reads of ribosomal, rRNA, depleted total RNA on a SOLiD 5500 instrument. They were able to map reads to the nuclear, mitochondrial, and chloroplast genomes, with 28–39% of reads deriving from the organelles. Reads were aligned with bowtie and tophat, normalized and compared in cufflinks, and annotated for GO term enrichment through blast2go and GOseq. The results showed 154 genes down-regulated during turion development, meaning that they were minimally four-fold less abundant, with roughly half being 0.2–0.1 the expression compared to control. These results had a false discovery rate, FDR, less than 0.01 thanks to the eight biological replicates. The 154 down-regulated genes were largely involved in carbon fixation, protein synthesis, DNA replication, and growth in general since turions no longer grow. For the 208 upregulated genes, the GO term enrichment showed that many of these upregulated genes functioned in starch and anthocyanin production, hormone response and signal transduction, cell wall synthesis, and seed dehydration. There were 13 genes in cell wall and anthocyanin production that were specific to turion induction. Similar to a desiccating seed of a terrestrial plant developing turions upregulated five and expressed two previously silent genes of the late embryogenesis abundant protein family. These LEA family proteins protect other proteins and confer resistance to dehydration, salinity, and cold stress. This transcriptomic study was properly timed to observe not only the structural changes of turion development, but the signaling pathway. They noticed upregulation of seven ABA-responsive, three ethylene-responsive, and two heat shock responsive transcription factors. There were also ABA transcription factor binding sites in 30 of the upregulated genes, while 119 had a bind site for ethylene-responsive transcription factors. This pathway matches the ABA or environment triggered, calcium-dependent signal pathway observed in maturing seeds, reinforcing the similarity of turions and seeds on a molecular, invisible level.

The same authors performed a second investigation of the rRNA depleted RNA-seq experiment in fronds and developing turions (Wang et al. 2015). Since 26% of the total RNA sequenced mapped to the chloroplast genome, they had 1000-fold coverage of most genes after stringent filtering. The PPR proteins are a massive family, characterized by the 35 amino acid pentatricopeptide repeat motif that specifically binds the 4th and 34th residues in the pfam model to an RNA base, creating a pattern of these motifs that bind to a specific RNA sequence (Barkan et al. 2012; Manna 2015). While these PPR proteins are found in prokaryotes and eukaryotes acting in splicing, processing, editing, stability, and translation of RNAs, this study focused on the DYW-type PPRs that correct certain missense mutations in the plastid genome by editing the mRNA from a cytosine to uracil residue, thereby creating a functional mRNA and protein product. Mapping the RNA-seq reads and detecting C to U SNPs with SAMtools revealed 66 sites of RNA editing with an average efficiency of 76% and a range of 6–100%. Comparison to developing turions showed very similar gene expression with no differentially expressed genes. There were, however, six over and five under edited sites (>two-fold difference, $p$ value <0.05) in seven genes during turion development compared to fronds. So while expression was constant, 1/6th of the sites were differentially edited, thereby altering the functional protein abundance of seven genes. These differences in editing efficiency even varied as much as 8–100% at multiple sites within the same gene due to the sequence-specific nature of PPR protein editing. A phylogenetic analysis with the Mega6 program revealed the 66 editing sites in *Spirodela* had an 81% overlap with the 75 in coconut. There was a 42 and 38% overlap with the 35 and 26 sites observed in the more evolutionarily distant rice and maize. This correlation confirms the hypothesis of a single origin of RNA editing PPR proteins in the early land plants like ferns, that have hundreds of edited sites that were gradually reduced and differentiated over time to ~80 in the basal monocots and 25–40 in the more recent angiosperm species.

Another indirect study of *Spirodela* mRNA expression was the degradome experiment found in Fourounjian et al. (2019). The primary purpose of this experiment was to confirm the cleavage activity of miRNAs on target mRNAs and observe regulatory differences between the biological triplicates of the control, 0 °C, 37 °C, ABA, kinetin, copper, nitrate, and sucrose conditions by sequencing 28–63 million uncapped mRNAs per library. While this degradome sequencing is not a perfect correlate to mRNA expression, the normalized read count of each gene (not kilobase normalized), its expression pattern can be viewed in this program hosted by the Myers laboratory of the Danforth Center https://mpss.danforthcenter.org/~private/dbs/index.php?SITE=messing_SPIRODELA_PARE. These patterns can even reveal unannotated exons, since all reads were polyadenylated. Finally, the miRNA cleavage study revealed that 15 genes, mainly well-conserved transcription factors were expressed and cleaved in four or more conditions, while 71% of the results were condition-specific targets, many of which had more structural and metabolic functions. Of these conditional specific changes, sucrose created the largest difference, followed by copper and heat exposure. This large transcriptomic and metabolic change of sucrose addition suggests that laboratory experiments modeling duckweeds in outdoor applications should avoid this often added media component.

The assembly of the *Spirodela* genome for strains 7498 and 9509 provided not only a scaffold for easy and accurate mapping of RNA-seq data, but a context for the gene expression. This is both in a physical sense for the chromatin modeling and DNA methylation studies (Cao et al. 2016; Michael et al. 2017), and in a physiological sense where the studies of turion development, for example, can be linked to the results (Kuehdorf et al. 2014). As it stands the research community can observe the Spirodela transcriptome in two or eight conditions as mRNA expression, editing, and degradation. It is expected that the transcriptional research will expand to include more stimuli exposure, tissue specific, life cycle, and microbe interaction experiments. These *Spirodela* genomes and transcriptome studies facilitate research across the family by providing a reference for the other genomes or transcriptomes. Even in cases of de novo assemblies and isoform sequencing of any other related species the *Spirodela* genome will provide annotated and characterized gene models.

## References


An D, Cao H, Li C et al (2018a) Isoform sequencing and state-of-art applications for unravelling complexity of plant transcriptomes. Genes (Basel) 9:43. https://doi.org/10.3390/genes9010043

An D, Li C, Zhou Y et al (2018b) Genomes and transcriptomes of duckweeds. Front Chem 6. https://doi.org/10.3389/fchem.2018.00230

Barkan A, Rojas M, Fujii S et al (2012) A combinatorial amino acid code for RNA recognition by pentatricopeptide repeat proteins. PLoS Genet 8:e1002910. https://doi.org/10.1371/journal.pgen.1002910

Cao HX, Vu GTH, Wang W et al (2016) The map-based genome sequence of *Spirodela polyrhiza* aligned with its chromosomes, a reference for karyotype evolution. New Phytol 209:354–363. https://doi.org/10.1111/nph.13592

Cieślik M, Chinnaiyan AM (2017) Cancer transcriptome profiling at the juncture of clinical translation. Nat Rev Genet 19:93–109. https://doi.org/10.1038/nrg.2017.96

Fourounjian P, Tang J, Tanyolac B, Feng Y, Gelfand B, Kakrana A, Tu M, Wakim C, Meyers BC, Ma J, Messing J (2019) Post-transcriptional adaptation of the aquatic plant under stress and hormonal stimuli. Plant J

Kuehdorf K, Jetschke G, Ballani L, Appenroth K-J (2014) The clonal dependence of turion formation in the duckweed *Spirodela polyrhiza*—an ecogeographical approach. Physiol Plant 150:46–54. https://doi.org/10.1111/ppl.12065

Manna S (2015) An overview of pentatricopeptide repeat proteins and their applications. Biochimie 113:93–99. https://doi.org/10.1016/J.BIOCHI.2015.04.004

Michael TP, Bryant D, Gutierrez R et al (2017) Comprehensive definition of genome features in *Spirodela polyrhiza* by high-depth physical mapping and short-read DNA sequencing strategies. Plant J 89:617–635. https://doi.org/10.1111/tpj.13400

Tao X, Fang Y, Xiao Y et al (2013) Comparative transcriptome analysis to investigate the high starch accumulation of duckweed (*Landoltia punctata*) under nutrient starvation. Biotechnol Biofuels 6:72. https://doi.org/10.1186/1754-6834-6-72

Van Hoeck A, Horemans N, Monsieurs P et al (2015) The first draft genome of the aquatic model plant Lemna minor opens the route for future stress physiology research and biotechnological applications. Biotechnol Biofuels 8:1–13. https://doi.org/10.1186/s13068-015-0381-1

Wang W, Haberer G, Gundlach H et al (2014a) The *Spirodela polyrhiza* genome reveals insights into its neotenous reduction fast growth and aquatic lifestyle. Nat Commun 5. https://doi.org/10.1038/ncomms4311

Wang W, Wu Y, Messing J (2014b) RNA-Seq transcriptome analysis of Spirodela dormancy without reproduction. BMC Genom 15:60. https://doi.org/10.1186/1471-2164-15-60

Wang W, Zhang W, Wu Y et al (2015) RNA editing in chloroplasts of *Spirodela polyrhiza*, an aquatic monocotelydonous species. PLoS ONE 10:e0140285. https://doi.org/10.1371/journal.pone.0140285

Wang W, Li R, Zhu Q et al (2016) Transcriptomic and physiological analysis of common duckweed Lemna minor responses to NH4+ toxicity. BMC Plant Biol 16:92. https://doi.org/10.1186/s12870-016-0774-8

Yu C, Zhao X, Qi G et al (2017) Integrated analysis of transcriptome and metabolites reveals an essential role of metabolic flux in starch accumulation under nitrogen starvation in duckweed. Biotechnol Biofuels 10:167. https://doi.org/10.1186/s13068-017-0851-8

# Proteomics in Duckweeds

# 14

Yang Fang, Anping Du, Li Tan, Kaize He, Yanling Jin, Xueping Tian, Yaliang Xu and Hai Zhao

**Abstract**

*Lemnaceae* (duckweed) is the smallest flowering plant, consisting of 5 genera and 37 species. It is a potential source for the production of biomass rich in starch, protein, flavonoids, and other high-value compounds. Also, it is deemed as a model for aquatic plants due to its small genome size. The biochemical and physiological features of duckweed have been studied for many years, while the molecular analysis at omics level was not studied until recently. To date, the genomes of three duckweed species, *Spirodela polyrhiza*, *Lemna minor*, and *Lemna gibba*, were sequenced and annotated. The expression profiles of duckweed were also studied at transcriptome level under various cultivation conditions. However, research on duckweed proteomics lags behind. The proteome analysis gives a precise estimate of gene expression and function. It plays an important role in mechanism elucidation of biochemical and physiological features of duckweed. Here, we review the progress of proteomic research of duckweed, to improve our insight into the mechanism of starch accumulation by duckweed at proteome level.

Y. Fang · A. Du · L. Tan · K. He · Y. Jin · X. Tian · Y. Xu · H. Zhao (✉)
Key Laboratory of Environmental and Applied Microbiology, Chengdu Institute of Biology, Chinese Academy of Sciences, Chengdu 610041, China
e-mail: zhaohai@cib.ac.cn

Y. Fang · A. Du · L. Tan · K. He · Y. Jin · X. Tian · Y. Xu · H. Zhao
Environmental Microbiology Key Laboratory of Sichuan Province, Chengdu 610041, China

X. Tian · Y. Xu
University of Chinese Academy of Sciences, Beijing 100049, China

## 14.1 Introduction

Duckweeds (the *Lemnaceae* family) are classified as the monocot order of Alismatales. There are five genera of *Spirodela*, *Landoltia*, *Lemna*, *Wolffiella*, and *Wolffia* within the *Lemnaceae* family. Each genus, except genus *Landoltia*, possesses multiple species. Genus *Landoltia* possesses only one species, naming *Landoltia punctata*. The natural ability to rapid biomass production and to thrive on anthropogenic wastewater makes the aquatic plants huge economic potential and extensive research interesting. It was reported that duckweed produces biomass faster than any other flowering plant (55 tons/ha/year dry weight) (Hillman and Culley 1978; Zhao et al. 2012). Duckweeds also have the potential to remediate wastewater and accumulate enormous amounts of starch for bioethanol fermentation (Zhao et al. 2014; Cui et al. 2011). Especially, duckweed biomass exhibits good characteristics for bioethanol production due to its relatively high starch and low lignin percentage.



X. H. Cao et al. (eds.), *The Duckweed Genomes*, Compendium of Plant Genomes,
https://doi.org/10.1007/978-3-030-11045-1_14

The starch content ranges from 3 to 75% (dry weight) depended on duckweed species and growing conditions (Cui et al. 2011). The starch produced in duckweed is mainly native small granular starch (with diameter lower than 10 μm) presenting higher gelatinization temperature, lower viscosity than large granular starch (with diameter greater than 10 μm), and B-type starch, which showing the potential utilization of starch from duckweed in food and non-food industries (Chen et al. 2016). Besides, duckweeds can be used for animal feed or fertilizer supplement for the high protein content varying from 15 to 45% of the dry weight (Cui et al. 2011).

The founding that DNA contains all of the genetic information that directs to construct an organism led to the principal dogma of molecular biology, which described a unidirectional flow of information from DNA to RNA to proteins (Crick 1970). The study of proteins, as key molecular entities, and the cell proteome, as a whole, is a challenging work, because of its complexity. Mass spectrometry-based proteomics is a current most effective way to study proteomes; because of high throughput, the raw data of mass spectrometry (MS) can result in up to 100,000-peptide spectra depended on the samples. So, proteomics, different from other omics, needs a dry lab, time-consuming in data mining. Quantitative protein expression profiling should identify the components of a proteome and compare the altered expression levels of two or more difference proteomes in response to a given treatment. iTRAQ, for isobaric tags for relative and absolute quantification, primary amino groups (the N-terminus and lysine side chains) of peptides are labeled chemically; there are eight different labels can be used to allow mixing of samples originating from different conditions or treatments for simultaneous analysis in a mass spectrometer (Angel et al. 2012). The iTRAQ technology had been used for proteome analysis during duckweed starch production recently (Huang et al. 2014, 2015).

Some transcriptome sequencing analysis had been carried out to address underlying physiological mechanism of duckweed response to changing environments. Although the protein concentrations are represented by the corresponding mRNA expression levels over a long period of time, a large number of studies are changing our understanding of protein expression regulation. Proteomics can directly reflect what plants have done responding to the external environment. There are fewer studies on duckweed proteome analysis than transcriptome analysis. At present, only two publications have reported proteomics study on high starch accumulation in *Landoltia punctata* 0202 (Table 14.1). In order to study the molecular mechanism of high starch accumulation at proteome level, proteomics technology combining transcriptome sequencing was adopted in *L. punctata* 0202 under nutritional stress and uniconazole treatment.

**Table 14.1** Summary of proteomics studies in duckweeds

| Species | Condition | Time points | Main findings | References |
|---|---|---|---|---|
| *Landoltia punctata* 0202 | Nutrient starvation | 0, 2, 5, 24 and 72 h | Directly and powerfully demonstrated that high starch and low lignin percentage resulted from regulated expression of enzymes and alternation of metabolism flux in the relevant pathways | Huang et al. (2014) |
| *Landoltia punctata* 0202 | Uniconazole treatment | 0, 2, 5, 72 and 240 h | Provided insights into the molecular mechanisms of uniconazole-induced hormone variation and starch accumulation at proteome level | Huang et al. (2015) |

## 14.2 Starch Production and Proteomics in *Landoltia punctata*

We previously determined *Landoltia punctata* 0202 was a candidate of highest biomass and starch percentage strain under nutrient starvation and uniconazole treatment through systematical screening. When growing under the nutrient starvation condition, high flavonoid and starch accumulation can be achieved simultaneously in *L. punctata* 0202. The flavonoids are a large class of secondary metabolites widely distributed in plants, which encompasses more than 10,000 structures, with different substituent groups, including chalcones, flavones, flavonols, flavandiols, anthocyanins, condensed tannins, and aurones (Winkel-Shirley 2001). The percentage of total flavonoid increases up to 5.56% following nutrient starvation for 168 h, of which seven components showed an obvious increase, accompanied abundant anthocyanin with purple coloration accumulated on the ventral side of fronds of *L. punctata* 0202 (Huang et al. 2014; Tao et al. 2017). Moreover, the cellulose and lignin contents of duckweed are 9.25% and 3.84%, respectively, which were four times lower than water hyacinth (39.93% and 10.15%, respectively), indicating that duckweed has more potential in animal food, fertilizer, and bioenergy production than water hyacinth (Zhao et al. 2014).

Uniconazole is another factor contributing to dry weight increase and high starch accumulation of duckweed, the dry weight in one flask can increase 3.1-fold compared to the control and the starch content can increase up to 48% (15.2-fold compared to the control) within 240 h after spraying frond with 800 mg/L uniconazole, and endogenous hormone content can be changed, the contents of abscisic acid (ABA), cytokinin (CK), and zeatin-riboside (ZR) increased, and on the contrary, the content of gibberellin (GA) decreased with uniconazole application. Besides, chlorophyll a and b content both increased compared with the control, resulting in the photosynthetic rate elevated (Liu et al. 2015a, b; Huang et al. 2015).

To elucidate the mechanisms of high starch accumulation, quantitative proteomics was firstly used to study the response of *L. punctata* 0202 to nutrient starvation with iTRAQ-LC-MS/MS technology. Duckweeds after expanding cultivation in sterile Hoagland nutrient solution for 14 days under stable condition were transferred into distilled water for an additional 7 days in the same condition, samples harvested on 11 time points (0, 0.5, 2, 5, 24, 48, 72, 96, 120, 144, and 168 h) in time course were used for composition characterization and enzymatic activity assay in three biological replicates, and five time point samples (0, 2, 5, 24, and 72 h) were subjected to iTRAQ proteomic analysis. A total of 2015 unique proteins were identified based on the duckweed protein sequence database using the mRNA transcripts predicted by RNA-seq results. In the identified proteins, 172 proteins were up-regulated and 43 proteins were down-regulated. Gene ontology (GO) categorization analysis revealed that the biological process was significantly enriched (76.7%) in the metabolic process. Notably, in starch metabolism, the expression levels of enzymes involved in starch biosynthesis were up-regulated, whereas those involved in starch degradation showed no significant difference. Importantly, in phenylpropanoid biosynthesis, the expression of several key enzymes involved in flavonoid biosynthesis showed up-regulated, but almost no enzyme related to the lignin biosynthetic branch exhibited sufficient expression abundance for detection. The proteomic analysis directly and powerfully demonstrated that high starch and low lignin percentage were regulated by the expression of enzymes and alteration of metabolic flux in the relevant pathways. This study helps us to understand the molecular mechanism of high starch accumulation and low lignin percentage in duckweed accurately, and promote the development of duckweed as a bioenergy crop (Huang et al. 2014).

Another proteomics research of duckweed is to investigate uniconazole-induced phytohormone variation and starch accumulation in *L. punctata* 0202. Duckweeds after expanding cultivation in standard 1/6 Hoagland nutrient solution for 3 days under stable condition were transferred into the same medium for an additional 10 days, but with uniconazole treatment in the homogeneous condition, samples were harvested at 13 time points (0, 1, 2, 3, 5, 7, 12, 24, 48, 72, 120, 168, and 240 h) in time course for composition characterization and enzymatic activity assay in three biological replicates. Five time point samples (0, 2, 5, 72, and 240 h) were used for iTRAQ proteomic analysis. A total of 3327 proteins were identified. Among these identified proteins, a large number of enzymes involved in endogenous hormone synthetic and starch metabolic pathways were affected. Notably, most of the enzymes involved in abscisic acid (ABA) biosynthesis showed up-regulated expression, which was consistent with the content variation. The increased endogenous ABA may up-regulate expression of ADP-glucose pyrophosphorylase to promote starch biosynthesis. Importantly, the up-regulated expression levels of several key enzymes in the starch biosynthetic pathway supported the enzymatic assay results and may explain why there is increased starch accumulation (Huang et al. 2015).

## 14.3 Flavonoids, Anthocyanin, Lignin Biosynthesis, and Proteomics in *Landoltia punctata*

The iTRAQ data described above was re-analyzed for flavonoid, anthocyanin, and lignin biosynthesis in phenylalanine metabolic networks based on another transcriptome as a new reference duckweed protein database described by Tao (Tao et al. 2017). The expression level of key enzymes that are responsible for flavonoid biosynthesis, such as phenylalanine ammonia-lyase (PAL), cinnamate 4-hydroxylase (C4H), 4-hydroxycinnamoyl-CoA ligase (4CL), chalcone synthase (CHS), chalcone isomerase (CHI), flavanone 3-hydroxylase (F3H), and anthocyanidin synthase (ANS), was improved, and the anthocyanin biosynthesis branch-related enzymes such as F3H, dihydroflavonol 4-reductase (DFR) and ANS expression were also increased under nutrient starvation in *L. punctata* 0202 based on the result of re-analysis for iTRAQ data, but PAL, C4H, and 4CL were suppressed immediately by uniconazole treatment. However, almost all the key enzymes of lignin biosynthetic branch in phenylalanine metabolic network and 25 laccases assembled by de novo RNA-Seq were not detected by re-analysis for iTRAQ data in duckweed under nutrient starvation and uniconazole treatment. This study supported previous omics research that the nutrient starvation treatment could improve the starch and flavonoid content simultaneously (Tao et al. 2013; Huang et al. 2014) and suggested that uniconazole treatment could induce starch accumulation and suppress the flavonoid content in duckweed (Tao et al. 2017).

## 14.4 Conclusion

Proteins are the direct undertakers of gene function. The high-throughput proteomic method facilitated our understanding of high starch accumulation mechanism by duckweed at proteome level. Also, it showed powerful in studying mechanism elucidation. However, few attention was paid to proteome analysis of duckweed. We encourage more researchers to focus on this field and thus promote our deepened understanding and wide application of duckweed.

## References

Angel TE, Aryal UK, Hengel SM et al (2012) Mass spectrometry-based proteomics: existing capabilities and future directions. Chem Soc Rev 41(10):3912–3928. https://doi.org/10.1039/c2cs15331a

Chen L, Yu C, Ma Y et al (2016) Insights into the structural and physicochemical properties of small granular starches from two hydrophyte duckweeds, *Spirodela oligorrhiza* and *Lemna minor*. Carbohydr Res 435:208–214. https://doi.org/10.1016/j.carres.2016.10.010

Crick F (1970) Central dogma of molecular biology. Nature 227(5258):561–563

Cui W, Xi J, Cheng JJ et al (2011) Starch accumulation in duckweed for bioethanol production. Biol Eng 3(4): 187–197

Hillman WS, Culley DD (1978) The uses of duckweed: the rapid growth, nutritional value, and high biomass productivity of these floating plants suggest their use in water treatment, as feed crops, and in energy-efficient farming. Am Sci 66(4):442–451

Huang M, Fang Y, Xiao Y et al (2014) Proteomic analysis to investigate the high starch accumulation of duckweed (*Landoltia punctata*) under nutrient starvation. Ind Crops Prod 59:299–308. https://doi.org/10.1016/j.indcrop.2014.05.029

Huang M, Fang Y, Liu Y et al (2015) Using proteomic analysis to investigate uniconazole-induced phytohormone variation and starch accumulation in duckweed (*Landoltia punctata*). BMC Biotechnol 15:81. https://doi.org/10.1186/s12896-015-0198-9

Liu Y, Fang Y, Huang M et al (2015a) Uniconazole-induced starch accumulation in the bioenergy crop duckweed (*Landoltia punctata*) I: transcriptome analysis of the effects of uniconazole on chlorophyll and endogenous hormone biosynthesis. Biotechnol Biofuels 8:57. https://doi.org/10.1186/s13068-015-0246-7

Liu Y, Fang Y, Huang M et al (2015b) Uniconazole-induced starch accumulation in the bioenergy crop duckweed (*Landoltia punctata*) II: transcriptome alterations of pathways involved in carbohydrate metabolism and endogenous hormone crosstalk. Biotechnol Biofuels 8:64. https://doi.org/10.1186/s13068-015-0245-8

Tao X, Fang Y, Xiao Y et al (2013) Comparative transcriptome analysis to investigate the high starch accumulation of duckweed (*Landoltia punctata*) under nutrient starvation. Biotechnol Biofuels 6(1):72

Tao X, Fang Y, Huang MJ et al (2017) High flavonoid accompanied with high starch accumulation triggered by nutrient starvation in bioenergy crop duckweed (*Landoltia punctata*). BMC Genom 18(1):166. https://doi.org/10.1186/s12864-017-3559-z

Winkel-Shirley B (2001) Flavonoid biosynthesis. A colorful model for genetics, biochemistry, cell biology, and biotechnology. Plant Physiol 126(2): 485–493

Zhao H, Appenroth K, Landesman L et al (2012) Duckweed rising at Chengdu: summary of the 1st international conference on duckweed application and research. Plant Mol Biol 78(6):627–632. https://doi.org/10.1007/s11103-012-9889-y

Zhao Y, Fang Y, Jin Y et al (2014) Potential of duckweed in the conversion of wastewater nutrients to valuable biomass: a pilot-scale comparison with water hyacinth. Bioresour Technol 163:82–91. https://doi.org/10.1016/j.biortech.2014.04.018

# Transformation Development in Duckweeds

**15**

Jingjing Yang, Shiqi Hu, Gaojie Li, Suliman Khan, Sunjeet Kumar, Lunguang Yao, Pengfei Duan and Hongwei Hou

**Abstract**

There are five genera (*Spirodela*, *Landoltia*, *Lemna*, *Wolffiella* and *Wolffia*) of duckweed species widely distributed in various freshwater habitats worldwide. Rapid growth rate, predominantly asexual reproduction and floating growth made them ideal for a plant model. The sensitivity to some toxicants and enrichment capacity also made duckweeds favorable in biomonitoring and bioremediation in contaminated water. Furthermore, duckweeds have increasingly been considered as alternative sources for bioenergy and food, due to their high biomass accumulation rate and nutritional contents. Both stable and transient transformation protocols have been established for some duckweed species. *Agrobacterium*-mediated method is the main approach in duckweeds genetic transformation, which could be affected by the type of explants, *Agrobacterium* strains, their densities, co-culture conditions and antibiotics and their concentrations. Particle treatment and other improvements such as vacuum infiltration can accelerate transient transformation efficiency by microprojectile bombardment method. Inadequacies are still present in genetic transformation of some duckweed species including low efficiency of transformation and long-time period especially using calli as infected materials. Therefore, more concentrated and persistent efforts to develop efficient approaches for genetic transformation of duckweeds are still needed. Furthermore, it is necessary to make an effort to express various types of genes so as to expand the development and utilization of duckweeds.

J. Yang · S. Hu · G. Li · S. Khan · S. Kumar · H. Hou (✉)
The State Key Laboratory of Freshwater Ecology and Biotechnology, The Key Laboratory of Aquatic Biodiversity and Conservation of Chinese Academy of Sciences, Institute of Hydrobiology, Chinese Academy of Sciences, Wuhan 430072, Hubei, People's Republic of China
e-mail: houhw@ihb.ac.cn

L. Yao · P. Duan
Collaborative Innovation Center of Water Security for Water Source Region of Mid-Line of South-to-North Diversion Project, Henan Province, Nanyang Normal University, Nanyang, China

Key Laboratory of Ecological Security for Water Source Region of Mid-Line of South-to-North Diversion Project of Henan Province, Nanyang Normal University, Nanyang, China

College of Agricultural Engineering, Nanyang Normal University, Nanyang, China

J. Yang · S. Hu · G. Li · S. Kumar · H. Hou
University of Chinese Academy of Sciences, Beijing 100049, People's Republic of China



X. H. Cao et al. (eds.), *The Duckweed Genomes*, Compendium of Plant Genomes,
https://doi.org/10.1007/978-3-030-11045-1_15

## 15.1 Introduction

It is obvious that duckweeds have captured more attention worldwide because of their special advantages: rapid growth rate, predominantly asexual reproduction, small growth space requirement and floating growth. Appenroth et al. (2015) pointed out that the physiological basis of the attractiveness of duckweeds as experimental organisms and for applications is mainly the rapid vegetative growth rate of the 37 species of this family. Duckweed clones represented the fastest growth of all flowering plants (Ziegler et al. 2015). The sensitivity to some toxicants and enrichment capacity made duckweeds widely used in biomonitoring and bioremediation in contaminated water. Furthermore, duckweeds have increasingly been considered as bioenergy and food sources alternatively due to their high biomass accumulation rate and nutrients content including carbohydrates and proteins as response to global resources and environmental crisis (Leng et al. 1995; Anderson et al. 2011; Qian et al. 2012; Xu et al. 2012; Zhao et al. 2012). Because of the economic and scientific potential, research on duckweeds in various aspects such as, morphology structure (White and Wise 1998; Lemon and Posluszny 2000), nutrient analysis (Appenroth et al. 2017), bioremediation (Karmakar et al. 2016; Teles et al. 2017) and genetic information (Hoeck et al. 2015; Cao et al. 2016) has increased exponentially in recent 30 years (Appenroth et al. 2013).

In comparison with traditional breeding, plant biotechnology has provided adequate opportunities for faster and directional introduction of beneficial traits into plants (Giri and Laxmi 2000; Sahrawat et al. 2003; Jones 2005). Duckweeds have also benefited from plant biotechnology, which has provided many potential for efficient selection of diverse plant genotypes and targeted modification of duckweeds. Introduction of exogenous genes into plant genome mainly contains the following processes: the introduction of genes into plant cells, regeneration of transgenic cells, selection and regeneration of transgenic plants. The conventional transformation methods include *Agrobacterium*-mediated, microinjection, electroporation, microprojectile bombardment, etc. *Agrobacterium*-mediated method has been considered as the most natural means because exogenous genes were integrated into plant through intergeneric transfer mechanism along with the growth of plants (Jouanin et al. 1993). Furthermore, this method has been widely applied to various plant species for obtaining stable transformation lines, due to their precise integration of exogenous genes into the plant genome and high stability (Dai et al. 2001). The other direct transformation techniques utilizing protoplasts or tissues were efficient in the regeneration of transiently transformed plant.

Transient transformation is a simple and rapid technology compared with stable transformation. When DNA is delivered into plant target cells, only a small portion (if any) will integrate into the host chromosomes. Transient expression does not depend on chromosomal integration of the heterologous DNA; therefore, analysis of gene expression cannot be confused by position effects (Jones et al. 2009). Expression of the heterologous DNA can be detected 3 h after DNA-delivery (Abel and Theologis 1994). It reaches a peak between 18 and 48 h and persists for 10 days (Werr and Lörz 1986; Abel and Theologis 1994). Although it lasts only for a few days, transient transformation occupies a useful niche in many aspects such as functional genomics, recombinant protein production, protein subcellular localization and protein–protein interactions, since it allows proteins to be highly expressed and effects to be seen in a short time (Gheysen et al. 1998).

Efficient methods for genetic transformation into different duckweed species have evolved significantly in last few years. Currently, both stable and transient transformation protocols have been successfully developed in some duckweed species (Table 15.1). In this paper, the corresponding work on the genetic transformation of duckweed species was reviewed.

**Table 15.1** Summary of genetic transformation of duckweeds

| Species | Stable/Transient | Transformation method | Gene and gene function | Selectable marker | References |
|---|---|---|---|---|---|
| *Lemna minor* | Stable | Lemna Expression System (LEX SystemSM, Biolex Therapeutics, Pittsboro, NC) | Hemagglutinin (*HA*) gene, against H5N1 HPAI viruses | Kanamycin | Bertran et al. (2015) |
| | | *Agrobacterium tumefaciens* (GV3101) | *gfp* gene, reporter gene | Phosphinothricin (PPT) | Cantó-Pastor et al. (2015) |
| | | *Agrobacterium tumefaciens* (CBE21) | M130-*β*-glucuronidase gene | Kanamycin | Firsov et al. (2015) |
| | | *Agrobacterium tumefaciens* (C58-z707) | Avian influenza hemagglutinin *HA* gene, against H5N1 HPAI viruses; threonine deaminase (TD) gene, expressing threonine deaminase | Kanamycin | Long et al. (2012) |
| | | *Agrobacterium tumefaciens* (EHA105) | Glyoxylate aminotransferase (SGAT) *AtAGT1*, involving in photorespiratory pathway in resistance to salt stress | Hygromycin | Yang, et al. (2013) |
| | | *Agrobacterium tumefaciens* (EHA105) | Porcine epidemic diarrhea virus (PEDV) spike protein-1 gene, producing a protective antigen for the PEDV spike protein 1 | Kanamycin | Ko et al. (2011) |
| | | *Agrobacterium tumefaciens* (C58-Z707) | *uidA* gene, reporter gene | Kanamycin | Yamamoto et al. (2001) |
| | | *Agrobacterium tumefaciens* (C58-z707) | E1 endoglucanase gene, expressing E1 protein | Kanamycin | Sun et al. (2007) |
| | | *Agrobacterium tumefaciens* (C58-z707) | Human monoclonal antibodies (mAbs) MDX-060 synthetic gene, an anti-CD30 antibody for the | Kanamycin | Cox et al. (2006) |

(continued)

**Table 15.1** (continued)

| Species | Stable/Transient | Transformation method | Gene and gene function | Selectable marker | References |
|---|---|---|---|---|---|
| | | | treatment of Hodgkin lymphoma and anaplastic large cell lymphoma | | |
| | Stable and transient | *Agrobacterium tumefaciens* (EHA105) | *uidA* gene, reporter gene | Kanamycin | Chhabra et al. (2011) |
| *Lemna gibba* | Stable | *Agrobacterium tumefaciens* (C58-Z707) | *uidA* gene, reporter gene | Kanamycin | Yamamoto et al. (2001) |
| *Lemna perpusilla* | Stable | *E. coli* plasmids infection | pMB9; pBR325 | Tetracycline | Frey et al. (1980) |
| *Spirodela punctata* | Stable | *Agrobacterium tumefaciens* (EHA105) | AntiTNFα-scFv (anti-tumor necrosis factor alpha single-chain variable fragment) gene, treatment for some pathophysiological conditions associated with TNFα | Hygromycin | Balaji et al. (2015) |
| *Spirodela oligorrhiza* | Stable | *Agrobacterium tumefaciens* (EHA105) | *gfp* gene, reporter gene | Kanamycin | Vunsh et al. (2007) |
| | Stable | *Agrobacterium tumefaciens* (EHA105) | Aprotinin synthetic gene, encoding the mature aprotinin sequence and a signal peptide for secretion | Kanamycin | Rival et al. (2008) |
| *Spirodela polyrhzia* | Stable | *Agrobacterium tumefaciens* (AGL1) | *HA1* gene, encoding the hemagglutinin antigen of the H5N1 virus | Hygromycin | Thu et al. (2015) |
| *Wolffia arrhiza* | Stable | *Agrobacterium tumefaciens* (EHA105) | *uidA* gene, reporter gene | Hygromycin | Khvatkov et al. (2015a, b) |
| | Transient | Biolistic transformation | *gfp* gene, reporter gene | Hygromycin | Khvatkov et al. (2015a, b) |
| *Wolffia columbiana* | Transient | *Agrobacterium tumefaciens* (LBA4404) | *uidA* gene, reporter gene | Kanamycin | Boehm et al. (2001) |
| | Transient | Microprojectile bombardment | *uidA* gene, reporter gene | | Kruse et al. (2002) |

## 15.2 Transformation Methods

### 15.2.1 Stable Transformation of Duckweeds

The first stable transformed duckweed was obtained by Frey et al. (1980) through incubating intact plant of *Lemna perpusilla* with the *Escherichia coli* plasmids pMB9 and pBR325 under optimized conditions. This research work confirmed that intact plant can directly absorb *E. coli* plasmids and these plasmids can be used as vectors for the introduction of exogenous genes into plants. Some factors including plasmid concentration, incubation time and temperature could affect the efficiency of transformation. In case of *Lemna* DNA, the highest transformation rate obtained per μg reported in this research was $1.9 \times 10^{-8}$ when 20 μg *Lemna* DNA was incubated with 57 μg/ml plasmid at 6 °C for 22 h. However, controversy suggested that exogenous plasmid may not enter plant cell even nucleus when using intact plants as receptors. There was even denial of the obtained transgenic lines.

*Agrobacterium*-mediated method has been a preferred technology in creating transgenic plants due to the crown gall development ability of *Agrobacterium* (Tsvetkov et al. 1997). The crown gall disease could transfer and integrate the T-DNA from Ti plasmid of the bacterium into the plant nuclear genome. Moreover, phenolic compounds released by the wound of dicotyledonous plants were essential for integration of *vir* genes (Razzaq et al. 2004). However, *Agrobacterium*-mediated transformation for monocotyledon plants has been lagging behind because monocotyledon plants cannot or can release minute phenolic compounds when wounded. Therefore, the addition of phenolic compounds has greatly promoted genetic transformation efficiency of monocotyledon plants (Cheng et al. 1997; Ali et al. 2007; Khan et al. 2013). Acetosyringone (AS), a phenolic compound, has been widely used in plant genetic transformation at the period of pre-culture or co-cultivation for inducing expression of *vir* genes and improving the transformation efficiency in monocotyledon plants. Other factors such as explant type, bacterial density, *Agrobacterium* strain and co-culture condition also affect the efficiency of *Agrobacterium*-mediated transformation (Shrawat and Horst Lörz 2006).

The stable transformation of *Lemna minor, Lemna gibba, Spirodela punctata, Spirodela oligorrhiza, Wolffia arrhiza* as well as the transient transformation of *Spirodela polyrhiza, Wolffia columbiana* have been obtained by *Agrobacterium*-mediated method. Yamamoto et al. (2001) were the first to establish efficient stable transformation protocols for *L. minor* and *L. gibba* mediated by *Agrobacterium tumefaciens* C58-Z707. Light green nodules instead of calli were used as infection receptor in their study. 100 μM AS played an important role and it was added in bacterial cultured plate, bacterial resuspension medium and co-culture medium to activate *A. tumefaciens*. Based on the protocol of Yamamoto et al. (2001), endoglucanase E1 from *Acidothermus cellulolyticus* was successfully expressed in *L. minor* 8627 without any obvious observable phenotypic effects on morphology or the rate of growth (Sun et al. 2007). Although the expression level of transgenic duckweed (up to 0.24% of total soluble protein) was lower than several other transgenic expression systems like tobacco (Ziegelhoffer et al. 2001), it is very encouraging for us to develop and improve duckweed expression system. Same transformation method was also successfully used in expressing monoclonal antibodies (mAbs) (Cox et al. 2006) as well as Avian influenza hemagglutinin HA gene (Long et al. 2012) in *L. minor*. To fully exploit the advantages of duckweed including rapid multiplication, secretion of recombinant proteins and high protein yields, Gasdaska et al. (2003) developed *Lemna* Expression System or "LEX SystemSM" for the production of recombinant proteins. The *Lemna* Expression System was also adopted to successfully express H5 hemagglutinin vaccine antigen by Bertran et al. (2015).

Chhabra et al. (2011) optimized the gene transformation method of *L. minor* and established both stable and transient transformation using *A. tumefaciens* strain EHA105, and 3.8%

stable transformation efficiency was obtained for the first time. In comparison with Yamamoto et al. (2001), the effect of different concentrations of *A. tumefaciens*, co-culture conditions and application of nonionic surfactants was studied. *Agrobacterium* at $10^7$ cell $ml^{-1}$ concentration, the presence of 100 μM AS in co-culture medium at pH 5.2, nodular calli inoculated with bacterial suspension for 60 min and co-culturing for 3 days at 25 °C under 16 h light/8 h dark photoperiod were significant for the transformation of *L. minor* and gave the highest percentage of calli showing GUS activity. In addition, nonionic surfactant (Tween 20) adopted in this study with relatively low concentration, i.e., 0.2% significantly elevated transient GUS expression. Following the transformation protocol described by Chhabra et al. (2011), an *Arabidopsis* photorespiratory pathway gene serine: glyoxylate aminotransferase (SGAT), named as *AtAGT1*, was successfully overexpressed in *L. minor* which provided an effective way to promote salt tolerance in duckweeds and solve the freshwater salinity problems (Yang et al. 2013).

Both of Firsov et al. (2015) and Cantó-Pastor et al. (2015) have established the stable transformation of *L. minor* using calli instead of nodular as infected object mediated by *A. tumefaciens* strains GV3101 and CBE21, respectively. Without AS application in their protocol, Firsov et al. (2015) obtained 20 different lines of duckweed with confirmed transgenic status. Unlike Firsov et al. (2015), 200 μM AS was added in *Agrobacterium* resuspension in the study of Cantó-Pastor et al. (2015). Furthermore, the selection and regeneration procedures were executed simultaneously in liquid media to reduce the overall transformation duration from 6–7 weeks to 5 weeks. 59% of GFP expressing was obtained in this study which was significantly higher than previous studies.

Rival et al. (2008) and Vunsh et al. (2007) successfully established the stable transformation of *S. oligorrhiza* using calli as infected object to express high levels of protein. Calli wounded with DNA-free tungsten particles using a PDS-1000/He System was adopted in the study of Rival et al. (2008). Then the wounded calli were co-cultured with *Agrobacterium* suspension adding 100 μM AS and a highly stable GFP expression level over 25% was obtained. The stable transformation of *S. polyrhiza* was established using the cut fronds co-cultured with prepared *A. tumefaciens* strain AGL1 suspension which has been shaken for 4–6 h with 200 μM AS. The above mixture was centrifuged and vacuumed before cultured for 3 days with photoperiod 12/24 h to promote the infection of *Agrobacterium* (Thu et al. 2015).

Balaji et al. (2015) put forward different opinions on the effect of the wound to improve transformation efficiency of duckweeds. Two-week old *S. punctata* fronds were used in plant transformation mediated by *A. tumefaciens* strain EHA105. Results showed that 95% of fronds without any wound were successfully transformed, whereas all of the wounded fronds gradually died during selection indicating that intact plants can be used for efficient transformation. Callus induction and regeneration of duckweeds always need a long period (Stefaniak et al. 2002; Li et al. 2004; Wang 2016), thus in plant transformation mediated by *A. tumefaciens* would be a good choice for saving time in stable transformation. The ability of the fronds of duckweeds to be transformed has also been found in other duckweed species (Ko et al. 2011). In the research of Ko et al. (2011), fronds of *L. minor* were immersed with the bacterial suspension of *A. tumefaciens* strain EHA105 harboring the PEDV spike protein 1 gene for 30 min after wounded with a pair of forceps and scalpel. The difference from previous studies was that a comparatively high concentration (200 mg/L) of kanamycin was used to select kanamycin-resistant fronds. The reason may be different sensitivity between fronds and calli to antibiotics as well as various geographic isolates of *Lemna* species. In addition, 100 μM AS was only added in prepared bacterial cells.

Comparatively, the stable transformation of *Wolffia* was indeed a time-consuming procedure because the period for inducing infected materials was at least 4 months (Khvatkov et al. 2015a, b). Only the stable transformation of *W. arrhiza* has been successfully established mediated by *A.*

*tumefaciens* strain EHA105. Both the cluster structures and calli obtained after 4 months cultivation were conducted to co-culture with *Agrobacterium* suspension. Results showed that the preferred type of explant for the transformation of *W. arrhiza* was cluster structure instead of calli due to their inefficiency in high necrotization. To promote transformation, plant materials immersed in *Agrobacterium* suspension were shaking first and then drying in an airflow laminar box. Furthermore, the addition of 2 mg/L 2,4-D and BA in the medium during the first 1–2 weeks and regulator-free after 2 weeks were found crucial for the successful transformation of *W. arrhiza.*

### 15.2.2 Transient Transformation of Duckweeds

Although *A. tumefaciens* has been widely used in the stable transformation of plants, *Agrobacterium*-mediated transient transformation has also attracted substantial attentions in recent years due to its time-saving property. However, the host-range restrictions and regeneration problems were considered to be limiting steps in *Agrobacterium*-mediated transformation. Particle bombardment, by using high-velocity microprojectiles for delivery of foreign DNA into intact plant tissues has also been demonstrated as an alternative method for transient transformation (Klein et al. 1987) and this method has been successfully applied on *W. Columbiana* and *W. arrhiza*. For *W. columbiana*, the plasmid pCAMBIA1301 containing *uidA* reporter construct under the control of constitutive CaMV35S promoter was transformed by biolistic approach (Kruse et al. 2002). When a gold particle with a size of 0.6 μm was accelerated at 1350 psi with a target distance of 60 mm, higher efficiency of transformation was obtained. *W. arrhiza* was also transiently transformed by particle bombardment (Khvatkov et al. 2015a, b). The vector pCamGFP containing CaMV35S-driven *m-gfp5-ER* (codon-optimized *gfp* with localization signal to the endoplasmic reticulum) was constructed and introduced into *W. arrhiza*. In this study, the optimal parameters of helium pressure and target distance were 1350 psi and 12 cm, respectively, showing high level of transient expression as 12%.

*Agrobacterium*-mediated transient transformation for *W. Columbiana* was also established (Boehm et al. 2001). *A. tumefaciens* strain LBA4404 harboring a binary vector p35SGUSINT (*uidA* gene under the control of CaMV35S promoter) was used for fast screening of transformation results. Since this plant has a compact structure and few stomata at the upper surface, other treatments were applied to increase the infection efficiency. In this study, particle treatment and vacuum infiltration were found to be more essential in increasing transformation efficiency than *Agrobacterium*-mediated method. Though, the average transformation efficiency was still low with 3.9% of the fronds showing GUS activity. Currently, only transient expression of marker genes was reported in the genus *Wolffia*, including species *W. australiana*, *W. globosa*, *W. columbiana* and *W. arrhiza* (Boehm et al. 2001; Kruse et al. 2002; Pham et al. 2010; Khvatkov et al. 2015a, b). Among those exogenous gene, the most commonly used reporter genes in transient transformation are the *β*-glucuronidase (*GUS*) gene and the green fluorescent protein GFP.

### 15.2.3 *Agrobacterium* Strains and Density

The ability of *Agrobacterium* to transfer T-DNA into plant genome varied in different strains and concentrations, therefore, produced different transformation effects. There are differences in the susceptibility among species even cultivars and genotypes of these species (Swarnapiria 2009). Almost all the stable transformation of *L. minor* was obtained using *Agrobacterium* strain c58-z707 followed by EHA105 (Yamamoto et al. 2001; Cox et al. 2006; Sun et al. 2007; Chhabra et al. 2011; Ko et al. 2011; Long et al. 2012; Yang et al. 2013). Other strains including GV3101 and CBE21 have also made considerable transformation effect in transferring the exogenous gene into *L. minor* genome (Cantó-Pastor et al. 2015; Firsov et al. 2015). The above revealed that *L. minor* specie was relatively easy to be transformed therefore it has more advantages in expressing some beneficial traits over other duckweed species. Furthermore, *Agrobacterium* strains EHA105 was found to respond best in the transformation experiments of *S. punctata* (Balaji et al. 2015), *S. oligorrhiza* (Rival et al. 2008; Vunsh et al. 2007) and *W. arrhiza* (Khvatkov et al. 2015a, b). While transgenic *S. polyrhiza* was only acquired under the effective infection of *Agrobacterium* strain AGL1 (Thu et al. 2015). In general, *Agrobacterium* strain EHA105 has made good effect in the stable transformation of monocotyledon plant of duckweeds.

High density of bacterial cells could cause the death of plants, meanwhile, low density of bacterial cells could lead to ineffective transformation. The effect of bacterial densities on transformation efficiency of duckweeds has not been reported in detail. Generally, the adoptive bacterial density (presented by OD600 value) in *L. minor* transformation mediated by *Agrobacterium* strains EHA105, C58-Z707 and GV3101 was about 1.0 (Yamamoto et al. 2001; Cox et al. 2006; Sun et al. 2007; Chhabra et al. 2011; Long et al. 2012; Yang et al. 2013; Cantó-Pastor et al. 2015). However, the setting value of OD600 in transformation experiment of *L. minor* mediated by *Agrobacterium* strain CBE21 was only 0.2 indicating distinctly difference in infecting capability of different *Agrobacterium* strains (Firsov et al. 2015). For three *Spirodela* species, the values of OD600 of *Agrobacterium* strain EHA105 varied from 0.5 to 1.5 (Vunsh et al. 2007; Thu et al. 2015; Balaji et al. 2015). Although the same *Agrobacterium* strain was used, the value of OD600 adopted in the transformation of *W. arrhiza* was only 0.4–0.6 (Khvatkov et al. 2015a, b).

## 15.3 Regeneration and Selection of Transformed Plants

A complete process of gene transformation includes explants cultivation, regeneration and selection of transgenic plants, as well as cultivation of transformed plants. Therefore, the successful establishment of genetic transformation system especially *Agrobacterium*-mediated and microprojectile bombardment method depends on one efficient and stable plant regeneration system. Almost all of the transformation experiments of duckweed species were conducted using calli, nodules or cluster structures as infected objects. Published researches on callus induction and regeneration of *Spirodela, Lemna* and *Wolffia* species laid the foundation for their gene transformation experiments (Stefaniak et al. 2002; Li et al. 2004; Wang 2016; Khvatkov et al. 2015a, b).

The regeneration and selection of transformed plants were very critical for obtaining single lines and always carried out simultaneously. Efficient selection depends on the kind of antibiotics employed which was determined by selectable marker genes and concentration applied (Rani et al. 2013). The most effective antibiotics are those which either inhibit regeneration of untransformed plants or slowly kill the untransformed plants while transformed plants survive in good conditions (Swarnapiria 2009). The neomycin phosphotransferase (*nptII*) gene encoding for kanamycin, neomycin, geneticin (G418) and paromomycin, and the hygromycin phosphotransferase (*hpt*) gene encoding for hygromycin have been used extensively in plant gene transformation. Different concentrations of kanamycin have been adopted in almost all the stable transformation of *L. minor, L. gibba, S. oligorrhiza* ranging from 10 to 200 mg/L (Yamamoto et al. 2001; Cox et al. 2006; Sun et al. 2007; Chhabra et al. 2011; Ko et al. 2011; Long et al. 2012; Bertran et al. 2015; Firsov et al. 2015). 10 mg/L phosphinothricin (PPT) was added in frond regeneration medium in the transformation experiment of *L. minor* by Cantó-Pastor et al. (2015), because the construct contained a selectable marker conferring resistance to PPT. Considerable results have been achieved in the transformation experiment of *S. polyrhiza* and *W. arrhiza* by adding 5 mg/L hygromycin (Khvatkov et al. 2015a, b; Thu et al. 2015). The conditions for regeneration and selection in plant gene transformation were not immutable and the optimal conditions should be evaluated with specific explants and plant vectors involved.

## 15.4 Applications of Genetic Transformation in Duckweeds

### 15.4.1 Fundamental Researches

The five genera (*Spirodela*, *Landoltia*, *Lemna*, *Wolffiella* and *Wolffia*) of duckweed species widely distributed in various freshwater habitats which were easy to be harvested (Appenroth et al. 2013). These small duckweeds (0.5–15 mm) propagated mostly or exclusively in a vegetative manner via budding of daughter fronds which arose from the primordia of mother fronds. Their organ constitution evolved from thalloid fronds and adventitious roots (*Spirodela*, *Landoltia*, *Lemna*) to thalloid fronds (*Wolffiella*, *Wolffia*) (Landolt 1986). The genome information of some duckweed species has been surveyed such as *S. polyrhiza* (Wang et al. 2011a, b), *L.* minor (Van et al. 2015). Some are being sequenced, such as *Landoltia punctata*, *Wolffiella neotropica*, *Wolffia brasiliensis* and *W. columbiana*. However, the genome sizes of duckweed species from *S. polyrhiza* (158 Mbp) to *W. arrhiza* (1881 Mbp) displayed a negative correlation with their body size and morphological structures (Landolt 1986; Wang et al. 2011a, b; Cao et al. 2015; Wang and Messing 2015). Furthermore, different degrees of interspecific genome size variation were also observed in five genera from little variation in *Spirodela* (150–167 Mbp) and *Landoltia* (372–427 Mbp) to 1.6- or 2-fold in *Wolffiella* (623–973 Mbp) or *Lemna* (323–760 Mbp) and up to 5.3-fold in the genus *Wolffia* (357–1881 Mbp) (Wang et al. 2011a, b; Bog et al. 2015). The above progress and advantages of duckweeds made them ideal to be used as a model plant for fundamental researches. Therefore, efficient transgenic methods of duckweed species are very important for future research (Zhao et al. 2012; Lam et al. 2014). There remain many questions to be answered in this family of plants, such as the relationship among duckweed genome, morphological structure analysis, evolution and development. More effective methods of molecular biology are to be developed.

### 15.4.2 Bioenergy and Wastewater Treatment

Bioenergy such as bioethanol and biobutanol is important energy alternative to reduce world dependence on fossil-based fuels (Cui and Cheng 2015). The utilization and popularization of bioenergy have created a large amount of economic, social and environmental benefits (Lynd et al. 1991). Materials from corn grain containing sugar, starch or cellulose are currently the dominant feedstock for bioethanol production which inevitably raised environmental concerns as well as competed for limited cropland (Pimentel 2003; Endo et al. 2008; Cheng 2010). Biobutanol, mainly produced by acetone-butanol-ethanol (ABE) fermentation, also needs new feedstock for the fermentation (Cheng 2010). Therefore, it is important to explore new materials that do not necessarily compete for cropland for production of bioenergy.

Duckweeds have the ability to double their biomass in every 16–24 h under appropriate conditions which is faster than most plants (Peng et al. 2007). The starch contents of duckweeds also varied from 3 to 75% (dry based) by manipulating culture conditions, such as temperature, light, pH, phosphate concentration or other nutrients (Reid and Bieleski 1970; McLaren and Smith 1976; Landolt and Kandeler 1987). In addition, duckweeds contain relatively low lignin content in comparison with other crops which enable us to utilize their feedstock more economically (Bai et al. 2008). The aquatic life of duckweeds is also farmland-free. Furthermore, duckweeds can convert nutrients to biomass by absorbing and purifying wastewater. Therefore, transgenic duckweeds by overexpressing the genes related to the synthesis of starch, sugars and cellulose are considered as a promising feedstock for biofuels as well as potential in wastewater treatment.

### 15.4.3 Bioreactor

Recently, plant bioreactor has emerged as an attractive area for its low cost, product safety and easy scale-up (Tiwari et al. 2009). Variety of products including vaccine antigens, medical diagnostics proteins, industrial and pharmaceutical proteins, nutritional supplements like minerals, vitamins, carbohydrates and biopolymers have been attempted to express in plant bioreactors based on transgenic plants systems (Sharma and Sharma 2009). As a safe and cost-effective alternative expression platform, plant bioreactor was of great importance in the area of animal and human health and diagnostics (Ma et al. 2005; Boehm 2007). Plant systems based on transgenic food crops such as tobacco, tomato, rice, potato, maize, carrot and soybean have already served as efficient bioreactors for expressing recombinant products as well as other non-food and/or non-crops (Cox et al. 2006; Tiwari et al. 2009; Rybicki 2010; Tremblay et al. 2010).

Selection of the host species is a critical step for the establishment of efficient bioreactor. Many factors such as the life cycle of plant species, reproductive rate, biomass yield and scale-up costs can affect the choice of suitable host. Therefore, duckweed species were ideal as plant bioreactor for their aquatic and short life cycle, rapid asexual reproduction, high biomass and protein yield, easy harvesting and cultivation, small living space, and easy to transform (Landolt 1986). Duckweed expression systems indeed promote rapid expansion of transgenic plants, secretion of recombinant proteins and high protein yield (Cox et al. 2006). The endoglucanase E1 gene from *Acidothermus cellulolyticus* has been expressed in transgenic *L. minor*. The duckweed-expressed enzyme was biologically active with expression level up to 0.24% of total soluble protein demonstrating possibilities for the expression of cellulolytic enzymes in transgenic duckweeds (Sun et al. 2007). Various proteins, such as mAbs, aprotinin, TNFα have also been expressed at high levels in *Lemna* and *Spirodela* species, which enables protein production by duckweed bioreactor in a robust and controllable format (Cox et al. 2006; Rival et al. 2008; Balaji et al. 2015).

Antigen protein porcine epidemic diarrhea virus (PEDV) expressed in transgenic *L. minor* was the first report on the expression of antigen vaccine against an animal infectious disease in duckweeds (Ko et al. 2011). A synthetic hemagglutinin (*HA*) gene from the highly pathogenic avian influenza (HPAI) virus A/chicken/Indonesia/7/2003 (H5N1) (Indo/03) was successfully expressed in *L. minor* (rLemnaHA) (Bertran et al. 2015). The transgenic duckweed derived HA produced high-quality antigen for an injectable vaccine against H5N1 HPAI viruses. *S. polyrhiza*, capable of growth and good biomass production, was also used to express *HA1* gene encoding hemagglutinin antigen of H5N1 virus for further generation of vaccine (Thu et al. 2015). In addition, the M2e peptide was expressed in nuclear-transformed duckweed plants with no noticeable impact on the plant morphology or growth rate, and the accumulation reached to 40 μg/g FW which was equivalent to levels obtained in transient virus-based systems (Firsov et al. 2015). The development of safe and effective vaccines against highly pathogenic influenza A virus subtype H5N1 has been recognized as an essential approach to decrease risk of transmission in poultry and humans, Furthermore, it opens the way to develop an edible plant vaccine against avian influenza virus (Bertran et al. 2015; Firsov et al. 2015).

## 15.5 Conclusion and Future Perspectives

Although protocols for gene transformation have been established in some duckweed species, inadequacies are still present. Low-transformation efficiency and long-period consumption were the main limitations. Therefore, more intense and persistent efforts are needed to develop efficient approaches for genetic transformation of duckweeds. In addition, most of the genes transferred into duckweeds are reported genes or vaccine related. More efforts are also needed to express various types of genes in order to expand the development and utilization of duckweeds.

## References


Abel S, Theologis A (1994) Transient transformation of *Arabidopsis* leaf protoplasts: a versatile experimental system to study gene expression. Plant J 5(3):421–427

Ali S, Zhang X, Xue Q, Hassan MJ (2007) Investigations for improved genetic transformation mediated by *Agrobacterium tumefaciens* in two rice cultivars. Biotechnology 6(1):138–147

Anderson KE, Lowman Z, Stomp AM, Chang J (2011) Duckweed as a feed ingredient in laying hen diets and its effect on egg production and composition. Int J Poult Sci 10(1):4–7

Appenroth KJ, Borisjuk N, Lam E (2013) Telling duckweed apart: genotyping technologies for the *Lemnaceae*. Chin J Appl Environ Biol 19(1):1–10

Appenroth KJ, Crawford DJ, Les DH (2015) After the genome sequencing of duckweed—how to proceed with research on the fastest growing angiosperm? Plant Biol 17(Suppl 1 s1):1–4

Appenroth KJ, Sree KS, Böhm V, Hammann S, Vetter W, Leiterer M et al (2017) Nutritional value of duckweeds (*Lemnaceae*) as human food. Food Chem 217: 266–273

Bai FW, Anderson WA, Mooyoung M (2008) Ethanol fermentation technologies from sugar and starch feedstocks. Biotechnol Adv 26(1):89–105

Balaji P, Satheeshkumar PK, Venkataraman K, Vijayalakshmi MA (2015) Expression of anti-tumor necrosis factor alpha (TNFα) single-chain variable fragment (scFv) in *Spirodela punctata* plants transformed with *Agrobacterium tumefaciens*. Biotechnol Appl Biochem 63(3):354

Bertran K, Thomas C, Guo X, Bublot M, Pritchard N, Regan JT et al (2015) Expression of H5 hemagglutinin vaccine antigen in common duckweed (*Lemna minor*) protects against H5N1 high pathogenicity avian influenza virus challenge in immunized chickens. Vaccine 33(30):3456–3462

Boehm R (2007) Bioproduction of therapeutic proteins in the 21st century and the role of plants and plant cells as production platforms. Ann N Y Acad Sci 1102 (1):121–134

Boehm R, Kruse CD, Barth S, Schnabl H (2001) A transient transformation system for duckweed (*Wolffia columbiana*) using *Agrobacterium*-mediated gene transfer. J Appl Bot 75(4):107–111

Bog M, Lautenschlager U, Landrock MF, Landolt E, Fuchs J, Sree KS et al (2015) Genetic characterization and barcoding of taxa in the genera *Landoltia* and *Spirodela*, (*Lemnaceae*) by three plastidic markers and amplified fragment length polymorphism (AFLP). Hydrobiologia 749(1):169–182

Cantó-Pastor A, Mollá-Morales A, Ernst E, Dahl W, Zhai J, Yan Y et al (2015) Efficient transformation and artificial miRNA gene silencing in *Lemna minor*. Plant Biol 17(Suppl 1 s1):59

Cao HX, Vu GT, Wang W, Messing J, Schubert I (2015) Chromatin organisation in duckweed interphase nuclei in relation to the nuclear DNA content. Plant Biol 17 (Suppl 1 s1):120

Cao HX, Vu GT, Wang W, Appenroth KJ, Messing J, Schubert I (2016) The map-based genome sequence of *Spirodela polyrhiza* aligned with its chromosomes, a reference for karyotype evolution. New Phytol 209 (1):354–363

Cheng J (2010) Biological process for ethanol production. In: Cheng J (ed) Biomass to renewable energy processes. CRC Press, Boca Raton, FL, USA, pp 209–270

Cheng M, Fry JE, Pang S, Zhou H, Hironaka CM, Duncan DR et al (1997) Genetic transformation of wheat mediated by *Agrobacterium tumefaciens*. Plant Physiol 115(3):971–980

Chhabra G, Chaudhary D, Sainger M, Jaiwal PK (2011) Genetic transformation of Indian isolate of *Lemna minor*, mediated by *Agrobacterium tumefaciens*, and recovery of transgenic plants. Physiol Mol Biol Plants Int J Funct Plant Biol 17(2):129–136

Cox KM, Sterling JD, Regan JT, Gasdaska JR, Frantz KK, Peele CG et al (2006) Glycan optimization of a human monoclonal antibody in the aquatic plant *Lemna minor*. Nat Biotechnol 24(12):1591–1597

Cui W, Cheng JJ (2015) Growing duckweed for biofuel production: a review. Plant Biol 17(Suppl 1 s1):16

Dai S, Zheng P, Marmey P, Zhang S, Tian W, Chen S et al (2001) Comparative analysis of transgenic rice plants obtained by *Agrobacterium*-mediated transformation and particle bombardment. Mol Breed 7(1): 25–33

Endo A, Nakamura T, Ando A, Tokuyasu K, Shima J (2008) Genome-wide screening of the genes required for tolerance to vanillin, which is a potential inhibitor of bioethanol fermentation, in *Saccharomyces cerevisiae*. Biotechnol Biofuels 1:1–6

Firsov A, Tarasenko I, Mitiouchkina T, Ismailova N, Shaloiko L, Vainstein A et al (2015) High-yield expression of M2e peptide of avian influenza virus H5N1 in transgenic duckweed plants. Mol Biotechnol 57(7):1–9

Frey M, Rall S, Roth A, Hemleben V (1980) Evidence for uptake of plasmid DNA into intact plants (*Lemna perpusilla*) proved by an *E. coli* transformation assay. Z Naturforsch 35(11–12):1104–1106

Gasdaska JR, Spencer D, Dickey L (2003) Advantages of therapeutic protein production in the aquatic plant *Lemna*. BioProcess J 2(2):49–56

Gheysen G, Angenon G, Van Montagu M (1998) *Agrobacterium*-mediated plant transformation: a scientifically intriguing story with significant applications. Transgenic Plant Res 1–33

Giri CC, Laxmi GV (2000) Production of transgenic rice with agronomically useful genes: an assessment. Biotechnol Adv 18(8):653–683

Hoeck AV, Horemans N, Monsieurs P, Cao HX, Vandenhove H, Blust R (2015) The first draft genome of the aquatic model plant *Lemna minor*, opens the route for future stress physiology research and biotechnological applications. Biotechnol Biofuels 8(1):188

Jones HD (2005) Wheat transformation: current technology and applications to grain development and composition. J Cereal Sci 41(2):137–147

Jones HD, Doherty A, Sparks CA (2009) Methods in molecular biology, vol 513: plant genomics. In: Transient transformation of plants, pp 131–152

Jouanin L, Brasileiro ACM, Leple JC, Pilate G, Cornu D (1993) Genetic transformation: a short review of methods and their applications, results and perspectives for forest trees [*Agrobacterium*]. Ann Sci Forestières 50:325–336

Karmakar S, Mukherjee J, Mukherjee S (2016) Removal of fluoride contamination in water by three aquatic plants. Int J Phytorem 18(3):222

Khan SA, Hanif Z, Irshad U, Ahmad R, Yasin M, Chaudhary MF et al (2013) Genetic transformation of sugarcane variety hsf-240 with marker gene *GUS*. Int J Agric Biol 15(6):1258–1264

Khvatkov P, Chernobrovkina M, Okuneva A, Pushin A, Dolgov S (2015a) Transformation of *Wolffia arrhiza* (L.) Horkel ex Wimm. Plant Cell Tissue Organ Cult 123(2):299–307

Khvatkov P, Chernobrovkina M, Okuneva A, Shvedova A, Chaban I, Dolgov S (2015b) Erratum to: callus induction and regeneration in *Wolffia, arrhiza* (L.) Horkel ex Wimm. Plant Cell Tissue Organ Cult 120(1):275–275

Klein TM, Wolf ED, Wu R, Sanford JC (1987) High-velocity microprojectiles for delivering nucleic acids into living cells. Nature 327(6117):70–73

Ko SM, Sun HJ, Oh MJ, Song IJ, Kim MJ, Sin HS et al (2011) Expression of the protective antigen for PEDV in transgenic duckweed, *Lemna minor*. Hortic Environ Biotechnol 52(5):511–515

Kruse C, Boehm R, Voeste D, Barth S, Schnabl H (2002) Transient transformation of *Wolffia columbiana* by particle bombardment. Aquat Bot 72(2):175–181

Lam E, Appenroth KJ, Mori K, Fakhoorian T (2014) Duckweed in bloom: the 2nd international conference on duckweed research and applications heralds the return of a plant model for plant biology. Plant Mol Biol 84(6):737–742

Landolt E (1986) The family of *Lemnaceae*—a monographic study, vol 1. Veroffentlichungen des Geobotanischen Institutes der Eidgenossischen Technischen Hochschule, Stiftung Rubel, Zurich, Switzerland

Landolt E, Kandeler R (1987) Biosystematic investigations in the family of duckweeds (*Lemnaceae*), vol 4: the family of *Lemnaceae*—a monographic study, vol 2 (phytochemistry, physiology, application, bibliography). Veroeffentlichungen Des Geobotanischen Instituts Der Eth Stiftung Ruebel

Lemon GD, Posluszny U (2000) Comparative shoot development and evolution in the *Lemnaceae*. Int J Plant Sci 161(5):733–748

Leng RA, Stambolie JH, Bell R (1995) Duckweed—a potential high-protein feed resource for domestic animals and fish. Livestock Res Rural Dev 7(1). http://www.lrrd.cipav.org.co/lrrd7/1/3.htm

Li J, Jain M, Vunsh R, Vishnevetsky J, Hanania U, Flaishman M et al (2004) Callus induction and regeneration in *Spirodela* and *Lemna*. Plant Cell Rep 22(7):457–464

Long VN, Cox KM, Ke JS, Peele CG, Dickey LF (2012) Genetic engineering of a *Lemna* isoleucine uxotroph. Transgenic Res 21(5):1071–1083

Lynd LR, Cushman JH, Nichols RJ, Wyman CE (1991) Fuel ethanol from cellulosic biomass. Science 251 (4999):1318–1323

Ma JK, Barros E, Bock R, Christou P, Dale PJ, Dix PJ et al (2005) Molecular farming for new drugs and vaccines. Current perspectives on the production of pharmaceuticals in transgenic plants. Embo Rep 6 (7):593

Mclaren JS, Smith H (1976) The effect of abscisic acid on growth, photosynthetic rate and carbohydrate metabolism in *Lemna minor* L. New Phytol 76:11–20

Peng J, Wang B, Song Y, Yuan P (2007) Model N transformation and removal in a duckweed pond: model application. Ecol Model 206:294–300

Pham TLT, Nguyen HA, Pham TH, Nguyen TH, Le HH (2010) Improvement of transformation procedure into duckweed (*Wolffia* sp.) via *Agrobacterium tumefaciens*. Tapchı Côngnghê Sinhhoc 8:53–60

Pimentel D (2003) Ethanol fuel: energy balance, economics, and environmental impacts are negative. Nat Resour Res 12:127–134

Qian C, Jin YL, Zhang GH, Yang F, Yao X, Hai Z (2012) Improving production of bioethanol from duckweed (*Landoltia punctata*) by pectinase pretreatment. Energies 5(8):3019–3032

Rani T, Yadav RC, Yadav NR, Rani A, Singh D (2013) Genetic transformation in oilseed brassicas—a review. Indian J Agric Sci 83(4)

Razzaq A, Ma Z, Wang H (2004) Genetic transformation of wheat (*Triticum aestivum*L): a review. Mol Plant Breed 2(4):457–464

Reid MS, Bieleski RL (1970) Response of *Spirodela oligorrhiza* to phosphorus deficiency. Plant Physiol 46 (4):609

Rival S, Wisniewski JP, Langlais A, Kaplan H, Freyssinet G, Vancanneyt G et al (2008) *Spirodela* (duckweed) as an alternative production system for

pharmaceuticals: a case study, aprotinin. Transgenic Res 17(4):503–513
Rybicki EP (2010) Plant-made vaccines for humans and animals. Plant Biotechnol J 8:620–637
Sahrawat AK, Becker D, Lütticke S, Lörz H (2003) Genetic improvement of wheat via alien gene transfer, an assessment. Plant Sci 165(5):1147–1168
Sharma AK, Sharma MK (2009) Plants as bioreactors: recent developments and emerging opportunities. Biotechnol Adv 27(6):811
Shrawat AK, Horst Lörz H (2006) *Agrobacterium*-mediated transformation of cereals: a promising approach crossing barriers. Plant Biotechnol J 4:575–603
Stefaniak B, Woźny A, Budna I (2002) Callus induction and plant regeneration in *Lemna minor* L. Biol Plant 45(3):469–472
Sun Y, Cheng JJ, Himmel ME, Skory CD, Adney WS, Thomas SR et al (2007) Expression and characterization of *Acidothermus cellulolyticus* E1 endoglucanase in transgenic duckweed *Lemna minor* 8627. Biores Technol 98(15):2866–2872
Swarnapiria R (2009) Genetic transformation in ornamentals—a review. Agric Rev 30(2):120–131
Teles CC, Mohedano RA, Tonon G, Filho PB, Rhr C (2017) Ecology of duckweed ponds used for nutrient recovery from wastewater. Water Sci Technol 75 (12):2926–2934
Thu PTL, Huong PT, Tien VV, Le HH, Khanh TD (2015) Regeneration and transformation of gene encoding the hemagglutinin antigen of the h5n1 virus in frond of duckweed (*Spirodela polyrhiza*L.). J Agric Stud 3(1):48
Tiwari S, Verma PC, Singh PK, Tuli R (2009) Plants as bioreactors for the production of vaccine antigens. Biotechnol Adv 27:449–467
Tremblay R, Wang D, Jevnikar AM, Ma S (2010) Tobacco, a highly efficient green bioreactor for production of therapeutic proteins. Biotechnol Adv 28:214–221
Tsvetkov I, Tsolova V, Atanassov A (1997) Genetic transformation of grape (review). Biotechnol Biotechnol Equip 11(1–2):23–28
Van HA, Horemans N, Monsieurs P, Cao HX, Vandenhove H, Blust R (2015) The first draft genome of the aquatic model plant *Lemna minor* opens the route for future stress physiology research and biotechnological applications. Biotechnol Biofuels 8:188
Vunsh R, Li J, Hanania U, Edelman M, Flaishman M, Perl A et al (2007) High expression of transgene protein in *Spirodela*. Plant Cell Rep 26(9):1511–1519
Wang Y (2016) Callus induction and frond regeneration in *Spirodela polyrhiza*. Czech J Genet Plant Breed 52 (3):114–119
Wang W, Messing J (2015) Status of duckweed genomics and transcriptomics. Plant Biol 17(Suppl 1 s1):10
Wang W, Haberer G, Gundlach H, Gläßer C, Nussbaumer T, Luo MC, Lomsadze A, Borodovsky M, Kerstetter RA, Shanklin J et al (2011a) The *Spirodela polyrhiza* genome reveals insights into its neotenous reduction fast growth and aquatic lifestyle. Nat Commun 5:3311
Wang W, Kerstetter RA, Michael TP (2011b) Evolution of genome size in duckweeds (*Lemnaceae*). J Botany 2011:9
Werr W, Lörz H (1986) Transient gene expression in a gramineae cell line. Mol Gen Genet MGG 202 (3):471–475
White SL, Wise RR (1998) Anatomy and ultrastructure of *Wolffia columbiana* and *Wolffia borealis*, two nonvascular aquatic angiosperms. Int J Plant Sci 159 (2):297–304
Xu JL, Cheng JJ, Stomp AM (2012) Growing *Spirodela polyrhiza* in swine wastewater for the production of animal feed and fuel ethanol: a pilot study. Clean-Soil Air Water 40(7):760–765
Yamamoto YT, Rajbhandari N, Lin X, Bergmann BA, Nishimura Y, Stomp AM (2001) Genetic transformation of duckweed *Lemna gibba*, and *Lemna minor*. Vitro Cell Dev Biol Plant 37(3):349–353
Yang L, Han H, Liu M, Zuo Z, Zhou K, Lü J et al (2013) Overexpression of the *Arabidopsis*, photorespiratory pathway gene, serine: glyoxylate aminotransferase (AtAGT1), leads to salt stress tolerance in transgenic duckweed (*Lemna minor*). Plant Cell Tissue Organ Cult 113(3):407–416
Zhao H, Appenroth K, Landesman L, Salmean AA, Lam E (2012) Duckweed rising at Chengdu: summary of the 1st international conference on duckweed application and research. Plant Mol Biol 78:627–632
Ziegelhoffer T, Raasch JA, Austin-Phillips S (2001) Dramatic effects of truncation and sub-cellular targeting on the accumulation of recombinant microbial cellulase in tobacco. Mol Breed 8(2):147–158
Ziegler P, Adelmann K, Zimmer S, Schmidt C, Appenroth KJ (2015) Relative in vitro growth rates of duckweeds (*Lemnaceae*)—the most rapidly growing higher plants. Plant Biol 17(Suppl 1 s1):33–41

# Small RNAs in Duckweeds

# 16

Paul Fourounjian

**Abstract**

Within eukaryotic genomes, there are several types of small RNAs including sn, sno, si, and miRNAs. With respect to the Lemnaceae family, the vast majority of the research to date has been conducted in *Spirodela polyrhiza*, focused mainly on the miRNAs. This research consists of three small RNA-sequencing experiments in strains from China, Germany, and the USA, with each experiment identifying conserved miRNAs and predicting novel miRNAs and targets. While the novel miRNAs and recently discovered miRNAs fluctuated, the family size and expression of well-known miRNA families was consistent between the three experiments. While miRBase likely contains many incorrect annotations, these miRNAs were annotated according to strict criteria and analyzed for the miRBase high confidence list. They were further characterized through degradome sequencing, which confirmed half of the conserved miRNAs and a third of the novel. Finally, *Spirodela polyrhiza* has a surprisingly low abundance of 24nt sRNAs, which are required to suppress transposon proliferation.

As scientists moved from sequencing the ϕX174 virus in 1977 to prokaryote genomes, simple eukaryotes, and then the first plant, *Arabidopsis thaliana* in the year 2000, they saw that these larger complex genomes were made of so much more than genes. We now know that eukaryotic genomes contain a host of structural repeats such as the centromere and telomere regions. There are also large stretches of tandem repeats, also called satellite DNA. Then, there are the virus-like transposable elements that are often copied and spread across the genome. Many of the transcribed RNA sequences are small RNAs like small interfering, micro, and small nucleolar RNAs (si, mi, and snoRNAs) that bind to protein complexes to regulate gene expression and assemble ribosomes. Larger RNA transcripts include long non-coding RNAs and the high copy number ribosomal and transfer RNAs (lnc, r, and tRNAs) that translate mRNAs to proteins. Each genome also contains plenty of pseudogenes, which are non-functional due to mutations. Finally, the genome contains the protein-coding genes themselves, with all their introns, exons, cis- and trans-regulatory elements and terminators, which are 2% of the human genome and roughly 20% of a typical angiosperm genome, with wide variation due to genome size differences.

Within this genome, there are several types of transcribed RNAs, with the longer varieties including m, r, t, and lncRNA. While the first three types are well characterized, long

P. Fourounjian (✉)
Waksman Institute of Microbiology, Rutgers University, Piscataway 08854, USA
e-mail: pjf99@scarletmail.rutgers.edu


X. H. Cao et al. (eds.), *The Duckweed Genomes*, Compendium of Plant Genomes,
https://doi.org/10.1007/978-3-030-11045-1_16

non-coding RNA (lncRNA) wasn't discovered until 1990 (Brannan et al. 1990). These spliced and polyadenylated RNAs function in epigenetic regulation, the generation and sequestration of miRNAs, and various other functions. While most of the studies have been run in animals, thousands of lncRNAs have been annotated in plant genomes, including IPS1, which sequesters miR399 with a non-cleavable target bulge in response to phosphate starvation across many plant species (Franco-Zorrilla et al. 2007).

The small RNAs in plants include sno, si, and miRNAs, with the snoRNAs evolutionarily conserved back to Archaea. They are produced from their own RNA precursors, or introns, which are cleaved by endonucleases and trimmed by exonucleases, until only the protein bound 60–250 bp snoRNA remains; they then guide the protein complex's methylation and pseudouridylation of rRNAs in the nucleolus. It is even hypothesized that snoRNAs gave rise to miRNAs based on their similarity in processing including some overlap of enzymes, their similar hairpin structure, and combination of function (Scott and Ono 2011). There have been reports of snoRNAs with miRNA-like characteristics, and vice versa, and even small RNAs with complete sno and miRNA function in animals, plants, and yeast. In plants, both miRNAs and siRNAs are cut to 22 and 21nt lengths by dicer-like proteins 1 and 4, respectively, and loaded onto Ago1 in the RISC, with the main difference being that an RNA hairpin is processed into a miRNA for mRNA gene suppression, while a dsRNA is diced into many siRNAs for pathogen gene silencing.

When the *Spirodela polyrhiza* genome was published in 2014, prediction programs were able to detect miRNA precursors through homologous sequences and RNA folding software (Wang et al. 2014a). In strain 7498, all miRBase plant mature sequences were mapped back to the genome, and flanking sequences analyzed by RNAfold and miRCheck (Denman 1993; Jones-Rhoades and Bartel 2004). The search predicted 413 miRNAs belonging to 93 families. This survey based on DNA sequencing aimed to provide all possible miRNA genes, for comparison to other plant genomes, with the eventual aim of detecting their activity in later RNA-seq experiments.

The earliest attempt at sequencing and analyzing *S. polyrhiza* miRNAs predated the published genome. This experiment, run at Peking University Shenzhen Graduate School, was run on strain LT5a, isolated from Lake Tai, using three populations grown in SH media for 1, 3, and 5 days under control conditions. Using 18-31nt sRNA on a HiSeq 2000 Illumina platform, they sequenced 24 million reads, 3.5 of which matched conserved miRNAs in miRBase, and 7.6 million that were not annotated in GenBank or Rfam. These 7.6 million reads were analyzed by the MIREAP program and validated by Mfold to identify 41 predicted novel miRNAs (Zuker 2003). A summary of this and the other small RNA-seq experiments is available in Table 16.1.

In strain 9509, conserved and novel miRNAs were identified through small RNA-sequencing and an analysis of read count and distribution (Michael et al. 2017). The study used 10 sRNA libraries from a SOLiD5500 sequencer, aligned to the genome allowing 1 mismatch, and then annotated if the candidate has a stable hairpin structure, sufficient miR reads, more than 1 miR* read, and a 2 or 3 nt 3′ overhang (Table 16.1). They identified conserved miRNAs by checking for a strong BLAST homology to not only the mature, but also hairpin structures in miRBase. Next they used the program TargetFinder with a cutoff score of 4 to identify the predicted targets (Fahlgren and Carrington 2010). These transcription and structural requirements lead to the prediction of 59 conserved miRNAs in 22 families, and 29 novel miRNAs, with 29 of the conserved and 25 of the novel miRNAs being predicted to regulate 991 mRNA targets.

Alongside the miRNA prediction, they were able to predict trans-acting siRNAs (tasiRNAs), from the sRNA library using previously established criteria (Howell et al. 2007; Johnson et al. 2009). Reads matching cDNA and the corresponding genomic regions had miRNA results filtered out, and then, 50nt candidate transcripts were required to have over 100 reads, with over

**Table 16.1** Summary of sRNA-sequencing experiments

| Strain | LT5a | 7498 | 9509 |
|---|---|---|---|
| Conditions | Control (SH media, 16 h days, 23 °C) | Control, heat, cold, abscisic acid, copper, kinetin, nitrate, sucrose | Control, abscisic acid |
| # reads | 25 million | 32 million | N/A |
| # conserved miRNAs | 158 | 58 | 59 |
| # novel miRNAs | 41 | 14 | 29 |
| # targets | N/A | 162 | 991 |
| # DE miRNAs | N/A | 15 | 12 |

70% being 21nt in length. These are sufficient to distinguish randomly degraded transcripts from mRNAs that had been transcribed into dsRNA and then diced into 21nt tasiRNAs. TargetFinder was then used with a cutoff of 6 and a requirement of two miRNA bind sites to identify the targeted genes. This search yielded two cleaved TAS3 genes, and the miR393 targeting another putative TAS gene that was also found in oil palm and banana.

The most recent miRNA survey started with strain 7498 grown in three replicate flasks of eight growth conditions: control, cold, heat, abscisic acid, copper, kinetin, nitrate, and sucrose stimuli. After harvest, RNA extraction, and size selection, 32 million reads of the 24 libraries were sequenced on the SOLiD5500 platform and mapped to the genome (Table 16.1). These results were filtered against *Brachypodium distachyon* non-coding RNAs, with miRNAs removed, and analyzed in miRPlant (An et al. 2014). Criteria required a miRPlant score greater than 3.0, over 20 miR reads and at least 1 miR* read. This yielded 58 conserved miRNAs and 14 novel miRNAs after the removal of those that had already been found in strain 9509. When consolidated with the results from strain LT5a and mapped back to the strain 7498 genome, these two showed a strong degree of overlap resulting in 63 conserved and 45 novel miRNAs. These miRNAs were then further judged by the stringent criteria for plant miRNA annotation by sRNA-seq indicating that only 30 were highly confident based on structure and read count (Axtell and Meyers 2018). These miRNAs were then used to predict 163 targets with a psRNATarget score better than 2 (Dai and Zhao 2011), with roughly half corresponding to novel miRNAs.

The first prediction of miRNAs based on genome sequence and hairpin structure saw 413 possible miRNAs, and this number dropped to 58 and 59 once the miRNAs were being predicted based on sequencing results (Table 16.1). Of the 413 miRNAs, many were from recently discovered families, with only 121 that corresponding to those 58 families sequenced in 7498 at 119 genomic loci. While numbers of miRNA loci within families mostly agree, the copy number of a few families based on expression data differs from the 7498 genome survey as shown in Table 16.2. Perhaps the 24 copies of miR156 include a number of unexpressed pseudogenes from duplication events. When the strain 7498 and 9509 conserved miRNA families were compared 20 overlapped, while two were only found in the 9509 genome, and the 7498 study included 11 less commonly conserved one-member miRNA families not observed in strain 9509. This overlap of family and sequence number of highly conserved families suggests we have robust identification of the expressed, heavily conserved miRNA families, while lower confidence previously reported and novel miRNAs require further investigation to characterize.

While much attention is always paid to proper identification and mapping of miRNAs in the first sequencing experiments of a genome, measuring miRNA abundance is also essential. Since miRNA families have high sequence homology and target the same family of gene targets, these results are grouped by expression of certain families. The three experiments studied strains LT5a, 9509, and 7498 which originated in China, USA, and Germany, providing a global perspective of the species. The control conditions were largely similar using Schenk & Hildebrandt medium at a pH of 5.8, with the known variations mainly being the 15 °C night time temperature, and relatively young cultures for LT5a and harvesting based on water surface coverage in 7498. While these expression results from strains across the world grown in control conditions vary in rank and abundance of miRNA families (Table 16.3), the same six families are within the top 5 in two of the three experiments demonstrating their prominent roles. As seen in Table 16.4, these miRNA families and their target gene families regulate growth, meristem development, and stress responses.

Strain 9509 was also exposed to 1uM ABA, which was shown to induce turion production irreversibly after 3 days (Wang et al. 2014b; Kuehdorf et al. 2014). At the 10 h time point, this hormonal stimuli changed the expression of 12 conserved miRNAs (over 100 reads in control, over twofold expression change in ABA), with the 169 and 396 families being underexpressed and the 159 and 168 families doubling in abundance (Michael et al. 2017). Then, at the 5-day time point, there were 28 miRNAs and targets with significant overexpression of the miRNA and underexpression of the mRNA compared to control and vice versa. Twelve of the miRNAs were novel miRNAs with relatively low expression, large fold change differences, and a wide variety of targets. Similar to the transcriptomic study at day 3, this experiment saw a decline in chloroplast proteins and an increase in polyphenol producing enzymes (Wang et al. 2014b; Michael et al. 2017).

The survey of miRNAs in strain 7498 in the control, cold, heat, abscisic acid, copper, kinetin, nitrate, and sucrose stimuli yielded a striking

**Table 16.2** Copy number variation of miRNA families between three publications

| miRNA family | 7498 genome survey | 9509 sRNA-seq | 7498 sRNA-seq |
|---|---|---|---|
| 156 | 24 | 6 | 9 |
| 159 | 1 | 3 | 4 |
| 169 | 9 | 5 | 7 |
| 396 | 11 | 5 | 9 |

**Table 16.3** miRNA expression of control conditions of three strains of *Spirodela polyrhiza*

| LT5a | 7498 | 9509 |
|---|---|---|
| 156 (47%) | 156 (41%) | 160 (68%) |
| 166 (24%) | 168 (18%) | 169 (7%) |
| 167 (20%) | 396 (16%) | 166 (6%) |
| 168 (5%) | 169 (6%) | 528 (5%) |
| 169 (1%) | 166 (4%) | 159 (3%) |

**Table 16.4** Biological roles of prominent miRNA families

| miRNA family | mRNA target family | Biological role |
|---|---|---|
| miR156 | SPLs | Maintains juvenile tissues |
| miR166 | HDZipIIIs | Regulates meristems |
| miR167 | IARs | Reduced by drought |
| miR168 | Ago1 | Viral defense |
| miR169 | NFYs | Drought and stress response |
| miR396 | GRFs | Regulates meristems |

result in that miR169c was between 33 and 82% of the reads in each condition, with large variability between the three biological replicates. This result was believed by the authors of the study to be an experimental artifact due to the lack of this expression in the other experiments, the only partial replication of the expression in the qPCR follow-up, and the current reputation of the SOLiD5500 sequencer. With this one sequence ignored and the dataset renormalized, we can accurately see the responses of other miRNAs to the various conditions. There were large increases in miR166 expression under the influence of cold and kinetin and miR168 in the heat and sucrose conditions. The meristem regulating 396 familiesy doubled expression in response to the heat, ABA, and copper stimuli. Finally, miR156, which maintains the juvenile, neotenous life cycle of the duckweed family, decreased over fourfold in response to sucrose, which was the condition responsible for 13 of the 19 instances of differential miRNA expression indicating that the mixotrophic lifestyle often used in laboratory experiments is quite different from duckweed grown in an outdoor setting.

Accurate miRNA annotation is quite difficult, since miRNAs are vastly outnumbered by similarly sized siRNAs in the genome, and even the more stringent miRNA prediction programs supply tens or hundreds of false predictions. An analysis in 2014 suggested that 75% of the land plant miRNA families in miRBase are questionable, especially those with only a single member (Taylor et al. 2014). In an attempt to manage the large number of submissions and false positives coming in, miRBase has established criteria for its high confidence miRNAs that analyze the structure of the hairpin, the read distribution along it, and the miR, and miR* read count. For plants in miRBase release 21, there are currently 6942 hairpins in 2408 distinct miRNA families, with only 587 from 227 families (9.7%) making the high confidence cutoff (Griffiths-Jones 2006). As an attempt to preserve miRNA annotation confidence, 21 of the leading minds of the field wrote the plant miRNA annotation criteria in 2008 that has since been updated by two of them thanks to new information and sequencing capabilities (Meyers et al. 2008; Axtell and Meyers 2018). The plant miRNA annotation criteria are generally more stringent than the high confidence criteria, except for the latter's requirement of 10 miR* reads, since plant miRNA biogenesis is quite specific. Both miRNA studies in *Spirodela* annotated miRNAs based on homology according to the 2008 criteria, with most of these being well-conserved, high-confidence miRNA families. The conserved miRNAs with family names above 535 are relatively likely to be based off of lower confidence annotations in previous reports. The novel miRNAs from strain 9509 were predicted in 2017 using cutoffs very similar to the 2018 criteria, demonstrating a high degree of confidence, while those predicted in the LT5a and 7498 study had a lower degree of confidence. These authors reviewed all their data, with the revised criteria finding that 30 of the 47 hairpin structures met the current standards.

In addition to applying the stringent structural and read distribution filters above, the authors of the 2018 study verified miRNAs through a method called degradome sequencing where uncapped mRNAs are sequenced and aligned to miRNA target sites to measure evidence of precise miRNA cleavage above random mRNA degradation. There were several methods available at the time, and the authors chose the GMUCT2.0 library for its read length and minimal PCR amplification and the sPARTA program for its accuracy in analysis of the degradome data (Kakrana et al. 2014; Willmann et al. 2014). Biological triplicate libraries of the same eight conditions observed in the miRNA-sequencing study were sequenced on the Illumina NextSeq 500, yielding 911 million total reads. When running the sPARTA program, the *Spirodela* 7498 gene models were extended 150nt upstream and 250nt downstream, since many of the degradome reads were from the UTRs of the mRNAs. The sequencing verified activity of 66 miRNAs on 149 targets. For the 42 conserved miRNAs, the targets were mainly the transcription factor families reported in other plant species. While these essential developmental transcription factors mostly made up the targets sequenced in over half of the conditions, 71% of the cleaved targets were sequence specific underscoring the importance of sampling a variety of post-transcriptional responses. Notably sucrose had the largest number of condition-specific results including metabolic and signaling proteins indicating a large shift in the mixotrophic lifestyle. This included a complete reversal where miR172 went from cleaving half as many targets as miR156 to twice as many despite being 0.4% of its expression. This suggested that sucrose may be inducing a less neotenous phenotype, and that highly expressed miRNAs are not necessarily highly active. Of the 81 novel miRNAs predicted within the three separate experiments of *Spirodela,* 24 were validated with 66 targets. This 30% validation rate, evenly spread between the three experiments, is consistent with similar surveys in other plant genomes thanks to the low expression and number of targets compared to conserved miRNAs, and the likelihood that novel miRNAs may be false predictions (Song et al. 2010; Li et al. 2010; Yang et al. 2013). While degradome evidence is a great way to confirm miRNAs, it does require co-expression and mRNA cleavage meaning that non-supported miRNAs may be found as active in later experiments with the right conditions and sequencing depth.

In order to provide other scientists easy access for further analysis, the raw data is available for LT5a results at GSE55208, 9509 at PRJNA308109, and 7498 at PRJNA473779 (SRP149336). As a second approach to increase transparency, ease replication, and enable further research, the data from the 2018 study and some of its analysis can be viewed in the Galaxy server as a history of the analysis, which includes the option of extracting the workflow and adapting it to analyze similar data Spirodela7498Galaxy-history (Afgan et al. 2016). Then, as a third method to make the data quick to review and useful to the community, the 7498 results are now displayed on an interactive viewer hosted by the Myers lab at the Danforth center https://mpss.danforthcenter.org/tools/mirna_apps/comPARE.php. Here the user can search for miRNAs, targets and sequences, see the expression across the 24 libraries, and download expression data (Fig. 16.1) (Nakano et al. 2006). The goal of this data accessibility was to enable other scientists to explore beyond the miRNAs, to the phased small interfering RNAs, the possible lncRNA intergenic targets in the degradome sequencing, or any other striking discovery within the datasets.

While the primary focus of both sRNA-seq experiments was to analyze miRNAs, Professor Jie Tang working with strain LT5a noted a surprising lack of 24nt RNAs typically found in plant genomes. These are often comparable in expression to the 22 and 21nt miRNAs, but they were rare as 7.3% of the small RNAs in strain

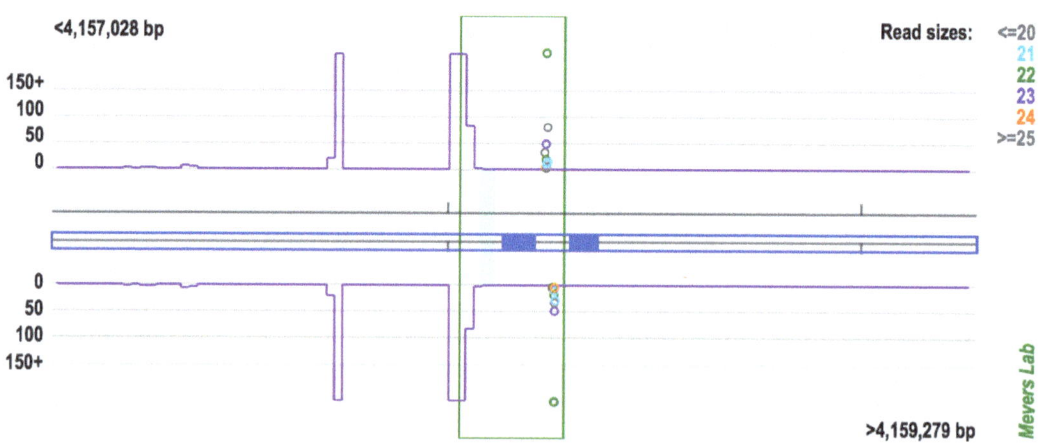


**Fig. 16.1** View of small RNA browser showing high expression of the 22nt miR396d in the intron of the unknown protein Spipo10G0052600 in the control 1 library

LT5a, and 1% in 7498. In other plant species, 24nt RNAs are a part of the RNA-directed DNA methylation pathway where transposons are transcribed into single-stranded and then double-stranded RNA, diced into 24nt heterochromatic small RNAs, and then used to guide a protein network that methylates matching sequences and then silences them as heterochromatin. Accordingly, Michael et al. also studied DNA methylation in the *Spirodela* 9509 genome and found it to be the least methylated plant sequenced! This DNA methylation pathway in duckweeds is a new and exciting field of study summarized in Chap. 5 that appears to be the cutting edge of small RNA research in the Lemnaceae.

**Acknowledgements** I would like to thank Professor Jie Tang of Chengdu University for providing data on his submitted and published work respectively and reviewing this manuscript to ensure its accuracy.

## References

Afgan E, Baker D, van den Beek M et al (2016) The galaxy platform for accessible, reproducible and collaborative biomedical analyses: 2016 update. Nucleic Acids Res 44:W3–W10. https://doi.org/10.1093/nar/gkw343

An J, Lai J, Sajjanhar A et al (2014) miRPlant: an integrated tool for identification of plant miRNA from RNA sequencing data. BMC Bioinform 15:275. https://doi.org/10.1186/1471-2105-15-275

Axtell MJ, Meyers BC (2018) Revisiting criteria for plant miRNA annotation in the era of big data. Plant Cell 30:tpc.00851.2017. https://doi.org/10.1105/tpc.17.00851

Brannan CI, Dees EC, Ingram RS, Tilghman SM (1990) The product of the H19 gene may function as an RNA. Mol Cell Biol 10:28–36. https://doi.org/10.1128/MCB.10.1.28

Dai X, Zhao PX (2011) PsRNATarget: A plant small RNA target analysis server. Nucleic Acids Res 39: W155–W159. https://doi.org/10.1093/nar/gkr319

Denman RB (1993) Using RNAFOLD to predict the activity of small catalytic RNAs. Biotechniques 15:1090–1095

Fahlgren N, Carrington JC (2010) miRNA target prediction in plants. Humana Press, pp 51–57

Franco-Zorrilla JM, Valli A, Todesco M et al (2007) Target mimicry provides a new mechanism for regulation of microRNA activity. Nat Genet 398 (39):1033. https://doi.org/10.1038/ng2079

Griffiths-Jones SM (2006) The microRNA sequence database. Methods Mol Biol 342:129–138

Howell MD, Fahlgren N, Chapman EJ et al (2007) Genome-wide analysis of the RNA-dependent RNA POLYMERASE6/DICER-LIKE4 pathway in Arabidopsis reveals dependency on miRNA- and tasiRNA-directed targeting. Plant Cell 19:926–942. https://doi.org/10.1105/tpc.107.050062

Johnson C, Kasprzewska A, Tennessen K et al (2009) Clusters and superclusters of phased small RNAs in the developing inflorescence of rice. Genome Res 19:1429–1440. https://doi.org/10.1101/gr.089854.108

Jones-Rhoades MW, Bartel DP (2004) Computational identification of plant MicroRNAs and their targets, including a stress-induced miRNA. Mol Cell 14:787–799. https://doi.org/10.1016/J.MOLCEL.2004.05.027

Kakrana A, Hammond R, Patel P et al (2014) SPARTA: a parallelized pipeline for integrated analysis of plant miRNA and cleaved mRNA data sets, including new miRNA target-identification software. Nucleic Acids Res 42:1–13. https://doi.org/10.1093/nar/gku693

Kuehdorf K, Jetschke G, Ballani L, Appenroth K-J (2014) The clonal dependence of turion formation in the duckweed *Spirodela polyrhiza*—an ecogeographical approach. Physiol Plant 150:46–54. https://doi.org/10.1111/ppl.12065

Li Y-F, Zheng Y, Addo-Quaye C et al (2010) Transcriptome-wide identification of microRNA targets in rice. Plant J 62:742–759. https://doi.org/10.1111/j.1365-313X.2010.04187.x

Meyers BC, Axtell MJ, Bartel B et al (2008) Criteria for annotation of plant MicroRNAs. Plant Cell 20:3186–3190. https://doi.org/10.1105/tpc.108.064311

Michael TP, Bryant D, Gutierrez R et al (2017) Comprehensive definition of genome features in *Spirodela polyrhiza* by high-depth physical mapping and short-read DNA sequencing strategies. Plant J 89:617–635. https://doi.org/10.1111/tpj.13400

Nakano M, Nobuta K, Vemaraju K et al (2006) Plant MPSS databases: signature-based transcriptional resources for analyses of mRNA and small RNA. Nucleic Acids Res 34:D731–D735. https://doi.org/10.1093/nar/gkj077

Scott MS, Ono M (2011) From snoRNA to miRNA: dual function regulatory non-coding RNAs. Biochimie 93:1987–1992. https://doi.org/10.1016/J.BIOCHI.2011.05.026

Song C, Wang C, Zhang C et al (2010) Deep sequencing discovery of novel and conserved microRNAs in trifoliate orange (citrus trifoliata). BMC Genom 11:431. https://doi.org/10.1186/1471-2164-11-431

Taylor RS, Tarver JE, Hiscock SJ, Donoghue PCJ (2014) Evolutionary history of plant microRNAs. Trends Plant Sci 19:175–182. https://doi.org/10.1016/J.TPLANTS.2013.11.008

Wang W, Haberer G, Gundlach H et al (2014a) The *Spirodela polyrhiza* genome reveals insights into its neotenous reduction fast growth and aquatic lifestyle. Nat Commun 5:1–13. https://doi.org/10.1038/ncomms4311

Wang W, Wu Y, Messing J (2014b) RNA-Seq transcriptome analysis of Spirodela dormancy without reproduction. BMC Genom 15:60. https://doi.org/10.1186/1471-2164-15-60

Willmann MR, Berkowitz ND, Gregory BD (2014) Improved genome-wide mapping of uncapped and cleaved transcripts in eukaryotes-GMUCT 2.0. Methods 67:64–73. https://doi.org/10.1016/j.ymeth.2013.07.003

Yang J, Liu X, Xu B et al (2013) Identification of miRNAs and their targets using high-throughput sequencing and degradome analysis in cytoplasmic male-sterile and its maintainer fertile lines of brassica juncea. BMC Genom 14:9. https://doi.org/10.1186/1471-2164-14-9

Zuker M (2003) Mfold web server for nucleic acid folding and hybridization prediction. Nucleic Acids Res 31:3406–3415. https://doi.org/10.1093/nar/gkg595

# Editing the Genome of *Wolffia australiana* 17

Thomas Reinard, Anke Londenberg, Merlin Brychcy, Kim Lühmann, Gerrich Behrendt and Maren Wichmann

**Abstract**

The genus *Wolffia* has not been the focus of biotechnological interest so far in contrast to *Lemna* and *Spirodela*. There are only a few publications on cultivation and regeneration and even fewer articles about transformation of species from this genus. However, *Wolffia* has great potential, because it can grow both floating and submerged, which allows for the production of a larger biomass compared to other genera. Furthermore, the *in planta* transformation is feasible, which may accelerate the development of new lines, since there is no need for time-consuming plant regeneration. The ground-breaking technology of genome editing allows the deletion, substitution, or insertion of genes at precise, predefined sites in the genome. Here, we present as proofs of principle the crucial steps needed to realize *Wolffia australiana* as a plant bioreactor. The procedures described include the cultivation and transformation of *W. australiana*, the identification of a suitable gene for genome editing, which can be used as a selectable marker as well, and the method of genome editing of duckweed itself, which resulted in knock-out *Wolffia* plants.

T. Reinard (✉) · A. Londenberg · M. Brychcy · K. Lühmann · G. Behrendt · M. Wichmann
Faculty of Natural Sciences, Institute of Plant Genetics, Leibniz University of Hannover, Herrenhäuser Str. 2, D-30419, Hannover, Germany
e-mail: reinard@genetik.uni-hannover.de

## 17.1 Introduction

*Wolffia* is one of the five genera of the worldwide aquatic family Lemnaceae commonly named duckweed. The genus consists of 11 species (Sree et al. 2016), categorized by their morphology and/or molecular barcoding (Table 17.1).

The taxonomic classification of Lemnaceae was rearranged in the past years. Here, we use the nomenclature according to Sree et al. (2016).

Among all Lemnaceae, the genus *Wolffia* is the most specialized with the simplest level of organization and the smallest flowering plants. The plant body of *Wolffia*, called the frond, shows an extreme morphological reduction, seen in the lack of roots and no differentiation into stem and leaves (Landolt 1986). Reproduction in *Wolffia* is one of the fastest among all angiosperms and occurs predominantly vegetatively as is true for all Lemnaceae. In contrast, sexual reproduction is very rare (Sree et al. 2015). New daughter fronds (DFs) bud from one basal cavity stay attached to the mother frond by a so-called stipe until maturation (Fig. 17.1a). Within the budding cavity of the mother frond and the connected daughter frond, the next generation will begin developing (Bernard et al. 1990). In contrast to most other Lemnaceae, *Wolffia* fronds are able to live submerged as well as floating (Thompson 1989), resulting in higher biomass per area (Fig. 17.1b). To survive unfavorable conditions like low temperature or starvation, it



X. H. Cao et al. (eds.), *The Duckweed Genomes*, Compendium of Plant Genomes,
https://doi.org/10.1007/978-3-030-11045-1_17

**Table 17.1** Genus *Wolffia* (distribution acc. (Bog et al. 2013), genome sizes acc. (Wang et al. 2011))

| Species | Distribution | Genome size (1C) (Mbp) | Comment |
|---|---|---|---|
| *W. angusta* Landolt | Australia, Malaysia | 1663 ± 34 | Own species since 1980 |
| *W. arrhiza* (L.) Horkel ex. Wimm. | Germany, Italy, Portugal, Morocco, Uganda, South Africa, Brazil | 1881 ± 83 | |
| *W. australiana* (Benth.) Hartog and Plas | South Australia, New Zealand, Tasmania | 375 ± 8/385[a] | Smallest genome of all *Wolffia* species, until 1972 variety of *W. arrhiza* |
| *W. borealis* (Engelm. Ex. Hegelm.) Landolt | USA | 889 ± 64 | |
| *W. brasiliensis* Wedd. | Brazil, Argentina, Venezuela, Dominican Rep., Bolivia | 776 ± 52 | |
| *W. columbiana* H. Karst | USA, Canada, Venezuela | 874 ± 69 | |
| *W. cylindracea* Hegelm. | Zimbabwe | 1076 ± 86 | |
| *W. elongasta* Landolt | Colombia | 847 ± 42 | |
| *W. globosa* (Roxb.) Hartog and Plas | USA, Japan, China, Vietnam, Indonesia, India | 1295 ± 42 | |
| *W. microscopia* (Griff.) | India | 1661 ± 12 | Rediscovered in 2013 |
| *W. neglecta* Landolt | Sri Lanka, Pakistan | 1176 ± 40 | |

[a]Our k-mer 17 analysis

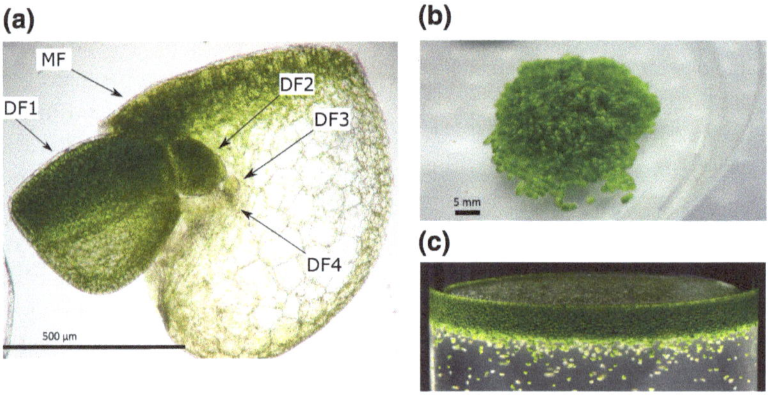


**Fig. 17.1** **a** *Wolffia australiana* with daughter frond at the left side. **b** Cultivation on solid SH medium. Each bundle consists of hundreds of plants. **c** Liquid culture of *W. australiana*, growing emerse and submerse in liquid SH medium

forms special "dormant fronds", named turions, with high starch content which sink to the bottom.

As is true with most duckweed species, these aquatic monocotyledons have high potential for biotechnological applications. Their high protein content amounts up to 30% of their dry weight, and their amino acid balance is of value for both food and feed (Appenroth et al. 1982; Landolt 1986; Cheng and Stomp 2009). Phytoremediation by growing *Wolffia* in wastewater allows for the recovering nutrients like phosphorous or nitrate and the accumulation of heavy metals (Lewis 1995; Dushenkov et al. 1995; Dhir et al. 2009). Finally, they are useful in the production of biofuels due to their high content of starch (Cheng and Stomp 2009; Cui and Cheng 2015; Xu et al. 2011).

*Wolffia australiana* was raised to the status of a species in 1972. Before this year, it was incorrectly regarded as a variety of *W. arrhiza* (den Hartog and van der Plas 1972). *W. australiana* originated from surfaces of calm waters of the temperate region of southern Australia and New Zealand (Landolt 1994). With a frond size of about 0.5–0.95 mm long, 0.30–0.52 mm wide, and 0.79–1.32 mm high, *W. australiana* represents one of the larger species among the genus *Wolffia* (den Hartog and van der Plas 1972; Landolt 1994). Figure 17.1a shows the elliptical dorsal surface and a bright green appearance, due to the high concentration of chloroplasts in the upper cell layer. The ventral bulge with its lighter green appearance is composed of larger, highly vacuolated cells with a lower concentration of chloroplasts. Usually, *W. australiana* appears as a two-plant colony consisting of one mother frond attached to a daughter frond. The average lifespan of one plant is about 17 days, during which approximately eleven daughter fronds develop by budding (Bernard et al. 1990). With doubling times of around 24–48 h, the growth rate is remarkably high (Ziegler et al. 2015). Because *Wolffia* can be grown on solid and on liquid media while both floating and submerged (Fig. 17.1c), the cultivation can be adapted to nearly all requirements (Kruse et al. 2002; Thompson 1989). Its cultivation in liquid can result in a higher biomass production compared to other duckweed species, which are limited to floating *modus vivendi*.

## 17.2 Culture and Transformation

As summarized by Khvatkov (Khvatkov et al. 2015a), different media have been used for the various duckweed species. We found Schenk and Hildebrandt (SH) medium adjusted to pH 6.0 to be the most suitable for *W. australiana* (Rechmann et al. 2007; Khvatkov et al. 2015a), but according to the literature, MS media is used as well (Boehm et al. 2001; Kruse et al. 2002). In our experiments, in vitro culture of the fronds occurred at 20 °C at a light intensity of 60–80 µmol $m^{-2}$ $s^{-1}$ during a 16-h day photoperiod (PAR-region of 400–700 nm). Solid media was used before and following the actual transformation of *W. australiana* with *A. tumefaciens*, as well as for long-term cultivation (Fig. 17.1b).

Several transformation approaches to obtain transgenic Lemnaceae have been described (Table 17.2). Most approaches rely on the transformation of induced structures like calli. For the genus, *Lemna* and *Spirodela* stable transformations have been shown by various groups, whereas all but one approach for *Wolffia* has been transient up to now.

We would like to quote that there are some patents as well (Edelman et al. 1998; Stomp and Rajbhandari 2000; Spencer et al. 2011) and two unpublished PhD theses from Bonn University (Friedrich 2005; Becker 2006), which are not listed in Table 17.2. Some groups used the transformation systems listed in Table 17.2 for their appropriate biotechnological applications of *Lemna* without further modification of the protocols (e.g., Woodard et al. 2009; Bertran et al. 2015).

For *Lemna minor, Lemna gibba*, as well as *Spirodela oligorrhiza* (Table 17.2), transformation protocols usually depend on the use of callus—respective nodule—culture as it was initially described by (Stomp and Rajbhandari 2000; Yamamoto et al. 2001). In contrast to *Spirodela* and *Lemna*, only very few approaches for culture and/or transformation of members of the genus *Wolffia* have been published. Kruse and co-worker used particle bombardment for the transient transformation of *Wolffia* (Kruse et al. 2002).

Another induced structure, obtained for *Wolffia arrhiza*, is the so-called cluster culture (Khvatkov et al. 2015a). The cluster structure develops directly from complete *W. arrhiza* fronds. Despite the cluster culture's success in the creation of a stable callus culture and regeneration of whole plants, only the novel

**Table 17.2** Overview of in vitro culture and transformation approaches of Lemnaceae in the literature

| Species | Publications | Method | Transformation: stable or transient |
|---|---|---|---|
| *Lemna minor* and/or *Lemna gibba* | Yamamoto et al. (2001), Cox et al. (2006), Chhabra et al. (2011), Ko et al. (2011), Nguyen et al. (2012), Bertran et al. (2015), Cantor–Pastor et al. (2015) | Co-cultivation of A. tumefaciens with nodule: Yamamoto, with frond: Ko; nodules with LEX system: Cox. All other authors used callus | Usually stable, either tested after >2 years of culture and/or Southern blot |
| *Spirodela* | Vunsh et al. (2007), Rival et al. (2008) | Vunsh: co-cultivation with nodule, Rival: with callus | |
| *W. columbiana* | Boehm et al. (2001) | Ballistic approach and A. tumefaciens-mediated | Transient |
| *W. australiana* | Rechmann et al. (2007) | No transformation | |
| *Wolffia arrhiza* | Khvatkov et al. (2015a, b) | Ballistic approach and A. tumefaciens-mediated transformation, callus culture | Claims stable by Southern analysis |
| *W. australiana* | Our data | Whole plant approach | Transient |

cluster culture allowed successful differentiation of transgenic *W. arrhiza* plants (Khvatkov et al. 2015b). To our knowledge, this is the only reported stable transformation for the genus *Wolffia*. Most approaches described above include a transgenic plant developing from a secondary structure, which has earlier been co-cultivated with *A. tumefaciens*.

In contrast, our approach focuses on carrying out the *Agrobacterium*-based transformation *in planta*, i.e., utilizing the full plant (Edelman et al. 1998; Stomp and Rajbhandari 2000). There is no need to cut the plant into small particles or to use tissue culture followed by in vitro regeneration for the transformation. Edelman (Edelman et al. 1998) co-cultivated complete *L. punctata* fronds with *A. tumefaciens* and obtained GUS-positive fronds. Prior to co-cultivation, existing daughter fronds of the to-be-transformed mother fronds were removed and vacuum infiltration was performed. Other reports examined the influence of injuring fronds before vacuum infiltration on transformation efficiency for *W. columbiana* (Boehm et al. 2001) or *L. minor* (Ko et al. 2011). Another less common option used for the larger *Lemna* or *Spirodela* species is the microinjection of DNA into the meristem (Edelman et al. 1998).

Our group also developed a highly efficient protocol for *in planta* transformation. Several parameters were tested; some of those resulted in an increase of transformation rate (Table 17.3).

- The frond injury before the transformation event was not performed manually, but by the use of an ultrasonic bath under white light;
- *A. tumefaciens* used for transformation were supplied with acetosyringone before inoculation (Stomp 2005);
- The inoculation itself was repeated three times using vacuum infiltration.

After 4–7 days of co-culture, *Wolffia* plants were washed at least two times for 20 min, in order to remove remaining Agrobacteria. SH media supplemented with 250 mg/l Timentin was used. After this procedure, plants were placed on solid SH medium with Timentin. If a selection with allyl alcohol (prop-2-en-1-ol) had to be performed (s. b.), the plants were placed in liquid SH medium with 20 μM allyl alcohol and incubated for one hour (Widholm and Kishinami 1988).

This *in planta* transformation approach led to transient transformed, chimeric plants. These chimeric plants could be used to develop the protocols needed for the transformation in a minimal time period (Fig. 17.2). A transient transformation is a prerequisite for performing the genome editing.

**Table 17.3** Detailed parameter for the transformation of *W. australiana*

| Step | Procedure |
|---|---|
| Step 1 | *A. tumefaciens* for the transformation of *W. australiana* were grown in YEB medium pH 7.2 containing 20 μM acetosyringone for preculture overnight |
| Step 2 | 100 μl of this preculture were used for the inoculation of the 50 ml main culture, growing on YEB medium with a pH 5.6 and 200 μM acetosyringone up to an $OD_{600}$ 1.0–1.5 |
| Step 3 | *W. australiana* plants, cultivated as described above from 2 to 3 weeks of cultures, were added to the 50 ml bacterial culture |
| The following two steps are repeated three times: | |
| Step 4 | Ultrasonic water bath for 30 s at 40–60% intensity |
| Step 5 | Vacuum infiltration (850 mbar) for 6 min |
| Step 6 | Only those plants, which sank down over the course of the treatment, were transferred to solid SH medium supplemented with 200 μM acetosyringone. Following this treatment, the plants were kept at 20 °C and a 16-h day photoperiod (intensity about 6080 μmol/sm$^2$) for culturing |

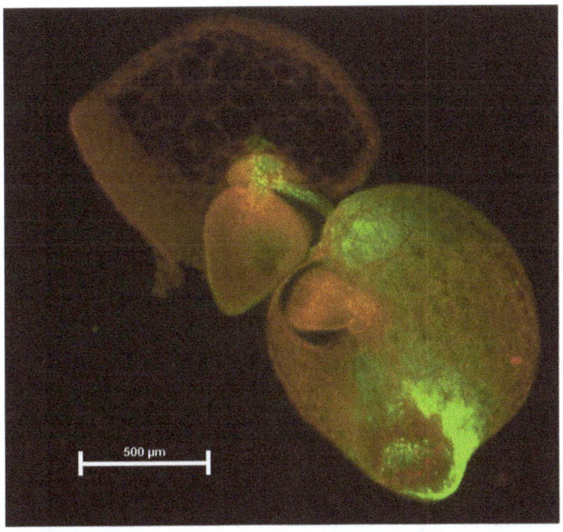


**Fig. 17.2** *W. australiana* plants under UV-light 21 days after transformation using *A. tumefaciens* strain GV2260: p1609 (This vector is a derivate of pBIN19 with mGFP5-gene (Acc: U87973) under the control of a 35S-promoter). The strongest GFP signals can be found in the area of the stipe and at the meristematic pocket of the mother and the daughter fronds. The mother frond tip area shows also a strong fluorescence but is interspersed by a ring of non-transformed tissue. We frequently observed this fluorescence pattern, suggesting that these plant parts are most amenable to *Agrobacterium*-mediated transformation

## 17.3 The Draft Genome of *W. australiana*

Because our final goal was the genome editing of *W. australiana*, we considered it appropriate to sequence its genome, in order to identify suitable target sequences after we established the transient transformation protocol.

The genome sizes of different species of the genus *Wolffia* differ almost fivefold (Wang et al. 2011). Compared to the other members of the genus, *W. australiana* has by far the smallest genome (375 Mbp measured by flow cytometry (Wang et al. 2011) to 385 Mbp, our data analyzed by k-mer 17 analysis), which has made it the preferred species for a genome sequencing approach. Thoroughly annotated duckweed genomes from other genera were already available, such as of *L. minor* (Van Hoeck et al. 2015) with an approximately equal genome size, as well as of *Spirodela polyrhiza* (Wang et al. 2014; Cao et al. 2016; Michael et al. 2017), with a genome half the size of that of *W. australiana*.

Therefore, a sequencing approach with 39× coverage of the *W. australiana* strain DWC304 using the Illumina next-generation sequencing platform (HiSeq PE150) appears sub-standardly. However, our focus was not on clarifying the genome sequence as precisely as possible, but our focus was on finding suitable target sequences for the genome editing experiments described below.

It is not surprising that the degree of heterozygosity is very low, which reflects the preferential vegetative propagation of all duckweeds. As is true for the sequenced genomes of the other genera (Van Hoeck et al. 2015; Wang et al. 2014; Cao et al. 2016; Michael et al. 2017), the percentage of repetitive sequences is quite high. This repetition complicates the assembly and annotation of the genome data. In contrast to the genome projects mentioned above, there was hardly any transcriptome data, besides an EST-library of about 1,988 sequences (SAMN00222771, ID: 222771) and the complete sequence of the plastome (Wang and Messing 2011). The assembly was performed using the well-established protocol described by Li et al. (2010), and k-mer 17 analysis revealed a genome size of 385 Mbp for *W. australiana* (Marçais and Kingsford 2011) and as described here, http://koke.asrc.kanazawa-u.ac.jp.

Due to the absence of the necessary computing capacity, a recently established, web-based pipeliner, the Genome Sequence Annotation Server, was chosen for the annotation (Humann et al. 2017). The GenSAS pipeliner (www.gensas.org) combines many tools and is precisely configurable. Because one may upload DNA evidence data, we added gene sets from Liliopsidae, ESTs from Lemnaceae, and 1,790 ESTs from a *W. australiana* EST project (JZ896467.1) in addition to the predefined plant reference genome dataset (Humann et al. 2017). After eliminating repetitive sequences (RepeatMasker and RepeatModeler), the alignment can be done using BLAT, nucleotide BLAST, and PASA. The data obtained was fine-tuned using gene modelers like Augustus and SNAP. A set of protein sequence-based annotation tools like BLASTp, InterProScan, Pfam, SignalP, and TargetP was applied, followed by the creation of the official gene set. This service allows even small workgroups without access to mainframes the annotation of genome data with a user-friendly interface.

The gene set obtained from GenSAS was used for further functional genome analysis on the protein level using the BLAST2GO software package (www.blast2go.com), which resulted in the identification of 18,617 protein sequences from *W. australiana*.

However, our main objective was to identify the genome context of the *adh1* gene (alcohol dehydrogenase 1), whose gene product can be used for selection using prop-2-en-1-ol, also known as allyl alcohol (Widholm and Kishinami 1988). Inactivation of ADH1 enzyme enables the plant to grow on allyl alcohol, which is otherwise toxic to the plant. It should be emphasized that there are typically several isoenzymes in a plant genome and this protein class shows only relatively weak homologies. Therefore, a PCR-based amplification of the gene from the genomic DNA was not possible; this is why genome sequencing was required. After identification of the *adh1* locus in *W. australiana*, we were able to use this locus as a selection system for targeted genome editing events without the need for other selection markers.

## 17.4 Genome Editing of *W. australiana*

Genome editing has become a major force in modern biotechnology. Its popularity is due to its underlying technologies which allow for the easy study of genes and their functions through knock-out/knock-ins or through the regulation of gene expression. Furthermore, genome editing allows for specific insertion of the gene of interest into a predefined site of the genome (knock-in). Because the knock-in scenario is somewhat tricky, we focused on a proof of principle approach using a knock-out strategy. We combined our knock-out strategy with the option to establish an *in planta* selection method, which would enable us to omit selection marker genes in the DNA sequences to be used for transformation. Since this is the first application of the CRISPR-Cas9 technology in duckweed, we would like to describe how genome editing works and how to apply genome editing in duckweed.

Nearly all available genome editing technologies rely on endonucleases, which create a double-strand break (DSB) in the target DNA (Osakabe and Osakabe 2015). DSBs, like any other DNA damage, are repaired in vivo through different repair mechanisms. The two major mechanisms are the non-homologous end joining (NHEJ) and the homology directed repair (HDR).

NHEJ can be described as an erroneous repair mechanism. There are two pathways in which NHEJ can repair a DSB. First, the proteins Ku70/Ku80 can lead a ligase and its cofactor to the DSB. This option usually leads to small base insertions or deletions (Deriano and Roth 2013). The second, often preferred pathway is based on microhomologies (Vu et al. 2014). Exonucleases digest the ends of the broken DNA until short homologies on both ends are available, which are then re-ligated (Crespan et al. 2012; Deriano and Roth 2013). Therefore, this pathway causes deletions instead of insertions. NHEJ is the main repair mechanism for DSBs in eukaryotic cells (Sonoda et al. 2006).

The other type of repair mechanism is available only in a small number of cell types and is named homology directed repair (HDR). HDR is based on homologous recombination (HR) and works in the same fashion. The process requires homologous DNA sequences as a repair template. This template DNA is used to repair the DSB. The HDR process allows the integration or substitution of bases if there are flanking, homologous sequences around the DSB. If the DNA to be integrated is flanked by homologous DNA sequences (each about 1000 bp), an integration at the site of the broken dsDNA strand will occur (Osakabe and Osakabe 2015). It was shown that whole genes can be inserted using this method in plants (Fauser et al. 2014; Schiml et al. 2014).

Today, two methods of genome editing are used: TALEN and CRISPR/Cas9 (Osakabe and Osakabe 2015; Singh et al. 2015; Samanta et al. 2016; Schiml and Puchta 2016). We have chosen the second procedure, the CRISPR/Cas9 system.

Originally, CRISPR (clustered regularly interspaced short palindromic repeat) was described as a defense mechanism in bacteria and archaea to fight phages (Wiedenheft et al. 2012). Genomes of these organisms contain a cluster of foreign (phage) DNA sequences, the so-called CRISPR array (Barrangou et al. 2007). Each repeat code for two RNAs: the crRNA represents the foreign DNA sequences, and the tracrRNA represents an integral element of the bacterial genome. The crRNA binds to foreign DNA, delivered by the phage in the case of an infection, whereas the tracrRNA binds to the crRNA. This binding causes a hairpin structure, which can be detected by the Cas9 protein, an endonuclease that creates a DSB in the target DNA (Garneau et al. 2010; Sternberg et al. 2014).

There is another prerequisite needed by the Cas9 nuclease; it is the so-called protospacer adjacent motif (PAM). The PAM consists of three bases (NGG) and must be located next to the sequence to be cut (Garneau et al. 2010; Sternberg et al. 2014). For practical reasons, the two RNAs involved in the CRISPR/Cas9 mechanism are combined into one so-called single guide RNA (sgRNA or gRNA) (Jinek et al. 2013), when used for genome editing.

CRISPR/Cas9 has successfully been applied to edit the genome of various organisms (e.g., Cong et al. 2013; Jiang et al. 2013a, 2013b; Mali et al. 2013; Dicarlo et al. 2013; Kim et al. 2014; Tang et al. 2017), but has yet to be demonstrated in Lemnaceae.

Cloning of sgRNAs can be performed relatively simply using the Golden Gate cloning technology (Engler et al. 2008). The tracrRNA section of the sgRNAs is already present in the plasmids used for genome editing, and the same is true for the Cas9 gene. Therefore, only the cloning of the homologous section of the sgRNA has to be cloned. These parts can be constructed from two 20 bp long oligonucleotides, which are annealed. Premade plasmids for CRISPR/Cas9 are available for plants and can be cloned using the Golden Gate system as well (Ordon et al. 2017). Our strategy to utilize CRISPR/Cas9 in *W. australiana* is founded on the MoClo system (Weber et al. 2011; Engler et al. 2014) and is outlined in Table 17.4.

To evaluate suitable sites for the sgRNA binding and to design and to score sgRNAs, we used the online sequence analysis tool Benchling (http://www.benchling.com, Doench et al. 2016). Our own research showed that sgRNAs should at least demonstrate a score of 70 to obtain a genome editing event later on. The chosen sgRNAs should enable the detection of the successful genome editing event by a shift in the band size in a gel compared to wild-type DNA. Two oligos containing the homologous part of the sgRNA and the specific overhangs for the entry vectors were designed, annealed, and cloned into the entry vectors. Then, multiple sgRNAs were cloned into the final vectors. Because the correct function of the designed sgRNAs is not completely assured, it is advisable to create several sgRNAs directed against different target sequences in the region to be edited. Therefore, we designed four different sgRNAs, which were all transferred into the plants in combination with the *Cas9* gene (Fig. 17.3).

In principle, two different entry vectors, pDGE and pMGE, are available. Since these vectors were created for the MoClo system (Ordon et al. 2017), they can be cloned from level to level with one of the type IIS restriction enzymes BpiI or BsaI. Furthermore, the selection gene changes with the respective level of the vector. The pDGE and pMGE systems differ mainly in the U6 promoter used. The promoter in the pDGE system originates from *Arabidopsis thaliana*, while the pMGE system uses the U6 promoter from the monocot *Oryza sativa*. However, only the pDGE system could successfully be applied for *W. australiana*. The U6 promoter is important for the expression of the sgRNAs. In contrast to other common promoters, such as the widely used CaMV35S promoter, the U6 promoter produces RNAs with defined transcription start, which leads to a precisely defined sgRNA.

The MoClo-based two-level system for cloning of the sgRNAs allows for the use of multiple sgRNAs in one backbone vector by the first cloning each single sgRNA into an entry vector. Following, the four entry vectors can all be sub-cloned in one reaction into the target (level 1) vector. This is enabled by the creation of different overhangs by BpiI or BsaI, respectively. Since the overhangs of successive sgRNA-modules are compatible with each other, all

**Table 17.4** Procedure of genome editing of *W. australiana*

| Step | Procedure |
|---|---|
| Step 1 | Finding the adh1 gene in the Wolffia genome sequence and producing primers for amplification of the genomic adh1 sequence |
| Step 2 | Identification of suitable target sequences for sgRNAs (taking into account neighboring PAM sites). Calculation of the on target scores was done using Benchling (www.benchling.com) |
| Step 3 | Cloning of four different sgRNAs with high score on Benchling into the pDGE001 vector using Golden Gate cloning (see Fig. 17.3) |
| Step 4 | Transformation of *E. coli* followed by transfer of the vector pDGE3_sg1-4ADH into *A. tumefaciens* |
| Step 5 | Transformation of *W. australiana* as described in Table 17.3 |
| Step 6 | Selection on tolerance against allyl alcohol |
| Step 7 | Verification of adh1-knock-outs by PCR and sequencing of the PCR product |

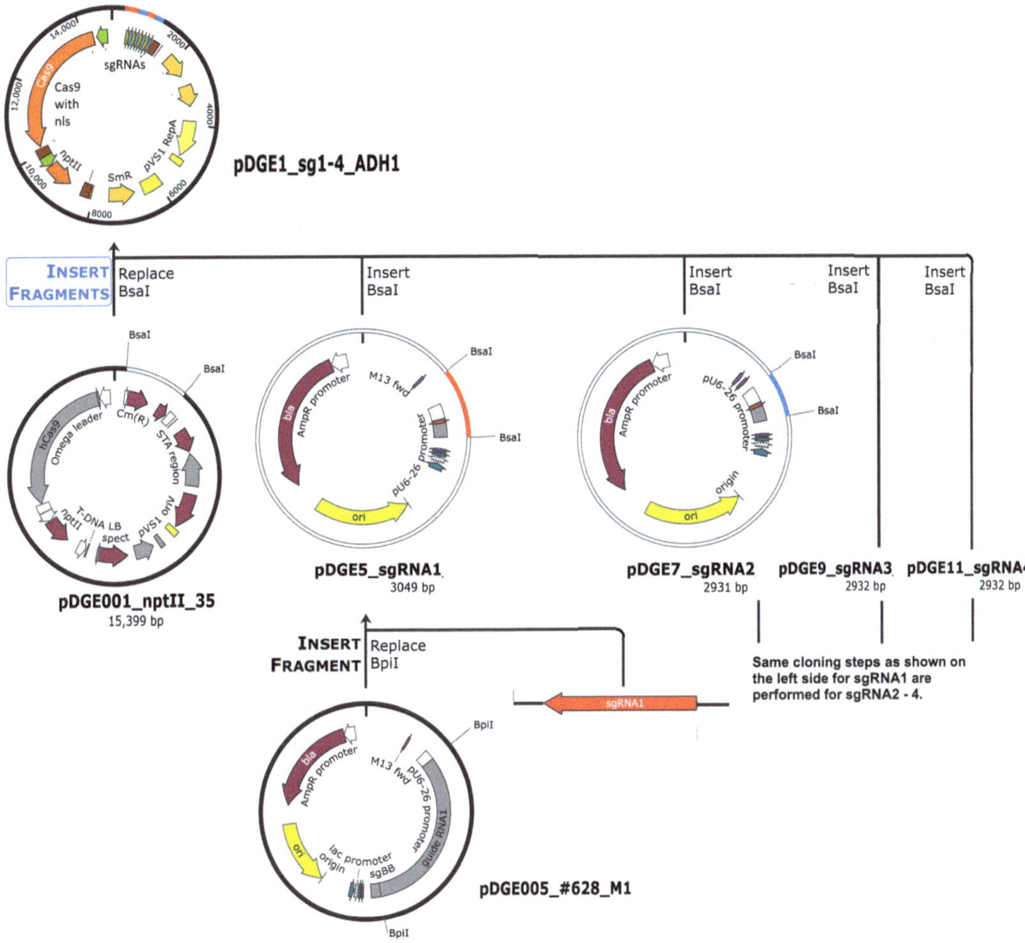


**Fig. 17.3** Cloning strategy for the use of four sgRNAs in parallel

elements are connected in the correct order. The Golden Gate system also demonstrates its superiority as the CutSmart® cloning method can also be used. In this procedure, restriction enzyme and ligase are combined with the DNA to be cloned in one reaction, in which both restriction and ligation occurred. Depending on the temperature (37 C or 16 C), either restriction or ligation occurs. The temperature is cycled for 8–50 times to achieve a very high efficiency compared to normal cloning. Because the recognition site of the restriction enzymes has been cut off in the resulting DNA molecules, cloned DNA molecules can no longer serve as substrate for the enzymes and are enriched in the pot.

The vector used for plant transformation is a typical binary vector. Its T–DNA contains the CDS of the Cas9 enzyme which is controlled by the ubiquitin promoter from *parsley* (pUbi4-2(parsley)). Our experiments revealed that this promoter showed the best efficiency in genome editing compared to other promoters like the CaMV35S promoter. As described above, the full construct contained both the Cas9 and four different sgRNAs. The plasmid was transformed via *A. tumefaciens* into *W. australiana* as described above (Table 17.3).

We were able to successfully knock-out the *adh1* gene of *W. australiana* and to verify that three of the four sgRNAs used led to a genome editing event. Typical observations, like the cut site being 3 bp upstream of the PAM site (Ran et al. 2013), were found as well.

Due to the successful genome editing event in *W. australiana*, the plants became tolerant to allyl alcohol treatment, through which they were easily selected. The addition of allyl alcohol to ADH1-expressing cells causes the production of the highly toxic acrolein. Genomic DNA of the tolerant plants was isolated and further analyzed. The *adh1* gene was amplified by PCR with primers neighboring the possible deletion in the *adh1* gene. The amplified DNA was separated during gel electrophoresis and showed small bands of approximately 150 bp which indicates a deletion in the *adh1* gene (Fig. 17.4). The small band appears in the samples of selected and edited *W. australiana*. The combination of sgRNA 1 and sgRNA 4 (lane 1–3), sgRNA 2 and sgRNA 4 (lane 4–6), and all four sgRNAs (lane 7–9) led to the small band. The samples in lane 7–9 also show the wild-type band of 2200 bp without a deletion. This lack of detection is due to the fact that the edited *W. australiana* is chimeric.

The small bands of 150 bp were cloned, sequenced, and aligned to the *adh1* gene of *W. australiana*. This alignment showed the deletion of 2051 bp of the adh1 gene. This deletion reaches from the binding sites of sgRNA1/2 and sgRNA4. The sgRNA3 never was found to result in a deletion, independently which other sgRNAs were used. Therefore, sgRNA3 is suggested to be not efficient.

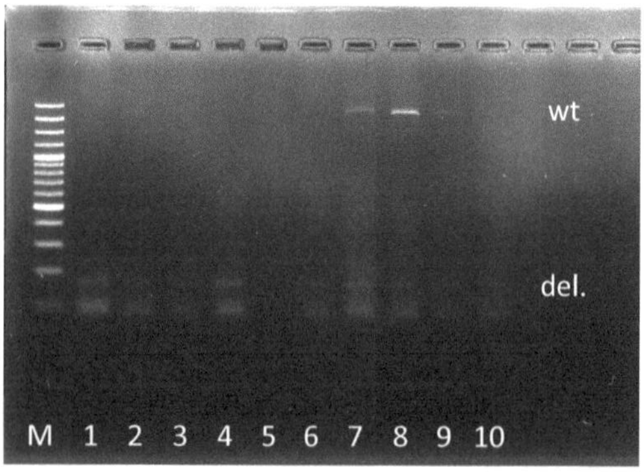


**Fig. 17.4** Agarose gel with PCR samples of several individual *W. australiana* plants after genome editing. M: GeneRuler™ 100 bp Plus (Thermo Fisher Scientific), lanes 1–10: different plants. The smaller bands indicate a deletion in the *adh1* gene (del), and bands at around 2,200 bp indicate the wild-type *adh1* gene (wt)

## 17.5 Conclusions and Perspectives for the Biotechnological Use of *W. australiana*

In this report, we showed the various steps of sterile culture, transformation including selection procedure, and genome editing, which are the prerequisites for the biotechnological use of *W. australiana*. However, the work presented here can only be regarded as proof of principle. Nevertheless, the economical use of *W. australiana* as a bioreactor for heterologous expressed proteins still requires further efforts. The two most crucial drawbacks which must be overcome are the chimeric status of the transformed plants and the undemonstrated stable and long-term transformation. Both problems may be solved by the use of protoplasts, which must be regenerated to whole plants. However, currently, no method for protoplast production in duckweed has been described. Although we have produced protoplasts, we have yet to perfect protoplast regeneration. Another important step toward a highly efficient bioreactor is the establishment of a knock-in protocol in *W. australiana* based on genome editing (Schiml et al. 2014; Schiml and Puchta 2016).

**Acknowledgements** Beate Meyer, Yvonne Koleczek, and Eva Plönnigs are acknowledged for excellent technical assistance and Shannon McCullan for her diligent proofreading of the manuscript. The authors do not declare any conflict of interest.

## References

Appenroth K-J, Augsten H, Liebermann B, Feist H (1982) Effects of light quality on amino acid composition of proteins in *Wolffia arrhiza* (L.) Wimm. using a specially modified bradford method. Biochem und Physiol der Pflanz 177:251–258. https://doi.org/10.1016/S0015-3796(82)80008-5

Barrangou R, Fremaux C, Deveau H et al (2007) CRISPR provides acquired resistance against vires in prokaryotes. Science 80(315):1709–1712. https://doi.org/10.1126/science.1138140

Becker S (2006) Immuntoxinproduktion in Pflanzen. Eine innovative Methode zur Herstellung neuartiger Therapeutika gegen das Hodgkin Lymphom. PHD Thesis

Bernard FA, Bernard JM, Denny P (1990) Flower structure, anatomy and life history of *Wolffia australiana* (Benth.) den Hartog and van der Plas. Bull Torrey Bot Club 117:18–26. https://doi.org/10.2307/2997125

Bertran K, Thomas C, Guo X et al (2015) Expression of H5 hemagglutinin vaccine antigen in common duckweed (*Lemna minor*) protects against H5N1 high pathogenicity avian influenza virus challenge in immunized chickens. Vaccine 33:3456–3462. https://doi.org/10.1016/j.vaccine.2015.05.076

Boehm R, Kruse C, Voeste D et al (2001) A transient transformation system for duckweed (*Wolffia columbiana*) using *Agrobacterium*-mediated gene transfer. J Appl Bot 75:107–111

Bog M, Schneider P, Hellwig F et al (2013) Genetic characterization and barcoding of taxa in the genus *Wolffia* Horkel ex Schleid (Lemnaceae) as revealed by two plastidic markers and amplified fragment length polymorphism (AFLP). Planta 237:1–13. https://doi.org/10.1007/s00425-012-1777-9

Cantor-Pastor A, Mollar-Morales A, Ernst E et al (2015) Efficient transformation and artificial miRNA gene silencing in *Lemna minor*. Plant Biol 17:59–65. https://doi.org/10.1111/plb.12215

Cao HX, Vu GTH, Wang W et al (2016) The map-based genome sequence of *Spirodela polyrhiza* aligned with its chromosomes, a reference for karyotype evolution. New Phytol 209:354–363. https://doi.org/10.1111/nph.13592

Cheng JJ, Stomp A-M (2009) Growing duckweed to recover nutrients from wastewaters and for production of fuel ethanol and animal feed. Clean-Soil Air Water 37:17–26. https://doi.org/10.1002/clen.200800210

Chhabra G, Chaudhary D, Sainger M, Jaiwal PK (2011) Genetic transformation of Indian isolate of *Lemna minor* mediated by *Agrobacterium tumefaciens* and recovery of transgenic plants. Physiol Mol Biol Plants 17:129–136. https://doi.org/10.1007/s12298-011-0059-5

Cong L, Ran FA, Cox D et al (2013) Multiplex genome engineering using CRISPR/Cas systems. Science 80 (339):819–823. https://doi.org/10.1126/science.1231143

Cox KM, Sterling JD, Regan JT et al (2006) Glycan optimization of a human monoclonal antibody in the aquatic plant *Lemna minor*. Nat Biotechnol 24:1591–1597. https://doi.org/10.1038/nbt1260

Crespan E, Czabany T, Maga G, Hübscher U (2012) Microhomology-mediated DNA strand annealing and elongation by human DNA polymerases λ and β on normal and repetitive DNA sequences. Nucleic Acids Res 40:5577–5590. https://doi.org/10.1093/nar/gks186

Cui W, Cheng JJ (2015) Growing duckweed for biofuel production: a review. Plant Biol 17:16–23. https://doi.org/10.1111/plb.12216

den Hartog C, van der Plas F (1972) The Australian species of *Wolffia* (Lemnaceae). Blumea—Biodiv Evol Biogeogr Plants 20:151–153

Deriano L, Roth DB (2013) Modernizing the nonhomologous end-joining repertoire: alternative and classical NHEJ share the stage. Annu Rev Genet 47:433–455. https://doi.org/10.1146/annurev-genet-110711-155540

Dhir B, Sharmila P, Saradhi PP (2009) Potential of aquatic macrophytes for removing contaminants from the environment. Crit Rev Environ Sci Technol 39:754–781

Dicarlo JE, Norville JE, Mali P et al (2013) Genome engineering in *Saccharomyces cerevisiae* using CRISPR-Cas systems. Nucleic Acids Res 41:4336–4343. https://doi.org/10.1093/nar/gkt135

Doench JG, Fusi N, Sullender M et al (2016) Optimized sgRNA design to maximize activity and minimize off-target effects of CRISPR-Cas9. Nat Biotechnol 34:184–191. https://doi.org/10.1038/nbt.3437

Dushenkov V, Kumar PBAN, Motto H, Raskin I (1995) Rhizofiltration: the use of plants to remove heavy metals from aqueous streams. Environ Sci Technol 29:1239–1245. https://doi.org/10.1021/es00005a015

Edelman M, Perl A, Flaishman M, Blumenthal A (1998) Patent US 7,176,352 B1: Transgenic Lemnaceen, pp 1–34

Engler C, Kandzia R, Marillonnet S (2008) A one pot, one step, precision cloning method with high throughput capability. PLoS ONE 3:e3647. https://doi.org/10.1371/journal.pone.0003647

Engler C, Youles M, Gruetzner R et al (2014) A golden gate modular cloning toolbox for plants. ACS Synth Biol 3:839–843. https://doi.org/10.1021/sb4001504

Fauser F, Schiml S, Puchta H (2014) Both CRISPR/Cas-based nucleases and nickases can be used efficiently for genome engineering in *Arabidopsis thaliana*. Plant J 79:348–359. https://doi.org/10.1111/tpj.12554

Friedrich AS (2005) Untersuchungen zu Kultivierung. Universität Bonn, Transformation und Fermentation von Wolffia spec

Garneau JE, Dupuis MÈ, Villion M et al (2010) The CRISPR/cas bacterial immune system cleaves bacteriophage and plasmid DNA. Nature 468:67–71. https://doi.org/10.1038/nature09523

Humann J, Ficklin S, Lee T et al (2017) Structural and functional annotation of model and non-model organisms with GenSAS v5.0, a web-based annotation platform. In: Bioinformatics Workshop, International Plant and Animal Genome Conference XXV. San Diego, CA, USA

Jiang W, Bikard D, Cox D et al (2013a) RNA-guided editing of bacterial genomes using CRISPR-Cas systems. Nat Biotechnol 31:233–239. https://doi.org/10.1038/nbt.2508

Jiang W, Zhou H, Bi H et al (2013b) Demonstration of CRISPR/Cas9/sgRNA-mediated targeted gene modification in *Arabidopsis*, tobacco, sorghum and rice.

Nucleic Acids Res 41:e188–e188. https://doi.org/10.1093/nar/gkt780

Jinek M, East A, Cheng A et al (2013) RNA-programmed genome editing in human cells. Elife 2013:e00471. https://doi.org/10.7554/eLife.00471

Khvatkov P, Chernobrovkina M, Okuneva A et al (2015a) Callus induction and regeneration in *Wolffia arrhiza* (L.) Horkel ex Wimm. Plant Cell, Tissue Organ Cult 120:263–273. https://doi.org/10.1007/s11240-014-0603-4

Khvatkov P, Chernobrovkina M, Okuneva A et al (2015b) Transformation of *Wolffia arrhiza* (L.) Horkel ex Wimm. Plant Cell, Tissue Organ Cul 123:299–307. https://doi.org/10.1007/s11240-015-0834-z

Kim S, Kim D, Cho SW et al (2014) Highly efficient RNA-guided genome editing in human cells via delivery of purified Cas9 ribonucleoproteins. Genome Res 24:1012–1019. https://doi.org/10.1101/gr.171322.113

Ko SM, Sun HJ, Oh MJ et al (2011) Expression of the protective antigen for PEDV in transgenic duckweed, *Lemna minor*. Hortic Environ Biotechnol 52:511–515. https://doi.org/10.1007/s13580-011-0007-x

Kruse C, Boehm R, Voeste D et al (2002) Transient transformation of *Wollfia columbiana* by particle bombardment. Vitro 72:175–181

Landolt E (1986) The family of Lemnacea: a monographic study. Veroffentlichungen des Geobot Institutes der ETH Zurich, Stift Rubel 71:566

Landolt E (1994) Taxonomy and ecology of the section wolffia of the genus wolffia (Lemnaceae). Veröffentlichungen des Geobot Institutes der ETH, Stift Rübel, Zürich 60:137–151

Lewis MA (1995) Use of freshwater plants for phytotoxicity testing: a review. Environ Pollut 87:319–336. https://doi.org/10.1016/0269-7491(94)P4164-J

Li R, Zhu H, Ruan J et al (2010) De novo assembly of human genomes with massively parallel short read sequencing. Genome Res 20:265–272. https://doi.org/10.1101/gr.097261.109

Mali P, Yang L, Esvelt KM et al (2013) RNA-guided human genome engineering via Cas9. Science 339:823–826. https://doi.org/10.1126/science.1232033

Marçais G, Kingsford C (2011) A fast, lock-free approach for efficient parallel counting of occurrences of k-mers. Bioinformatics 27:764–770. https://doi.org/10.1093/bioinformatics/btr011)

Michael TP, Bryant D, Gutierrez R et al (2017) Comprehensive definition of genome features in Spirodela polyrhiza by high-depth physical mapping and short-read DNA sequencing strategies. Plant J 89:617–635. https://doi.org/10.1111/tpj.13400

Nguyen LV, Cox KM, Ke JS et al (2012) Genetic engineering of a Lemna isoleucine auxotroph. Transgenic Res 21:1071–1083. https://doi.org/10.1007/s11248-012-9594-2

Ordon J, Gantner J, Kemna J et al (2017) Generation of chromosomal deletions in dicotyledonous plants employing a user-friendly genome editing toolkit. Plant J 89:155–168. https://doi.org/10.1111/tpj.13319

Osakabe Y, Osakabe K (2015) Genome editing with engineered nucleases in plants. Plant Cell Physiol 56:389–400. https://doi.org/10.1093/pcp/pcu170

Ran FA, Hsu PD, Wright J et al (2013) Genome engineering using the CRISPR-Cas9 system. Nat Protoc 8:2281–2308. https://doi.org/10.1038/nprot.2013.143

Rechmann H, Friedrich A, Forouzan D et al (2007) Characterization of photosynthetically active duckweed (*Wolffia australiana*) in vitro culture by respiration activity monitoring system (RAMOS). Biotechnol Lett 29:971–977. https://doi.org/10.1007/s10529-007-9330-9

Rival S, Wisniewski JP, Langlais A et al (2008) Spirodela (duckweed) as an alternative production system for pharmaceuticals: a case study, aprotinin. Transgenic Res 17:503–513. https://doi.org/10.1007/s11248-007-9123-x

Samanta MK, Dey A, Gayen S (2016) CRISPR/Cas9: an advanced tool for editing plant genomes. Transgenic Res 25:561–573

Schiml S, Fauser F, Puchta H (2014) The CRISPR/Cas system can be used as nuclease for in planta gene targeting and as paired nickases for directed mutagenesis in Arabidopsis resulting in heritable progeny. Plant J 80:1139–1150. https://doi.org/10.1111/tpj.12704

Schiml S, Puchta H (2016) Revolutionizing plant biology: multiple ways of genome engineering by CRISPR/Cas. Plant Methods 12:8. https://doi.org/10.1186/s13007-016-0103-0

Singh R, Kuscu C, Quinlan A et al (2015) Cas9-chromatin binding information enables more accurate CRISPR off-target prediction. Nucleic Acids Res 43:e118. https://doi.org/10.1093/nar/gkv575

Sonoda E, Hochegger H, Saberi A et al (2006) Differential usage of non-homologous end-joining and homologous recombination in double strand break repair. DNA Repair (Amst) 5:1021–1029. https://doi.org/10.1016/j.dnarep.2006.05.022

Spencer D, Dickey LF, Gasdaska JR et al (2011) Expression of Plasminogen and Microplasminogen in Duckweed

Sree KS, Bog M, Appenroth KJ (2016) Taxonomy of duckweeds (Lemnaceae), potential new crop plants. Emirates J Food Agric 28:291–302. https://doi.org/10.9755/ejfa.2016-01-038

Sree KS, Sudakaran S, Appenroth K-JJ (2015) How fast can angiosperms grow? Species and clonal diversity of growth rates in the genus *Wolffia* (Lemnaceae). Acta Physiol Plant 37:1–7. https://doi.org/10.1007/s11738-015-1951-3

Sternberg SH, Redding S, Jinek M et al (2014) DNA interrogation by the CRISPR RNA-guided endonuclease Cas9. Nature 507:62–67. https://doi.org/10.1038/nature13011

Stomp A-M, Rajbhandari N (2000) Patent US 6,040,498: Genetically engineered duckweed 66:37

Stomp AM (2005) The duckweeds: a valuable plant for biomanufacturing. Biotechnol Annu Rev 11:69–99. https://doi.org/10.1016/S1387-2656(05)11002-3

Tang L, Zeng Y, Du H et al (2017) CRISPR/Cas9-mediated gene editing in human zygotes using Cas9 protein. Mol Genet Genomics 292:525–533. https://doi.org/10.1007/s00438-017-1299-z

Thompson BG (1989) The maximization of the productivity of aquatic plants for use in controlled ecological life support systems (CELSS). Acta Astronaut 19:269–273. https://doi.org/10.1016/0094-5765(89)90039-8

Van Hoeck A, Horemans N, Monsieurs P et al (2015) The first draft genome of the aquatic model plant *Lemna minor* opens the route for future stress physiology research and biotechnological applications. Biotechnol Biofuels 8:188. https://doi.org/10.1186/s13068-015-0381-1

Vu GTH, Cao HX, Watanabe K et al (2014) Repair of site-specific DNA double-strand breaks in barley occurs via diverse pathways primarily involving the sister chromatid. Plant Cell 26:2156–2167. https://doi.org/10.1105/tpc.114.126607

Vunsh R, Li J, Hanania U et al (2007) High expression of transgene protein in *Spirodela*. Plant Cell Rep 26:1511–1519. https://doi.org/10.1007/s00299-007-0361-4

Wang W, Haberer G, Gundlach H et al (2014) The *Spirodela polyrhiza* genome reveals insights into its neotenous reduction fast growth and aquatic lifestyle. Nat Commun 5:3311. https://doi.org/10.1038/ncomms4311

Wang W, Kerstetter RA, Michael TP (2011) Evolution of genome size in duckweeds (Lemnaceae). J Bot 2011:1–9. https://doi.org/10.1155/2011/570319

Wang W, Messing J (2011) High-throughput sequencing of three Lemnoideae (duckweeds) chloroplast genomes from total DNA. PLoS ONE 6:e24670. https://doi.org/10.1371/journal.pone.0024670

Weber E, Engler C, Gruetzner R et al (2011) A modular cloning system for standardized assembly of multigene constructs. PLoS ONE 6:e16765. https://doi.org/10.1371/journal.pone.0016765

Widholm JM, Kishinami I (1988) Allyl alcohol selection for lower alcohol dehydrogenase activity in *Nicotiana plumbaginifolia* cultured cells. Plant Physiol 86:266–269. https://doi.org/10.1104/PP.86.1.266

Wiedenheft B, Sternberg SH, Doudna JA (2012) RNA-guided genetic silencing systems in bacteria and archaea. Nature 482:331–338. https://doi.org/10.1038/nature10886

Woodard SL, Wilken LR, Barros GOF et al (2009) Evaluation of monoclonal antibody and phenolic extraction from transgenic *Lemna* for purification process development. Biotechnol Bioeng 104:562–571. https://doi.org/10.1002/bit.22428

Xu J, Cui W, Cheng JJ, Stomp AM (2011) Production of high-starch duckweed and its conversion to bioethanol. Biosyst Eng 110:67–72. https://doi.org/10.1016/j.biosystemseng.2011.06.007

Yamamoto YT, Rajbhandari N, Lin X et al (2001) Genetic transformation of duckweed *Lemna gibba* and *Lemna minor*. Vitr Cell Dev Biol - Plant 37:349–353. https://doi.org/10.1007/s11627-001-0062-6

Ziegler P, Adelmann K, Zimmer S et al (2015) Relative invitro growth rates of duckweeds (Lemnaceae)—the most rapidly growing higher plants. Plant Biol 17:33–41. https://doi.org/10.1111/plb.12184

# 18 Future Prospects of Duckweed Research and Applications

Giang T. H. Vu, Paul Fourounjian, Wenqin Wang and Xuan Hieu Cao

**Abstract**

The duckweeds have fascinated many generations of biologists since 1960s because of their simplicity of body structure and extremely fast vegetative reproduction. Since the beginning of the era of plant molecular biology, duckweeds have emerged as model plants with important findings in basic research as well as plant-based applications. Recent advances in the omic-tools and techniques have rejuvenated research on duckweeds, resulting in a fast accumulation of genomic, transcriptomic, and epigenomic resources to the field. This recent progress enables us to use duckweeds to address biological questions related to stress physiology, phytotoxicity, adaptive strategies of aquatic plants and the molecular interactions between aquatic microorganisms and plants. Throughout the past decades, unimaginably wide-range applications of duckweeds like wastewater purification, use of protein-rich biomass for animal feed and human nutrition, use of starch-rich biomass as source of renewable energy and as factories for producing plant-derived pharmaceuticals, to name a few, have been in development, or laboratory scale. We look forward to a deeper understanding of these wonderful plants and many new functional technologies on the horizon.

Giang T. H. Vu, Paul Fourounjian—These authors contributed equally to this work.

G. T. H. Vu
Leibniz Institute of Plant Genetics and Crop Plant Research (IPK), Corrensstrasse 3, 06166 Gatersleben, Germany

P. Fourounjian
Waksman Institute of Microbiology, Rutgers University, Piscataway 08854, USA

W. Wang
School of Agriculture and Biology, Shanghai Jiao Tong University, 800 Dong Chuan Road, Shanghai 200240, China

X. H. Cao (✉)
Institute of Biology/Plant Physiology, Martin-Luther-University of Halle-Wittenberg, Weinbergweg 10, 06120 Halle (Saale), Germany
e-mail: xuan.cao@biologie.uni-halle.de

## 18.1 Introduction

Not only arable fields are suitable for robust cultivation of crops. With a fast-growing global population, the challenges of expanding demand for resources need to be addressed by developing new and alternative agricultural systems for production of food, feed, and raw materials. Fortunately, duckweeds (Lemnaceae) which are productive and versatile aquatic monocot plants offer tremendous and unique opportunities for floating cultures under a wide range of conditions



X. H. Cao et al. (eds.), *The Duckweed Genomes*, Compendium of Plant Genomes,
https://doi.org/10.1007/978-3-030-11045-1_18

(except in nutrient-poor or highly acidic water). In addition to its nutritional value and fast biomass production, these plants can purify highly concentrated waste streams into safe drinking water. Importantly, duckweed production can possibly be scalable from households (e.g., 5 L tanks), small garden or village ponds to several-hectare-sized wastewater lagoons. Provided that duckweed is globally distributed, duckweed-based solutions can be available on all continents but Antarctica (for review, see Cao et al. 2018). Fascinatingly, duckweed was additionally described as one of the most attractive higher plants for long-duration supporting human life in space (Yuan and Xu 2017). However, the implementation of these multi-purpose plants has still been underrated partly due to currently incomplete knowledge on duckweed biology and agriculture.

The availability of high-quality duckweed reference genome sequences has ushered in new and detailed insights into the mechanisms underpinning biological processes of these aquatic, highly morphologically reduced plant models. The first duckweed genome *Spirodela polyrhiza* (aka., the Greater Duckweed), chosen due to its basal phylogenetic position and smallest genome size, was sequenced by an international consortium led by scientists from Rutgers University (Wang et al. 2014). This high-quality reference genome was improved to the chromosome-level assembly by using multi-color fluorescence in situ hybridization (Cao et al. 2016a), optical mapping (Michael et al. 2017), and Oxford Nanopore Technology sequencing of another ecotype (aka., clone 9509) of the same species (Hoang et al. 2018). Since publication in 2014, the Spirodela genome has been cited more than 100 times on the Google Scholar indexing platform (dated July 2019), serving as a duckweed reference genome not only for greatly facilitating duckweed genomic investigation (e.g., MAD box genes by Gramzow and Theissen 2015; pentatricopeptide-repeat proteins by Wang et al. 2016) but also for comparative genome analysis of other monocot and aquatic plants (e.g., seagrass for angiosperm adaptation to the sea by Olsen et al. 2016).

## 18.2 Future Prospects in Duckweed Genome Research

Living in a very specialized aquatic environment, duckweeds do not need all essential genes of terrestrial flowering plants, such as water transport, lignin biosynthesis, and cell wall-loosening proteins. In fact, the Spirodela genome was streamlined to be small (as the genome of *Arabidopsis thaliana*) with less than 20,000 protein-coding genes (three-quarters of *A. thaliana* genes). However, there are certain functional categories of genes being enriched (e.g., ammonia assimilation, defense-related processes) or specific (e.g., underlining mechanisms for fast growth, aquatic life style or highly neotenous morphology) to duckweed genomes (Wang et al. 2014; Van Hoeck et al. 2015). In this sense, comparative genome analysis among duckweed genomes, ideally with representative species of all five genera, will help advance our knowledge of duckweed genome structure and evolution. So far, in addition to published genomes of *S. polyrhiza* and *Lemna minor*, there have been several on-going sequencing projects working on *Spirodela intermedia*, *Landoltia punctata*, *Lemna gibba*, and *Wolffia australiana*. With the advent of fast and cost-effective next-generation sequencing platforms together with recent and continued advances in third generation of long, single-molecule sequencing platforms, it is imaginable that further development of multiple technologies for effectively constructing long-range, chromosome-size scaffolds will enable the complete assemblies of all 37 duckweed species ranging from 158 Mbp to almost 2 Gbp.

In addition to the goal of gapless, chromosome-scale genome assemblies, future advances in technology, and growing interest of research community may overcome the remaining hurdles on comprehensive understanding duckweed gene functions, networks, and metabolic pathways. Although in many cases function and biochemical activity of gene products may be inferable from the presence of the common and conserved protein domains, precise biological role and detailed functional information of genes, especially duckweed-specific genes, requires intensive experimental analysis on functional characterization. The future functional genomics in duckweeds will be in line with new robust, precise and versatile genome editing tools (for review, see Cao et al. 2016b) which not only knocks out unwanted genes but also precisely replace or insert single nucleotides, single gene or multiple genes of a specific agronomic trait ("trait stacking").

In recent years, the field of duckweed research and applications has entered the post-genomic era with a growing number of large-scale-omics datasets generated from different duckweed species, for example, whole genome resequencing of 68 *S. polyrhiza* ecotypes/genotypes representing the global distribution of the species illustrated that the low genetic variation was associated with low mutation rate in duckweeds (Xu et al. 2019). Examination of growing genomic, epigenomic, transcriptomic, proteomic, and metabolomic resources would facilitate dissecting the genetics underlying the fast growth rate, turion formation, protein content and other important physiological and agronomic traits of duckweeds (An et al. 2018). Future work should critically enable the systematic and comprehensive interpretation of these resources, in an ideal model, from an online, open and evolving catalog.

Given the understanding that incremental gains in conventional agriculture will not meet resource demand in 2050, most experts in this field preach the importance of novel breakthrough technologies. Large-scale research grants from the world's great nations aimed at long-term funding of multiple academic laboratories and companies across the world in a collaborative way, similar to the duckweed genome initiative, would be a great way to foster cooperation and accelerate research in understanding and utilizing duckweed, as well as other promising new crops. To successfully implement the manifold potential use of duckweed and finally to create a new industry, we anticipate the importance of collaborations between academic and commercial partners, of which the involvement of commercial stakeholders starts from the very beginning.

## 18.3 Future Prospects in Duckweed-Based Applications

While duckweeds have been used as a livestock feed for centuries and harvested for human consumption in Southeast Asia so long, it has been incorporated into traditional cuisine, they have been overlooked and under-utilized for most of history. There was a spike in commercial interest during the oil crisis of the 1970s, including an analysis of traditional duckweed harvesting (Bhanthumnavin and Mcgarry 1971), followed by a lull. At the same time from 1950 to 1990, there was strong academic interest in the duckweed family as a simple model plant to understand the physiology of crops of the day, yet the *Lemnaceae* was not seen as a crop at the time. Then, the 1990s provided both an economic model for duckweed water treatment and farming in the form of the book "Duckweed Aquaculture" (Skillicorn et al. 1993) and the first attempt to produce plantibodiesTM in duckweed by Biolex, which unfortunately went out of business. Then, there was a sustained academic and commercial interest in this plant family since the late 2000s, likely fueled by the 2008 oil price that has lasted till the present day 2019. This decade of duckweed research including multiple genome sequences and transcriptomic studies, physiology research, nutritional and animal feed studies, and advances in genetic engineering across the family has led to a much deeper understanding of these plants. Combined with a renewed interest in biofuels, more toxicological testing, greater demand for water treatment, the need for alternative food and protein sources, and a medical interest in "biologics" aka therapeutic proteins, we have seen a recent revolution in duckweed applications and companies thoroughly reviewed in Chap. 1 and summarized or restated here. Considering that there is virtually no duckweed focused company more than 10 years old this is an extremely new industry which probably has more future ahead of it than past behind it (Table 18.1).

**Table 18.1** Summary of the duckweed applications in use or development and the major companies working on them

| Application | Company (if blank academic) | Genera |
|---|---|---|
| Human food | Hinoman, Green Onyx, Parabel | Wolffia, Lemna |
| Protein isolate | Plantible, Parabel, CAIS | Lemna |
| Livestock | Many small-scale farmers | Lemna, Spirodela, others |
| Conversion chemicals | MamaGrande | Lemna |
| Wastewater treatment | MamaGrande, CAIS | Mixture |
| Space life support | Space Lab Technologies | Lemna, Wolffia |
| Isolation chemicals | CAIS | Mixture |
| Transformation | | |
| Specialty (cosmetics, pets, tea) | | |
| Biofuels or energy | Greenbelt Resources | |

Duckweed as a food and livestock feed source is likely the oldest human application and an easy one to project into the future. Due to rising incomes and rising populations, global demand for non-fish animal protein, particularly poultry is expected to increase at 1.3% per year till 2050, with the largest growth of 4.2% in South Asia, (Alexandratos and Bruinsma 2012; OECD/FAO 2017). Additionally, the largest increase in animal protein supply will be aquaculture, which grew largely in Asia between 4 and 10% per year since 1990 and is forecasted to exceed the global catch in 2020 (OECD/FAO 2017). Considering that the duckweed species can clean wastewater from agricultural operations, provide a feed for fish, chicken, pigs, goats, and other livestock, generally as ~20% of their diet, grow year round in the tropics and subtropics, and raise farmer income (reviewed in Cao et al. 2018), duckweed seems wonderfully positioned to clean wastewaters from and provide feed to livestock, especially fish and poultry in the tropics where the largest growth in the livestock sector will occur. In terms of human nutrition, Hinoman and GreenOnyx were able to achieve the GRAS (generally recognized as safe) status for the *Wolffia* species *arrhiza* and *globosa* in the USA in 2015 and 2016, respectively, thanks to supporting data from academic laboratories, records of historical consumption, and thorough testing of their product for harmful metals and oxalates. Thanks to this breakthrough, the recent legal recognition of select Wolffia species as food in Israel and traditional consumption in Thailand, Laos, and Cambodia the duckweed production as a food crop is set to expand. Likewise, the whole US market in plant-based protein has been growing at 12%/year and is expected to grow at 6.7% for the near future (Henchion et al. 2017). There will also be an expansion of duckweed food options in the future as traditional cuisine recipes are expanded upon, and whole vegetables or isolates enter the processed foods market.

Another sector where duckweed species will likely play an expanding role is supplying clean water. In 2018, the Duckweed Forum issue 22 described 23 companies in 9 countries, with four each working in water quality testing and water treatment (Shoham 2018). Provided the perpetual rise of water pollution and increased testing, and the roughly 50% lower capital and operating costs of duckweed (Skillicorn 2013) and constructed wetlands (Zhang et al. 2014) treatment systems compared to their bacterial counterparts these industries are expected to grow, likely more so in developing countries. Sadly, 14 years of satellite observations reveal decreasing water availability in heavily populated areas like California, The Middle East, Northern India, and Northern China where groundwater is being depleted (Rodell et al. 2018). Duckweed treatment systems to reclaim water, as well as water efficient duckweed crops, with many other measures, might be utilized in these and other regions to increase supply. Similar to water reclamation there is a lesser known need for phosphorous reclamation, since our current practice is to mine and refine phosphorous deposits, fertilize our crops, and then let the phosphorous run directly off of fields and into the ocean, or through our wastewater treatment systems into the ocean where it causes eutrophication damage. Economically mineable phosphorous is expected to be scarce by 2050 or 2100, and production might decline by 2030 raising its price possibly beyond the reach of poorer farmers (Childers et al. 2011). Fortunately, phosphorous can be recycled by better farming practices or by using more aquatic plants and other methods to recapture more than the current rate of 50% from human wastes. Given the water and fertilizer scarcities, this century will likely pose to billions of people we sincerely hope that duckweed-based water treatment systems, and many other water and nutrient reclamation technologies will be applied at larger scale to "close the loop" and avoid scarcity.

Today there is even the option to genetically engineer many different species of the duckweed family. There have been over 20 transgenic therapeutic proteins that could improve human health, reaching as high as 7% of total soluble protein (Balaji et al. 2016), and the door is open for many other types of transgenes like industrial enzymes, those that improve nutritional content of animal feed (Ghosh et al. 2018), and edible vaccines expressed in duckweed for livestock feed (Firsov et al. 2018). While some of these transgenic duckweeds could be used whole, others will go through an extraction process to isolate the protein, especially the therapeutic proteins. These extraction products may also isolate other high value products like vitamin E and omega-3 fatty acids before proceeding to further processing. As described in Chap. 1, modern agriculture often leads into a feedstock being processed into a wide range of chemical compounds that can be used as ingredients in other finished goods, aka biorefining. This has occurred with corn and soy finding their way into a wide variety of products like emulsifiers and gums, soy-based inks, and baby powder. For example, duckweed protein and therapeutic proteins specifically, can be isolated, perhaps secondary metabolites could be alcohol extracted, and then the starch can be digested into simple sugars that can be isolated or fermented into ethanol, butanol, or polylactic acid. Additionally hydrothermal liquefaction could turn duckweed biomass at any point in that process to into natural gas, biochar, and a bio-crude oil that could be fractionated into countless chemical compounds. Maybe in the future, if duckweed becomes a cheap and common crop, it will find its way into tea, cosmetics, pet foods, garden products, and other diverse finished goods. In some cases in the future people may look at duckweed thinking that the sum of its parts are greater than the whole.

Thanks to their ability to clean wastewater while providing food and fresh air, duckweeds can be arguably seen as not only a crop species, but also a life support system. In order to create a life support system with low resupply needs for long-term space travel, NASA has been investigating plant-based life support for decades and specifically duckweed-based life support since 1966 (Landolt and Kandeler 1987). Therefore, Space Lab Technologies, LLC is currently collaborating with the University of Colorado at Boulder on a Phase 2 grant from NASA to develop the μG-LilyPondTM growth chamber as part of a plant-based life support system (Escobar and Escobar 2017). Thanks to their high growth rate, ability to grow in shallow trays, preference for ammonia, simple aquatic lifestyle, ability to grow and thrive in microgravity, high carbon dioxide tolerance, and entirely edible nutritious biomass duckweed is currently the prime candidate for the job. Presently, it is designed to provide fresh food, and oxygen, with the eventual goal of converting urine to clean water. Part of their project is studying how bursts of high-intensity light can stimulate production of carotenoids, vitamin E, and other nutritious secondary metabolites (Demmig-Adams and Adams 2002), since these and other vitamins have limited shelf stability meaning they must be produced onboard to enable longer flights. This intimate reliance on duckweed in a closed-loop system provides both a technical and symbolic example of how humans and duckweed complement each other, and how we can use the smallest plants to solve the largest challenges.

**Acknowledgements** We would like to acknowledge Paul Skillicorn from Agriquatics, Ron Salpeter from Hinoman, and Tsipi and Ben Shoham from Green Onyx, and Christine Escobar from Space Lab Technologies for sharing data on their companies and research and reviewing this document to ensure its accuracy.

## References

Alexandratos N, Bruinsma J (2012) World agriculture towards 2030/2050: the 2012 revision. Food and Agriculture Organization of the United Nations (FAO), Rome

An D, Li C, Zhou Y, Wu Y, Wang W (2018) Genomes and transcriptomes of duckweeds. Front Chem 6:230. https://doi.org/10.3389/fchem.2018.00230

Balaji P, Satheeshkumar PK, Venkataraman K, Vijayalakshmi MA (2016) Expression of anti-tumor necrosis factor alpha (TNFalpha) single-chain variable fragment (scFv) in *Spirodela punctata* plants transformed with agrobacterium tumefaciens. Biotechnol Appl Biochem 63(3):354–361. https://doi.org/10.1002/bab.1373

Bhanthumnavin K, Mcgarry MG (1971) Wolffia arrhiza as a possible source of inexpensive protein. Nature 232(5311):495

Cao HX, Fourounjian P, Wang W (2018) The importance and potential of duckweeds as a model and crop plant for biomass-based applications and beyond. In: Hussain CM (ed) Handbook of environmental materials management. Springer International Publishing, Cham, pp 1–16. https://doi.org/10.1007/978-3-319-58538-3_67-1

Cao HX, Vu GT, Wang W, Appenroth KJ, Messing J, Schubert I (2016a) The map-based genome sequence of *Spirodela polyrhiza* aligned with its chromosomes, a reference for karyotype evolution. New Phytol 209 (1):354–363. https://doi.org/10.1111/nph.13592

Cao HX, Wang W, Le HTT, Vu GTH (2016b) The power of CRISPR-Cas9-induced genome editing to speed up plant breeding. Int J Genom 2016:10. https://doi.org/10.1155/2016/5078796

Childers DL, Corman J, Edwards M, Elser JJ (2011) Sustainability challenges of phosphorus and food: solutions from closing the human phosphorus cycle. Bioscience 61(2):117–124. https://doi.org/10.1525/bio.2011.61.2.6

Demmig-Adams B, Adams WW 3rd (2002) Antioxidants in photosynthesis and human nutrition. Science 298 (5601):2149–2153. https://doi.org/10.1126/science.1078002

Escobar CM, Escobar AC (2017) Duckweed: a tiny aquatic plant with enormous potential for bioregenerative life support systems. In: 47th international conference on environmental systems, Charleston, South Carolina, 16–20 July 2017 2017. ICES, p 281

Firsov A, Tarasenko I, Mitiouchkina T, Shaloiko L, Kozlov O, Vinokurov L, Rasskazova E, Murashev A, Vainstein A, Dolgov S (2018) Expression and immunogenicity of M2e peptide of avian influenza virus H5N1 fused to ricin toxin b chain produced in duckweed plants. Front Chem 6:22

Ghosh M, Sharma N, Gera M, Kim N, Huynh D, Zhang J, Min T, Sodhi SS, Kim MB, Rekha V (2018) Insights into phytase-containing transgenic *Lemna minor* (L) as a novel feed additive. Transgen Res 27 (2):211–224

Gramzow L, Theissen G (2015) Phylogenomics reveals surprising sets of essential and dispensable clades of MIKC(c)-group MADS-box genes in flowering plants. J Exp Zool B Mol Dev Evol 324(4):353–362. https://doi.org/10.1002/jez.b.22598

Henchion M, Hayes M, Mullen AM, Fenelon M, Tiwari B (2017) Future protein supply and demand: strategies and factors influencing a sustainable equilibrium. Foods 6(7):53. https://doi.org/10.3390/foods6070053

Hoang PNT, Michael TP, Gilbert S, Chu P, Motley TS, Appenroth KJ, Schubert I, Lam E (2018) Generating a high-confidence reference genome map of the greater duckweed by integration of cytogenomic, optical mapping and oxford nanopore technologies. Plant J. https://doi.org/10.1111/tpj.14049

Landolt E, Kandeler R (1987) Biosystematic investigations in the family of duckweeds (Lemnaceae), vol 4: the family of Lemnaceae-a monographic study, vol 2 (phytochemistry, physiology, application, bibliography). Veroeffentlichungen des Geobotanischen Instituts der ETH, Stiftung Ruebel (Switzerland)

Michael TP, Bryant D, Gutierrez R, Borisjuk N, Chu P, Zhang H, Xia J, Zhou J, Peng H, El Baidouri M, Ten Hallers B, Hastie AR, Liang T, Acosta K, Gilbert S, McEntee C, Jackson SA, Mockler TC, Zhang W, Lam E (2017) Comprehensive definition of genome features in *Spirodela polyrhiza* by high-depth physical mapping and short-read DNA sequencing strategies. Plant J 89 (3):617–635. https://doi.org/10.1111/tpj.13400

OECD, FAO (2017) OECD-FAO agricultural outlook 2017–2026. OECD Publishing, Paris. https://doi.org/10.1787/19991142

Olsen JL, Rouze P, Verhelst B, Lin YC, Bayer T, Collen J, Dattolo E, De Paoli E, Dittami S, Maumus F, Michel G, Kersting A, Lauritano C, Lohaus R, Topel M, Tonon T, Vanneste K, Amirebrahimi M, Brakel J, Bostrom C, Chovatia M, Grimwood J, Jenkins JW, Jueterbock A, Mraz A, Stam WT, Tice H, Bornberg-Bauer E, Green PJ, Pearson GA, Procaccini G, Duarte CM, Schmutz J, Reusch TB, Van de Peer Y (2016) The genome of the seagrass *Zostera marina* reveals angiosperm adaptation to the sea. Nature. https://doi.org/10.1038/nature16548

Rodell M, Famiglietti JS, Wiese DN, Reager JT, Beaudoing HK, Landerer FW, Lo MH (2018) Emerging trends in global freshwater availability. Nature 557 (7707):651–659. https://doi.org/10.1038/s41586-018-0123-1

Shoham T (2018) On a mission to introduce duckweed as the twenty first century cash crop: current active industrial efforts. Duckweed Forum 6(3):92–97

Skillicorn P (2013) The agriquatics system versus activated sludge: a comparative analysis. https://paulskillicorn.wordpress.com/2013/04/17/the-agriquatics-system-vs-activated-sludge-a-comparative-analysis/. Accessed 09.09 2019

Skillicorn P, Spira W, Journey W (1993) Duckweed aquaculture: a new aquatic farming system for developing countries. World Bank

Van Hoeck A, Horemans N, Monsieurs P, Cao HX, Vandenhove H, Blust R (2015) The first draft genome of the aquatic model plant Lemna minor opens the route for future stress physiology research and biotechnological applications. Biotechnol Biofuels 8 (1):1–13. https://doi.org/10.1186/s13068-015-0381-1

Wang W, Haberer G, Gundlach H, Glasser C, Nussbaumer T, Luo MC, Lomsadze A, Borodovsky M, Kerstetter RA, Shanklin J, Byrant DW, Mockler TC, Appenroth KJ, Grimwood J, Jenkins J, Chow J, Choi C, Adam C, Cao XH, Fuchs J, Schubert I, Rokhsar D, Schmutz J, Michael TP, Mayer KF, Messing J (2014) The *Spirodela polyrhiza* genome reveals insights into its neotenous reduction fast growth and aquatic lifestyle. Nat Commun 5:3311. https://doi.org/10.1038/ncomms4311

Wang W, Wu Y, Messing J (2016) Genome-wide analysis of pentatricopeptide-repeat proteins of an aquatic plant. Planta. https://doi.org/10.1007/s00425-016-2555-x

Xu S, Stapley J, Gablenz S, Boyer J, Appenroth KJ, Sree KS, Gershenzon J, Widmer A, Huber M (2019) Low genetic variation is associated with low mutation rate in the giant duckweed. Nat Commun 10(1):1243. https://doi.org/10.1038/s41467-019-09235-5

Yuan J, Xu K (2017) Effects of simulated microgravity on the performance of the duckweeds Lemna aequinoctialis and Wolffia globosa. Aquat Bot 137:65–71. https://doi.org/10.1016/j.aquabot.2016.11.010

Zhang DQ, Jinadasa K, Gersberg RM, Liu Y, Ng WJ, Tan SK (2014) Application of constructed wetlands for wastewater treatment in developing countries–a review of recent developments (2000–2013). J Environ Manag 141:116–131

Kolibri 360 GmbH
Ferdinand-Jühlke-Straße 7
99095 Erfurt, Deutschland
produktsicherheit@kolibri360.de